Highlights in Colloid Science

Edited by
Dimo Platikanov and Dotchi Exerowa

Related Titles

Tadros, Th. F. (ed.)

Topics in Colloid and Interface Science

Volume 2

2008

ISBN: 978-3-527-31991-6

Tadros, Th. F. (ed.)

Colloids and Interface Science Series

6 Volume Set

2008

ISBN: 978-3-527-31461-4

Tadros, Th. F. (ed.)

Topics in Colloid and Interface Science

Volume 1

2008

ISBN: 978-3-527-31990-9

Wilkinson, K. J.

Environmental Colloids and Particles - Behaviour, Separation and Characterisation

2007

ISBN: 978-0-470-02432-4

Schramm, L. L.

Emulsions, Foams, and Suspensions

Fundamentals and Applications

2005

ISBN: 978-3-527-30743-2

Tadros, Th. F.

Applied Surfactants

Principles and Applications

2005

ISBN: 978-3-527-30629-9

Pashley, R., Karaman, M.

Applied Colloid and Surface Chemistry

2004

ISBN: 978-0-470-86882-9

Rosen, M. J.

Surfactants and Interfacial Phenomena

2004

ISBN: 978-0-471-47818-8

Goodwin, J.

Colloids and Interfaces with Surfactants and Polymers - An Introduction

2004

ISBN: 978-0-470-84143-3

Highlights in Colloid Science

Edited by
Dimo Platikanov and Dotchi Exerowa

WILEY-VCH Verlag GmbH & Co. KGaA

The Editors

Professor Dimo Platikanov
University of Sofia
Department of Physical Chemistry
Blvd James Bourchier 1
1164 Sofia
Bulgaria

Professor Dotchi Exerowa
Bulgarian Academy of Sciences
Institute of Physical Chemistry
1113 Sofia
Bulgaria

All books published by **Wiley-VCH** are carefully produced. Nevertheless, authors, editors, and publisher do not warrant the information contained in these books, including this book, to be free of errors. Readers are advised to keep in mind that statements, data, illustrations, procedural details or other items may inadvertently be inaccurate.

Library of Congress Card No.: applied for

British Library Cataloguing-in-Publication Data
A catalogue record for this book is available from the British Library.

Bibliographic information published by the Deutsche Nationalbibliothek
Die Deutsche Nationalbibliothek lists this publication in the Deutsche Nationalbibliografie; detailed bibliographic data are available on the Internet at http://dnb.d-nb.de.

© 2009 WILEY-VCH Verlag GmbH & Co. KGaA, Weinheim

All rights reserved (including those of translation into other languages). No part of this book may be reproduced in any form – by photoprinting, microfilm, or any other means – nor transmitted or translated into a machine language without written permission from the publishers. Registered names, trademarks, etc. used in this book, even when not specifically marked as such, are not to be considered unprotected by law.

Composition   Thomson Digital, Noida, India
Printing   Strauss GmbH, Mörlenbach
Bookbinding   Litges & Dopf GmbH, Heppenheim
Cover Design   Adam Design, Weinheim

Printed in the Federal Republic of Germany
Printed on acid-free paper

ISBN: 978-3-527-32037-0

Contents

Preface *XI*
Tharwat F. Tadros *XIII*
List of Contributors *XVII*

1	**Orthokinetic Heteroflocculation in Papermaking** *1*	
	Theo G.M. van de Ven	
1.1	Introduction *1*	
1.2	Polymer-Induced Orthokinetic Heteroflocculation *2*	
1.2.1	Polymer Adsorption and Desorption on Fibers and Colloids *3*	
1.2.2	Deposition of Colloids on Fibers Subjected to Shear *5*	
1.2.3	Polymer Transfer *10*	
1.2.4	Time Dependence of Deposition and Detachment Rate Coefficients *12*	
1.3	Heteroflocculation Among Colloids *13*	
1.4	Heteroflocculation of Fines and Colloids *17*	
1.5	Concluding Remarks *18*	
	References *19*	
2	**Uptake and Release of Active Species into and from Microgel Particles** *21*	
	Melanie Bradley, Paul Davies, and Brian Vincent	
2.1	Introduction to Microgel Particles *21*	
2.2	Absorption of Small Molecules *23*	
2.3	Absorption of Surfactants *25*	
2.4	Absorption of Polymers and Proteins *29*	
2.5	Absorption of Nanoparticles *34*	
	References *38*	
3	**Stability of Fluorinated Systems: Structure-Mechanical Barrier as a Factor of Strong Stabilization** *41*	
	Eugene D. Shchukin, Elena A. Amelina[†], and Aksana M. Parfenova	
3.1	Introduction *41*	

Highlights in Colloid Science. Edited by Dimo Platikanov and Dotchi Exerowa
Copyright © 2009 WILEY-VCH Verlag GmbH & Co. KGaA, Weinheim
ISBN: 978-3-527-32037-0

3.2	Rheological Studies of Interfacial Adsorption Layers in Fluorinated Systems	42
3.3	Studies of the Rupture and Coalescence of Individual Droplets	44
3.4	Studies of the Interaction of Hydrophobized Solid Surfaces in Nonpolar Liquids	47
3.5	Discussion	49
3.6	Conclusion	51
	References	51

4 Particle Characterization Using Electro-Acoustic Spectroscopy 55
Richard W. O'Brien, James K. Beattie, and Robert J. Hunter

4.1	Introduction	55
4.2	Understanding the ESA Effect	55
4.3	The Dynamic Mobility	59
4.4	The Dynamic Mobility for Thin Double Layer Systems	63
4.5	Particles with Adsorbed Polymer Layers	65
4.6	Surface Conductance	70
4.7	Nanoparticles	73
	References	76

5 Modeling the Structure and Stability of Charged Hemi-Micelles at the Air–Water Interface 79
Johannes Lyklema, Ana B. Jódar-Reyes, and Frans A.M. Leermakers

5.1	Introduction	79
5.2	Thermodynamics	80
5.3	Fundamentals of SCF Theory and the Molecular Model	82
5.3.1	The Lattice	82
5.3.2	From Volume Fractions to Potentials	83
5.3.3	From Potentials to Volume Fractions	86
5.3.4	Grand Potential	88
5.4	Results	90
5.4.1	Stability Analysis	91
5.4.2	Structural Analysis	94
5.5	Conclusions	95
	References	96

6 Foam, Emulsion and Wetting Films Stabilized by Polymeric Surfactants 97
Dotchi Exerowa and Dimo Platikanov

6.1	Introduction	97
6.2	Microinterferometric Method for Investigation of Thin Liquid Films	98
6.3	Intercation Forces in Foam Films	100
6.3.1	Foam Films Stabilized by A-B-A Block Copolymers. Brush-to-Brush Interaction	100

6.3.2	Foam Films Stabilized by Hydrophobically Modified Inulin Polymeric Surfactants. Loop-to-Loop Interaction *103*	
6.4	Interaction Forces in Emulsion Films *106*	
6.4.1	Emulsion Films Stabilized by A–B–A Block Copolymers: Brush-to-Brush Interaction and Transition to the Newton Black Film *106*	
6.4.2	Emulsion Films Stabilized by Hydrophobically Modified Inulin: Loop-to-Loop Interaction and Transition to the Newton Black Film *108*	
6.4.3	Comparison of Film Stability and the Stability of a Real Emulsion *112*	
6.5	Wetting Films Stabilized by Hydrophobically Modified Inulin Polymeric Surfactant *113*	
6.6	Conclusion *115*	
	References *116*	
7	**Conditions for the Existence of a Stable Colloidal Liquid** *119*	
	Gerard J. Fleer and Remco Tuinier	
7.1	Introduction *119*	
7.2	Theory *120*	
7.2.1	Free Energy *120*	
7.2.2	Yukawa Attraction *122*	
7.2.3	FVT ("Fix") *122*	
7.2.4	GFVT ("Var") *124*	
7.3	Phase Diagrams $\varepsilon(\eta)$ *125*	
7.4	Phase Diagrams pv/kT Versus ε/kT *128*	
7.5	Phase Diagrams pv/ε Versus kT/ε *131*	
7.6	Concluding Remarks *132*	
	References *132*	
8	**Preparation, Properties and Chemical Modification of Nanosized Cellulose Fibrils** *135*	
	Per Stenius and Martin Andresen	
8.1	Introduction *135*	
8.2	Microfibrillar Cellulose *135*	
8.3	Preparation of Microfibrillar and Nanocrystalline Cellulose *137*	
8.3.1	Acid Hydrolysis *137*	
8.3.2	High Shear Mechanical Treatment *138*	
8.3.3	Other Routes to Cellulose Microfibrils *140*	
8.4	Methods Used to Characterize Cellulose Microfibrils *140*	
8.4.1	Fibril Morphology and Structure *140*	
8.4.2	Fibril Surface Chemistry *141*	
8.5	Modification of Microfibril Surfaces *141*	
8.5.1	Esterification Reactions *142*	
8.5.1.1	Acetylation *142*	

8.5.1.2	Reaction with Anhydrides	*142*
8.5.1.3	Carboxymethylation	*143*
8.5.1.4	Isocyanate Grafting	*143*
8.5.2	Etherification	*144*
8.5.2.1	Silylation	*144*
8.5.3	Nitration	*145*
8.5.4	Oxidation Reactions	*145*
8.5.4.1	TEMPO-Mediated Oxidation	*145*
8.5.4.2	Cerium Induced Grafting	*146*
8.5.5	Coating with Surfactant	*146*
8.6	Applications of Nanofibrillar Cellulose	*147*
8.6.1	Rheology	*147*
8.6.2	Nanocomposites	*148*
8.6.3	Thin Films	*148*
8.6.4	Dispersion Stabilizers	*149*
8.6.5	Biochemical and Biomedical Applications	*149*
8.6.5.1	Enzymatic Assay	*150*
8.6.5.2	Grafting of an Antimicrobial	*150*
8.6.6	Paper Products	*151*
8.7	Concluding Remarks	*151*
	References	*152*

9 Melting/Freezing Phase Transitions in Confined Systems *155*
Ludmila Boinovich and Alexandre Emelyanenko

9.1	Introduction	*155*
9.2	Surface Phase Transitions at the Plane Interface	*158*
9.3	Confinement by Curved Interfaces	*166*
9.3.1	Phase Transitions at the Surface and in the Interior of Small Particles	*169*
9.3.2	Phase Transitions at the Interfaces and in the Interior of the Substance Condensed in a Porous Matrix	*172*
9.4	Concluding Remarks	*175*
	References	*175*

10 Manipulation of DNA by Surfactants *179*
Björn Lindman, Rita S. Dias, M. Graça Miguel, M. Carmen Morán, and Diana Costa

10.1	Introduction	*179*
10.2	Surfactants Bind to ds-DNA and Induce Compaction	*181*
10.3	Surfactant Addition Can Lead to Phase Separation of DNA	*184*
10.3.1	Effect of Salt	*187*
10.3.2	Effect of Temperature	*188*
10.4	DNA is an Amphiphilic Polyelectrolyte	*189*
10.5	Phase Separation Phenomena Underlie the Preparation of Novel Particles	*192*

10.6	DNA Can be Crosslinked into Gels	*196*
10.7	Perspectives	*199*
	References	*200*

11 Deposition of Colloid Particles at Heterogeneous Surfaces *203*
Zbigniew Adamczyk, Jakub Barbasz, and Małgorzata Nattich

11.1	Introduction	*203*
11.2	Theoretical Models	*204*
11.2.1	Random Sequential Adsorption Approach	*205*
11.3	Illustrative Theoretical Results	*207*
11.3.1	Deposition at Quasi-Continuous Surfaces	*207*
11.3.2	Deposition at Random Site Surfaces	*207*
11.3.3	Particle Deposition at Surface Features	*212*
11.4	Comparison with Experimental Results	*214*
11.4.1	Deposition at Surface Features and Patterns	*222*
11.5	Concluding Remarks	*224*
	References	*225*

12 Effect of the Interaction Between Heavy Crude Oil Components and Stabilizing Solids with Different Wetting Properties *229*
Simone Less, Andreas Hannisdal, Heléne Magnusson, and Johan Sjöblom

12.1	Introduction	*229*
12.2	Experimental	*231*
12.2.1	Extraction of Asphaltenes from the Crude Oils	*231*
12.2.2	Silica Particles: Characterization and Properties	*231*
12.2.3	Preparation of the Emulsions	*232*
12.2.4	Emulsion Stability Measurements and Drop Size Determination	*233*
12.3	Results and Discussion	*234*
12.3.1	Droplet Size Distributions	*234*
12.3.2	Viscosity Observations	*234*
12.3.3	Stability Measurements	*236*
12.4	Conclusions	*242*
	References	*244*

13 Impact of Micellar Kinetics on Dynamic Interfacial Properties of Surfactant Solutions *247*
Reinhard Miller, Boris A. Noskov, Valentin B. Fainerman, and Jordon T. Petkov

13.1	Introduction	*247*
13.2	Micellization Kinetics Mechanisms	*249*
13.3	Impact of Micelles on Adsorption Kinetics	*250*
13.4	Impact of Micelle Kinetics on Interfacial Dilational Visco-Elasticity	*254*
13.5	Summary	*256*
	References	*257*

14	**Aggregation of Colloids: Recent Developments in Population Balance Modeling** *261*	
	Ponisseril Somasundaran and Venkataramana Runkana	
14.1	Introduction *261*	
14.2	Aggregation in Quiescent Environments *262*	
14.2.1	Models Incorporating Surface Forces *263*	
14.2.1.1	Aggregation in the Presence of Inorganic Electrolytes *264*	
14.2.1.2	Aggregation in the Presence of Polymers *266*	
14.3	Aggregation in Shear Environments *269*	
14.3.1	Models Incorporating Surface Forces *270*	
14.3.2	Models Incorporating Evolution of Aggregate Structure *271*	
14.3.3	Coupled Population Balance – Fluid Flow Models *272*	
14.4	Summary and Suggestions for Future Research *274*	
14.4.1	Multidimensional Population Balances *274*	
14.4.2	Polymer Adsorption Dynamics *275*	
14.4.3	Computationally Efficient Population Balance–Fluid Flow Models *275*	
14.4.4	Depletion Flocculation *275*	
	References *276*	
15	**Cubosomes as Delivery Vehicles** *279*	
	Nissim Garti, Idit Amar-Yuli, Dima Libster, and Abraham Aserin	
15.1	Introduction *279*	
15.2	Preparation Techniques *281*	
15.3	Drug Delivery Applications *282*	
15.4	Summary *288*	
	References *288*	
16	**Highly Concentrated (Gel) Emulsions as Reaction Media for the Preparation of Advanced Materials** *291*	
	Conxita Solans and Jordi Esquena	
16.1	Introduction *291*	
16.2	Highly Concentrated Emulsions as Templates for Low-Density Macroporous Materials *294*	
16.3	Materials with Dual Meso- and Macroporous Structure Templated in Macroporous Foams Obtained From Highly Concentrated Emulsions *295*	
16.4	Conclusions *296*	
	References *297*	

Index *299*

Preface

The year 2007 saw two remarkable anniversaries: the 200th year since the publishing house Wiley was established in Manhattan, New York, and the 70th birthday of the distinguished scientist in the field of colloid and interface science Tharwat Tadros. It seemed, therefore, fitting that the publishers Wiley-VCH accepted the idea of preparing a book dedicated to Tharwat Tadros. As a result of invitations sent to well-known colloid and interface scientists the 16 review articles were received. Together with the publishers we entitled the book *Highlights in Colloid Science* – a title that may be a little pretentious but one that reflects well the content of this book. A great variety of topics is presented – colloid particles: aggregation, deposition and characterization; surfactant solutions: micelles, interfacial properties and the manipulation of DNA; delivery agents: microgel particles and cubosomes; thin liquid films; colloidal liquids; structure-mechanical barrier; phase transitions in confined systems; papermaking: heteroflocculation and cellulose fibrils; wetting phenomena in crude oil; highly concentrated emulsions, and so on.

The reviews, with different scopes, offered here are written by leading scientists from all over the world. They are comprehensive, with many references, and they should be very useful for those engaged (both in academia and in industry) in fundamental and applied studies in the area of colloid and interface science. As it proved difficult to arrange the reviews in a particular order, they are presented here according to the time the manuscript was received. Before the reviews, a short biography and summary of selected scientific achievements of Tharwat Tadros is given by Brian Vincent, the immediate past president of the International Association of Colloid and Interface Scientists.

We express our gratitude to the invited authors and their co-authors for the preparation of these high level review articles, as well as to the staff of Wiley-VCH for producing this excellent volume

Sofia, February 2008

Dimo Platikanov and
Dotchi Exerowa

Highlights in Colloid Science. Edited by Dimo Platikanov and Dotchi Exerowa
Copyright © 2009 WILEY-VCH Verlag GmbH & Co. KGaA, Weinheim
ISBN: 978-3-527-32037-0

Tharwat F. Tadros

Tharwat F. Tadros was born on 29 July 1937 near Luxor in Upper Egypt, where he grew up and went to school. He subsequently attended the University of Alexandria, where he obtained a first class honors BSc degree in 1956 (aged 19), followed by an MSc degree in 1959. He then pursued his PhD studies in electrochemistry at Alexandria with Professor Sadek. Afterwards, Tharwat was appointed to a lectureship in chemistry at that university. In 1966 he sought a position overseas and began a two-year postdoctoral position with Hans Lyklema in the Agricultural University in Wageningen, The Netherlands. There he met Wikie Buter, a Dutch lady who was working in plant protection at the University. They married in the summer of 1969 in Wassenaar, near Den Haag. Before this happy event, Tharwat had intended returning to his lectureship position in Alexandria after his period in Wageningen, but now he sort a position in Holland, and went to TNO in Delft for two years (1968–1969), where he worked on electrochemical machining. After that he applied to, and was awarded a position with, ICI Plant Protection Division, in Jealotts Hill, Berkshire, in the UK.

During his early days at Jealotts Hill, as well as running a strong product development group, Tharwat was encouraged to make contacts with academics. As a result, Tharwat started his lifelong collaborations with many universities worldwide, initially in Bristol with Ron Ottewill and myself, soon followed by collaborations at Imperial College, with Anita Bailey and Paul Luckham. Subsequently, he worked with academics from many others institutions, including Barcelona (Conxita Solans), Crackow (Piotr Warszynski and Maria Zembala), Groningen (H. Busscher), Liverpool Polytechnic (Alec Smith), Nottingham (Mike Hey and Bob Davies), Reading (Thelma Herrington), Sofia (Dotchi Exerowa and Dimo Platikanov) and of course further collaborations with people from Wageningen.

Over the years, Tharwat built up a strong team of coworkers at Jealotts Hill, including Chris Hart, Peter Wynn, David Heath, Phil Taylor and many others. He has also had many visitors and students working with him in his laboratory. He stayed through the transition to Zeneca Agrochemicals, and, although Tharwat formally retired in 1994, he is still employed as a consultant to this day by Syngenta, who now occupy the site. Indeed his ongoing consulting roles and short courses for a wide range of companies and organizations are legendary.

Highlights in Colloid Science. Edited by Dimo Platikanov and Dotchi Exerowa
Copyright © 2009 WILEY-VCH Verlag GmbH & Co. KGaA, Weinheim
ISBN: 978-3-527-32037-0

Tharwat is the author, or a co-author, of around 270 published papers, eight books (see later) and six patents (as at the end of 2007). Many of these are a result of his ongoing collaborations with academics around the world. Tharwat's PhD work in Alexandria with Professor Sadek was largely concerned with the conduction properties of electrolyte solutions, in particular solutions of acids and their mixtures. As mentioned earlier, his first papers in surface and colloid science, as such, were with Hans Lyklema on the electrical double layer properties of the porous silica/aqueous electrolyte interface. After his move to ICI, one of his major interests became the adsorption of polymers at solid/solution interfaces. My group collaborated extensively with Tharwat in this area at that time. In the 1960s industrial researchers, in particular, had started to study this topic in a systematic manner, using carefully designed, synthetic polymers, especially block or graft copolymers; this early work included researchers at ICI, in particular at the Paints Division. Moreover, at that time, people were beginning to build theories of polymers at interfaces and the steric interaction between polymer-coated particles.

Tharwat and I decided to test some of these emerging ideas using a model aqueous-based system, namely polystyrene particles plus, what is effectively a block copolymer, namely poly(vinyl alcohol-*co*-vinyl acetate) [PVA]. We had our first, ICI-funded, joint PhD student working on this topic, Mike Garvey. Because it was clear even then that polymer molecular weight is an important parameter in determining the properties of the adsorbed polymer layer, such as the adsorbed amount and the adsorbed layer thickness, as well as in steric stabilization, we decided to fractionate a commercial sample of PVA. This Mike achieved using a GPC column. Those early studies by Mike Garvey were followed up by a succession of students, working with Tharwat and myself. Moreover, around 1980 we were joined in this research in Bristol by Terry Cosgrove, and a mathematically gifted young PhD student, seconded from ICI (Runcorn), Trevor Crowley, and we stared to apply the newer techniques of small-angle neutron scattering (SANS) and solid-state NMR to studying polymers at interfaces. Using SANS, thanks largely to Trevor, who was able to solve the complex Laplace transforms involved, we were able to determine the segment density profile of adsorbed polymers normal to the particle surface. Terry's interests in NMR enabled us to determine the bound fraction of segments (i.e. in "trains" as opposed to tails or loops). Much of this early work on polymer adsorption and steric stabilization, inspired by Tharwat, led to a much better practical understanding of how polymers could be used in commercial formulations, such as agrochemicals, to control dispersion stability and flocculation.

The other main field of study that Tharwat and I explored together in those early days was the adsorption of small positive latex particles on much larger negative particles, and how the adsorbed particles could act as particle "bridging" flocculants at low coverage, by analogy with polymer flocculants. It was felt that this might lead to a more "robust" method of achieving controlled, reversible flocculation in particle suspensions – a very important topic for agrochemical dispersions, such as pesticides, herbicides and fungicides, where it is necessary to prevent "claying," that is, the hard bed of closed-packed particles that forms when particles, stable with respect to aggregation, settle slowly with time on standing.

Tharwat's collaborations with Ron Ottewill at Bristol have led to papers on topics as diverse as understanding settling in Newtonian and non-Newtonian media (also with Jim Goodwin and Richard Buscall), and to fundamental studies on microemulsions (also with A.T. Florence).

For the last 20 years or so, however, Tharwat's principle academic collaboration has been with Imperial College. His earlier work with Anita Bailey was focused on electro-spraying, a very important application being the spraying of agrochemical formulations. The work with Paul Luckham, on the other hand, has mostly been concerned with combining Paul's expertise in measuring the forces between surfaces (using the surface forces apparatus or the atomic force microscope) with Tharwat's expertise in rheology. Indeed, Tharwat has established himself as one of the leading experts on what one might call "the practical rheology of industrial formulations."

Two of the main European collaborations Tharwat has had in recent years have been the groups of Dotchi Exerowa in Sofia and that of Conxita Solans in Barcelona. With Dotchi the work has focused on foam and emulsion films and with Coxita (plus Paul Luckham) on the stabilization of latex particles. Indeed in much of this latter work on foams, emulsions, and particle dispersions, very efficient stabilizers, based on hydrophobically-modified, sugar-based polymers, have been studied. It was observed, for example, that, when these are added, polymer latex dispersions are stable in Na_2SO_4 solutions of up to 1.5 mol dm^{-3}.

I am aware that this selection from Tharwat's large research activity has been very selective, so I apologies to anyone where no mention of them, or their work with Tharwat, has been made!

Tharwat has not only contributed widely to the research-base in colloid science, but he has also made major contributions to the broader education not only of younger colloid scientists in universities but also to colloid scientists of all ages, through the very many industrial consultancies, courses and seminars he has given over the years.

His more formal university connections include appointments as a visiting professor at three universities in the UK: Imperial College (1988), Bristol (1988) and Reading (1994). As well as co-supervising PhD students with academics from various institutions, where ICI provided at least partial funding, Tharwat has been an external examiner for PhD students in many places, as well as for MSc courses. For example, he was the external examiner for the MSc advanced course in surface chemistry and colloids at Bristol – twice (1983–1985 and then 1992–1994).

Tharwat has received many invitations to give plenary or keynote lectures. He has given (and is still giving!) specialist courses on rheology, surfactants, emulsions, the dispersion of powders in liquids, wetting, spreading and adhesion, amongst other topics. He has edited six books, and authored two other books: *Surfactants in Agrochemicals* (1994) and *Applied Surfactants* (2005). He has been a senior editor for *Colloids and Surfaces* and for *Advances in Colloid and Interface Science* (both for Elsevier). He is also currently editor of two series published by Wiley-VCH: *Colloids and Interface Science Series* (two volumes published in 2006 and others in the pipeline), and *Topics in Colloid and Interface Science* (commencing 2007). Tharwat has acted as a reviewer and evaluator for many organizations, including the UK government funding body (the Engineering and Physical Science Research Council, EPSRC), the NSF in the USA and the Consultative Board of the University of Alexandria.

Tharwat served as Chair of the UK Society of Chemical Industry (SCI) Surface and Colloid Chemistry section committee (1987–1989) and was elected President of IACIS (1990–1992). All this work, in and for colloid science, has been recognized by the award of a number of medals and named lectureships. These include the UK Royal Society of Chemistry's (RSC) Colloid and Surface Chemistry Medal (1989), the RSC's silver medal and industrial lectureship (1990), and the SCI's Founder's (now the Rideal) lecture (1991).

Tharwat and Wikie live in Wokingham in the South of England, and are very proud parents and grandparents. Wikie has had a very successful life in the art world, with many exhibitions of her work. I know that Tharwat is very proud of all his family, both in the UK and back in Egypt. May he long continue into the future as one of colloid science's "father figures" and a good friend to many of us.

Bristol, December 2007 *Brian Vincent*

List of Contributors

Zbigniew Adamczyk
Institute of Catalysis and Surface
Chemistry
Polish Academy of Sciences,
ul. Niezapominajek 8
30-239 Cracow
Poland

Idit Amar-Yuli
Casali Institute of Applied Chemistry
The Institute of Chemistry
The Hebrew University of Jerusalem
91904 Jerusalem
Israel

Elena A. Amelina[†]
Moscow State University
Department of Chemistry
119899 Moscow
Russia

Martin Andresen
Ugelstad Laboratory
Department of Chemical Engineering
Norwegian University of Science and
Technology
7491 Trondheim
Norway

Abraham Aserin
Casali Institute of Applied Chemistry
The Institute of Chemistry
The Hebrew University of Jerusalem
91904 Jerusalem
Israel

Jakub Barbasz
Institute of Catalysis and Surface
Chemistry
Polish Academy of Sciences,
ul. Niezapominajek 8
30-239 Cracow
Poland

James K. Beattie
School of Chemistry
University of Sydney
Sydney NSW 2006
Australia

Ludmila Boinovich
Russian Academy of Sciences
A. N. Frumkin Institute of Physical
Chemistry and Electrochemistry
31 Leninsky prosp
119991 Moscow
Russia

Highlights in Colloid Science. Edited by Dimo Platikanov and Dotchi Exerowa
Copyright © 2009 WILEY-VCH Verlag GmbH & Co. KGaA, Weinheim
ISBN: 978-3-527-32037-0

Melanie Bradley
School of Chemistry
University of Bristol
Bristol BS8 1TS
UK

Diana Costa
Physical Chemistry 1
Centre for Chemistry and Chemical Engineering
Lund University
P.O. Box 124
22100 Lund
Sweden
and
Department of Chemistry
University of Coimbra
3004-535 Coimbra
Portugal

Paul Davies
School of Chemistry
University of Bristol
Bristol BS8 1TS
UK

Rita S. Dias
Physical Chemistry 1
Centre for Chemistry and Chemical Engineering
Lund University
P.O. Box 124
22100 Lund
Sweden

Alexandre Emelyanenko
A. N. Frumkin Institute of Physical Chemistry and Electrochemistry
Russian Academy of Sciences
31 Leninsky prosp
119991 Moscow
Russia

Jordi Esquena
Institut de Química Avançada de Catalunya (IQAC)
Consejo Superior de Investigaciones Científicas (CSIC)
Jordi Girona 18-26
08034-Barcelona
Spain

Dotchi Exerowa
Institute of Physical Chemistry
Bulgarian Academy of Sciences
1113 Sofia
Bulgaria

Valentin B. Fainerman
Medical Physicochemical Centre
Donetsk Medical University
83003 Donetsk
Ukraine

Gerard J. Fleer
Laboratory of Physical Chemistry and Colloid Science
Wageningen University
6703 HB Wageningen
The Netherlands

Nissim Garti
Casali Institute of Applied Chemistry
The Institute of Chemistry
The Hebrew University of Jerusalem
91904 Jerusalem
Israel

Andreas Hannisdal
Aibel AS
Technology and Products
1375 Billingstad
Norway

List of Contributors | XIX

Robert J. Hunter
School of Chemistry
University of Sydney
Sydney NSW 2006
Australia

Ana B. Jódar-Reyes
Universidad de Extremadura
Dpto. de Física Aplicada
Facultad de Veterinaria
Avda. de la Universidad s/n
10071 Cáceres
Spain

Frans A.M. Leermakers
Wageningen University
Lab. of Physical Chemistry and
Colloid Science
Dreijenplein 6
6703 HB Wageningen
The Netherlands

Simone Less
Ugelstad Laboratory
Department of Chemical Engineering
Norwegian University of Science and
Technology (NTNU)
7491 Trondheim
Norway

Dima Libster
Casali Institute of Applied Chemistry
The Institute of Chemistry
The Hebrew University of Jerusalem
91904 Jerusalem
Israel

Björn Lindman
Physical Chemistry 1
Centre for Chemistry and
Chemical Engineering
Lund University
P.O. Box 124
22100 Lund
Sweden
and
Department of Chemistry
University of Coimbra
3004-535 Coimbra
Portugal

Johannes Lyklema
Wageningen University
Lab. of Physical Chemistry and
Colloid Science
Dreijenplein 6
6703 HB Wageningen
The Netherlands

Heléne Magnusson
Ugelstad Laboratory
Department of Chemical Engineering
Norwegian University of Science and
Technology (NTNU)
7491 Trondheim
Norway

M. Graça Miguel
Department of Chemistry
University of Coimbra
3004-535 Coimbra
Portugal

Reinhard Miller
Max-Planck-Institut für Kolloid- und
Grenzflächenforschung
14424 Potsdam
Germany

M. Carmen Morán
Department of Chemistry
University of Coimbra
3004-535 Coimbra
Portugal

Małgorzata Nattich
Institute of Catalysis and Surface
Chemistry
Polish Academy of Sciences,
ul. Niezapominajek 8
30-239 Cracow
Poland

Boris A. Noskov
St. Petersburg State University
Chemical Faculty
198904 St. Petersburg
Russia

Richard W. O'Brien
School of Chemistry
University of Sydney
Sydney NSW 2006
Australia

Aksana M. Parfenova
Moscow State University
Department of Chemistry
119899 Moscow
Russia

Jordon T. Petkov
Unilever R&D Port Sunlight
Quarry Road East
Bebington CH63 3JW
UK

Dimo Platikanov
Department of Physical Chemistry
University of Sofia
1164 Sofia
Bulgaria

Venkataramana Runkana
Tata Research Development and
Design Centre
A Division of Tata Consultancy
Services Limited
54-B, Hadapsar Industrial Estate
Pune 411013
India

Eugene D. Shchukin
Moscow State University
Department of Chemistry
119899 Moscow
Russia
and
Institute for Physical Chemistry and
Electrochemistry
Russian Academy of Sciences
117915 Moscow
Russia
and
Department of Geography
Johns Hopkins University
2704 Hanson Avenue
Baltimore MD 21218
USA

Johan Sjöblom
Ugelstad Laboratory
Department of Chemical Engineering
Norwegian University of Science and
Technology (NTNU)
7491 Trondheim
Norway

Conxita Solans
Institut de Química Avançada
de Catalunya (IQAC)
Consejo Superior de Investigaciones
Científicas (CSIC)
Jordi Girona 18-26
08034-Barcelona
Spain

Ponisseril Somasundaran
NSF Industry/University Cooperative
Research Center for Advanced Studies
in Novel Surfactants
School of Engineering and Applied
Science
Columbia University
New York NY 10027
USA

Per Stenius
Ugelstad Laboratory
Department of Chemical Engineering
Norwegian University of Science and
Technology
7491 Trondheim
Norway
and
Laboratory of Forest Products
Chemistry
Helsinki University of Technology
P.O. Box 6300
02015 TKK
Finland

Remco Tuinier
Forschungszentrum Jülich
IFF-Soft Matter group
52425 Jülich
Germany
Current affiliation:
DSM Research
ACES
P.O. Box 18
6160 MD Geleen
The Netherlands

Theo G.M. van de Ven
NSERC/Paprican Industrial Research
Chair in "Colloid and Papermaking
Chemistry"
Pulp & Paper Research Centre
Department of Chemistry
McGill University
Montreal
Canada

Brian Vincent
School of Chemistry
University of Bristol
Bristol BS8 1TS
UK

1
Orthokinetic Heteroflocculation in Papermaking
Theo G.M. van de Ven

1.1
Introduction

In modern papermaking many colloidal particles are added to a papermaking suspension consisting of pulp fibers and fines. Examples of such particles are fillers (clays, calcium carbonates, titanium dioxide, etc.), internal sizing agents, which modify the hydrophobicity of paper [wood resins or solidified emulsion droplets, consisting of reactive sizing agents such as alkyl ketene dimer (AKD) or alkenyl succinic acid (ASA)], or color pigments [1, 2]. Other colloids that might be present are colloidal pitch, liberated from the wood in the pulping stage, or residual ink particles, arising from incomplete deinking of recycled paper. To incorporate such particles in paper requires retention agents that in essence act as glue, sticking the particles to fibers or flocculating them into large aggregates that are captured in the pores of a forming sheet. Examples of polymeric retention aids are polyethylene imine (PEI), cationic polyacrylamides (cPAM), poly(ethylene oxide)/cofactor (PEO/CF) systems, poly(diethyl-dimethylammonium chloride) (PDADMAC) and starch [3]. Often, additional polymers are added that function as dry or wet strength agents. All these colloids and polymers interact with the pulp fibers and fines, resulting in homoflocculation, heteroflocculation and coflocculation. Certain types of flocculation are desirable, others are not. For instance, fiber flocculation (an example of homoflocculation) is undesirable as this leads to poor paper uniformity, whereas deposition of fillers on fibers (an example of heteroflocculation) is desirable, since fillers in paper improve paper properties and reduce costs. Consequently, papermaking requires selective flocculation among the many components that are present [4].

In addition, colloids can be added in paper surface treatments, such as surface sizing or coating. In coating, often, mixtures of pigments and binders [e.g. polystyrene butadiene (PSB) latex] are used, together with polymers, such as rheology modifiers, which can also lead to flocculation. Clearly, from these examples, flocculation is an important phenomenon in papermaking, especially heteroflocculation, since most ingredients are added to interact with the pulp fibers. In this chapter we highlight some examples of heteroflocculation occurring in papermaking, namely the deposition of fillers on fibers [5–8], the deposition of solidified sizing droplets on fibers [9], the heteroflocculation of fillers with such droplets [10], the heteroflocculation of clay and latex, which could either happen on a paper machine when coated broke is returned to the papermaking process or in coating applications, and, finally, the heteroflocculation of fines and clay [11]. Papermaking is a fast process, with speeds typically in the range $10–30\,\mathrm{m\,s^{-1}}$. This implies that all interactions must take place in a very short time and are subject to the high shear forces operating in a papermaking suspension traveling through a paper machine. Thus, hydrodynamic particle interactions play a crucial role in papermaking. Because, in most interactions, polymers or polyelectrolytes are involved as well in papermaking, we are mainly dealing with polymer-induced orthokinetic heteroflocculation.

1.2
Polymer-Induced Orthokinetic Heteroflocculation

As an example let us consider the orthokinetic (shear-induced) deposition of colloids on fibers, induced by polymers or polyelectrolytes. Usually, the colloids are stable prior to the addition of polymer. The pulp fibers form flocs also in the absence of polymers due to mechanical entanglements. These flocs can be broken up by high shear, resulting in a dynamic equilibrium between floc formation and break-up. The higher the shear, the fewer is the number of fibers in flocs. Polymers can shift this dynamic equilibrium towards an increased flocculation [12]. Usually at various locations on a paper machine the shear is sufficiently high to break up most fiber flocs, which reform in regions of lower shear. Here we will assume that fibers are well-dispersed due to high shear.

When polymers are added to a mixture of fibers and colloids subjected to high shear, several processes can take place:

1. Polymer adsorption on fiber;
2. Polymer adsorption on colloids;
3. Polymer-induced flocculation of fibers;
4. Polymer-induced flocculation of colloids;
5. Polymer-induced deposition of colloids on fibers;
6. Polymer transfer from fibers to colloids;
7. Polymer transfer from colloids to fibers.

Some of these processes can possibly be reversible, resulting in polymer desorption from fibers and colloids, break-up of fiber flocs and colloidal aggregates and detachment of colloids from fibers. Which of these processes actually occurs depends on the characteristic times of these processes. To be able to make estimates of these times, we will discuss some of these processes in more detail.

1.2.1
Polymer Adsorption and Desorption on Fibers and Colloids

The initial polymer adsorption on surfaces can be modeled reasonably well by considering the polymer coils as tiny spheres. If the kinetics of adsorption are faster than polymer rearrangement (reconformation) on the surface, the surface will quickly be covered by a monolayer of random coils (provided there is enough polymer). Thus an apparent plateau in polymer adsorption will be reached, which is transient because more polymer can adsorb when the polymers rearrange, and for fibers that have a porous wall the polymer can adsorb in the interior of the fibers. However, since papermaking is a fast process, considering the initial adsorption is usually sufficient. For times shorter than the polymer rearrangement time on the surface, the kinetics of adsorption can be fairly well described by Langmuir kinetics [13]:

$$\frac{d\theta_p}{dt} = k_{ads}(n_o - \theta_p)(1 - \theta_p) - k_{des}\theta_p \tag{1.1}$$

Here t is time; $\theta_p = \Gamma/\Gamma_{max}$ is the fractional coverage of polymer on the surface of the particles, with Γ being the amount of polymer adsorbed and Γ_{max} that corresponding to (apparent) monolayer coverage; n_o is the initial polymer concentration, relative to monolayer coverage; and k_{ads} and k_{des} are the adsorption and desorption rate constants, respectively. Equation 1.1 shows that polymer adsorption is initially linear in time and reaches a plateau at long times. The characteristic adsorption time, τ_{ads}, defined as the time where the linear increase intersects the plateau value, and the characteristic desorption time, τ_{des}, the average time a polymer spends on the surface, are:

$$\tau_{ads} = \frac{1}{k_{ads}n_o}; \quad \tau_{des} = \frac{1}{k_{des}} \tag{1.2}$$

The dimensionless polymer concentration can be expressed, for random adsorption of spherical Gaussian coils of radius a_p on fibers, as:

$$n_o = \frac{10^3 \pi a_p^2 N_{Av}}{0.55 M A_F} c_p \tag{1.3}$$

Here c_p is the polymer concentration (kg m^{-3}), N_{Av} is Avogadro's number, M the molecular weight of the polymer and A_F is the area of fiber per unit volume. In papermaking, fibers are typically 1–3 mm long with a diameter of 20–30 µm and their specific surface is about 1 m^{-1}, with fiber concentrations typically 10 kg m^{-3}, and thus A_F is of order 10^4 m^2 m^{-1}. Notice that for ideal coils n_o is independent of the molecular weight. For typical polymer concentrations of 1 g m^{-3}, and for $M = 5$ MDa and $a_p = 100$ nm, $n_o = 0.7$.

Since we are considering shear-induced adsorption, the adsorption rate constant can be approximated from Smoluchowski's theory [14] for the collision between two

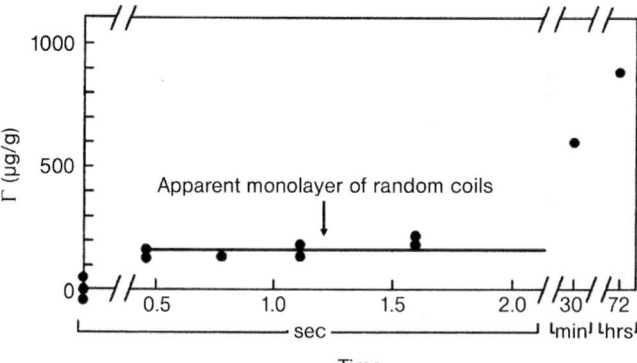

Figure 1.1 Adsorption kinetics of PDADMAC adsorption on pulp fibers subjected to high shear (adapted from Ref. [16]).

unequal-sized spheres in shear, modified by Petlicki and van de Ven [15] for collisions between a large spheroid and a small sphere:

$$k_{ads} = \frac{\gamma G \varphi}{\pi} \tag{1.4}$$

Fibers can be modeled as long slender spheroids. Here G is the effective shear rate the suspension is subjected to, γ the fast collision efficiency and φ is the volume fraction of particles. The factor γ accounts for differences between the approximate theory of Smoluchowski (in which $\gamma = 1$) and the full hydrodynamic theory. Equation 1.4 is valid when the polymer is much smaller than the particle. Filler levels in papermaking are usually less than 30% by weight (although there is a trend towards higher filled papers), and because the density of fillers is about twice that of fibers and the fibers are hollow, which increases their effective volume fraction, the volume fraction of fibers is typically an order of magnitude larger than that of fillers. From Equations 1.4 and 1.2 we can conclude that most polymers first adsorb on fibers, prior to adsorption on fillers. Polymer dosages are usually below a monolayer coverage on fibers (with the exception of strength agents), which implies that at the end of the adsorption process most polymer is on the fibers and little is on the fillers.

The validity of Equation 1.1 can be judged from experimental polymer adsorption studies. Figure 1.1 shows results from fast PDADMAC adsorption on pulp fibers, subjected to high shear in a flow loop [16]. An apparent plateau is reached at times less than 0.5 s. This plateau was independent of polymer dosage and corresponds to a monolayer of random coils on the fiber surface. At much longer times additional polymer adsorption takes place, mainly due to polymer penetration in pores.

Figure 1.2 gives another example of the kinetics of polymer adsorption – this time the adsorption of PEI on pulp fibers [17]. The curve for the lowest PEI concentration is fitted to Equation 1.1, with $n_0 = 3.5$, $k_{ads} = 0.2 \, \text{min}^{-1}$ and $k_{des} = 0$. The curves for higher PEI concentrations were modeled by assuming that a fraction of the poly disperse PEI is small enough to penetrate the pores or the lumen. At higher PEI concentrations, more low-molecular weight PEI is available for pore penetration [18].

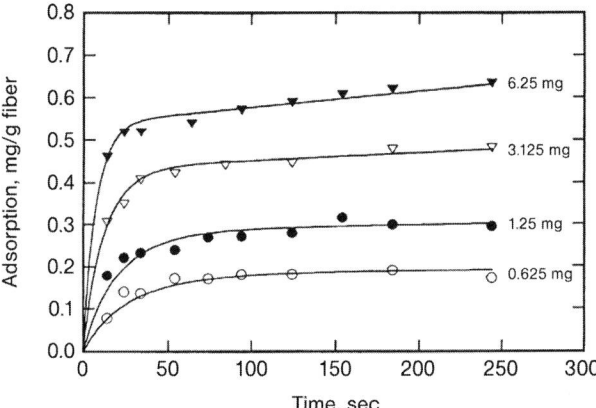

Figure 1.2 PEI adsorption on pulp fibers subjected to shear in a beaker, for various PEI dosages shown in the figure (mg g^{-1} fiber); pH 6 (after Ref. [17]).

Polymer desorption is often negligible, especially for high molecular weight retention aids. Globular polymers have a larger tendency to desorb: PEI was found to desorb from pulp fibers after increasing the ionic strength [18]; cationic starch was found to be present in the form of clusters, which were easier to desorb from the fibers than well-dispersed starch, likely because of large hydrodynamic forces acting on the large clusters [19].

Some polymers, such as PEO, do not adsorb on pulp fibers. PEO adsorption on fibers can be triggered by a cofactor, by a process we call association-induced adsorption [20].

1.2.2
Deposition of Colloids on Fibers Subjected to Shear

As soon as polymer has adsorbed on the fibers, creating patches of polymer on the fibers, the colloidal particles can deposit on these polymer patches. The kinetics of colloid deposition on fibers are very similar to those of polymer adsorption on fibers, discussed above. It can be described by [13]:

$$\frac{d\theta}{dt} = k_1(n_o^c - \theta)(1-\theta) - k_2\theta \quad (1.5)$$

Here θ is the fractional coverage of fibers by colloids, n_o^c the initial colloid concentration relative to monolayer coverage on fibers, and k_1 and k_2 are the deposition and detachment rate coefficients, respectively. In Equation 1.5 it is assumed that blocking by deposited particles can simply be described by the probability $(1-\theta)$ that a particle deposits on a bare area on the fiber. More complex blocking functions have been formulated [21], but for particle deposition on fibers this simple model adequately describes the experimental data (see below). The deposition rate coefficient can be expressed as:

$$k_1 = \frac{\alpha \gamma G \varphi}{\pi} \tag{1.6}$$

For fibers the collision efficiency γ depends on the size ratio of colloid and fiber diameter and on the ratio of colloidal (van der Waals) to hydrodynamic forces, but does not depend on fiber length [15]. The term α is the deposition efficiency, given according to the classical polymer bridging model [22] by:

$$\alpha = \theta_1(1-\theta_2) + \theta_2(1-\theta_1) \simeq \theta_p \tag{1.7}$$

Here θ_i are the fractional coverages of the polymer on the fibers and the colloids, and since most polymer adsorbs on the fibers $\theta_2 \approx 0$ and $\theta_1 = \theta_p$. Equation 1.7 is only valid when the retention aid functions by polymer bridging, as is usually the case. It does not apply for highly charged cationic polyelectrolytes, which function by charge modification, in which case α is determined by the height of the energy barrier, and reaches 1 when the fibers reach their point of zero charge.

Assumptions made in writing Equation 1.7 are that collisions between a bare patch and a polymer patch lead to deposition and those between polymer patches or between bare patches do not. This is plausible when, during a collision, the particle trajectory is determined mainly by the attractive force between a bare patch and a polymer patch. Collisions in shear are more complicated, as rotating particles are surrounded by fluid that rotates with it, leading to trajectories that can orbit a fiber several times before colloid–fiber contact occurs [23]. When particles are partially coated by polymer, they will experience fluctuating colloidal interaction forces, depending on where the patches are located. Nevertheless, also in such cases $\alpha = 0$ when polymer is absent and $\alpha = 1$ when fibers are fully coated (and colloids not). Thus, also in this case Equation 1.7 is probably not a bad approximation.

The respective characteristic times of deposition and detachment are:

$$\tau_{dep} = \frac{1}{k_1 n_o^c}; \quad \tau_{det} = \frac{1}{k_2} \tag{1.8}$$

The dimensionless colloid concentration can be expressed, for random deposition of spheres of radius a, or disks of thickness h, on fibers, as:

$$n_o^c = \frac{3c}{2.2 a A_F \rho}; \quad n_o^c = \frac{c}{0.55 h A_F \rho} \tag{1.9}$$

Here c is the concentration of colloids (kg m^{-3}) and ρ is their density. For a typical loading level of 20% (wt filler/wt fibers), n_o^c is about 0.1 for a 1 micron radius particle and 1.3 for plate-like clay particles 100 nm thick. For close packing on the surface, the factor 0.55 in Equation 1.9 has to be replaced by 0.9 [24], and the right-hand side of Equation 1.9 must be multiplied by 0.61.

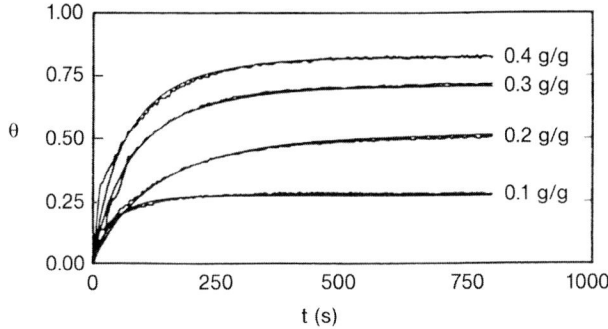

Figure 1.3 Deposition kinetics of $CaCO_3$ particles on pulp fibers subjected to shear for various dosages (indicated in the figure). Wiggly lines are experimental observations and smooth curves are best fits to Equation 1.5, with $k_1 = 1.4 \times 10^{-2}\,s^{-1}$ and $k_2 = 1.1 \times 10^{-3}\,s^{-1}$ (after Ref. [4]).

Because the polymer adsorption is close to the particle deposition time (cf. Equations 1.2 and 1.8), both processes take place simultaneously. However, colloids can only deposit on polymer patches on the fibers, that is, after polymer adsorption has occurred. Notice that Equation 1.7, combined with Equation 1.1 for polymer adsorption predicts that at short times the deposition increases quadratically ($\theta \propto t^2$). When polymers are added before the fillers, θ_p and thus the deposition efficiency is constant and does not depend on time; in such cases the initial kinetics are linear in time.

Equation 1.7 has been validated for several particles depositing on fibers, such as $CaCO_3$ particles, either non-treated or coated by cPAM [4] or by poly(propylene imine) dendrimers [6], for positive clay particles depositing on negative fibers, for negative clay particles depositing on positive (PEI-treated) fibers [25, 26], for the deposition of fines on fibers [27], for the deposition of latex particles and for the deposition of cationic and anionic AKD particles on cPAM-coated fibers [9]. Figure 1.3 shows the results for the deposition of positively charged $CaCO_3$ particles, in the absence of any retention aids, together with comparisons with Equation 1.5. In the calculation for n_0, close packing was assumed, since particle detachment will lead to denser than random packing.

For $CaCO_3$ particles, particle detachment leads to a dynamic equilibrium between deposition and detachment. In the absence of retention aids the bond between colloids and fibers is weak and retention on a paper machine is low.

Figure 1.4 presents results for cationic latex particles for various particle sizes. Because the addition level was the same for all latexes, this results in different dimensionless latex concentrations, which are expressed relative to close packed monolayers. The figure shows that the deposition is a strong function of particle size – in agreement with theoretical predictions for model systems. This is because the hydrodynamic collision efficiency γ depends strongly on the size ratio of colloid to fiber diameter. Small particles tend to go around fibers in a large orbit, resulting in large minimum separation distances, for which colloidal forces are weak. Large

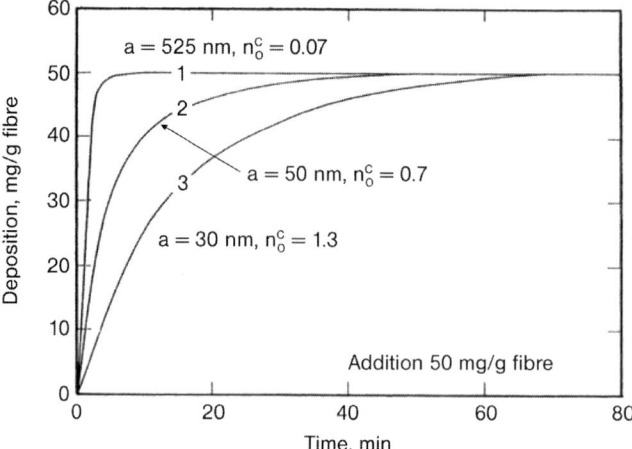

Figure 1.4 Deposition of cationic latex particles of various sizes on pulp fibers subjected to low shear in a beaker (adapted from Ref. [28]).

particles can approach the fibers much closer, resulting in much larger colloidal attractive forces. For the data in Figure 1.4, $k_1 = 0.37$, 0.15 and 0.040 min^{-1} for the three particle sizes used ($k_2 = 0$). Because all experiments were performed at the same shear conditions, γ is more than nine times larger for the largest particle compared to the smallest one.

Notice that for the smallest latex $n_o^c = 1.3$, whereas all particles deposit. This implies that some of the particles entered the macropores of the fiber wall (estimated to have pores of about 80 nm), which have an accessible surface area about 15 times that of the external surface [29]. Interestingly, particle deposition on fibers provides a method to fractionate particles of different sizes. Adding a mixture of small and large particles to fibers and subjecting the system to shear for a short time (a few minutes) will result in the deposition of most or all large particles, with little deposition of small particles. The fibers with the large particles can be readily separated from the mixture, leaving the small particles suspended in the liquid.

A SEM of a fiber fully coated by polystyrene latex particles is shown in Figure 1.5a [30]. By annealing the fibers with latex above the glass transition temperature of the latex, a smooth latex film surrounding the fibers is obtained (Figure 1.5b). Such treatment leads to fibers fully surrounded by latex. The thickness of the latex layer can be controlled by controlling the latex particle size during deposition. Such fibers might be useful, for example, in fiber–polymer composites.

Figure 1.6 gives another example of colloid deposition on fibers, for negative AKD particles (stabilized by CMC, carboxymethyl cellulose), using cPAM as a retention aid [9]. Little deposition occurs in the absence of cPAM, and the initial deposition rate increases with increasing cPAM concentrations, as predicted by Equation 1.7, till $\alpha = \theta_p = 1$, above which the deposition no longer depends on cPAM dosage. At excess cPAM dosages, the colloids become coated as well but, since the fibers are coated by cPAM before the colloids, no more deposition takes place.

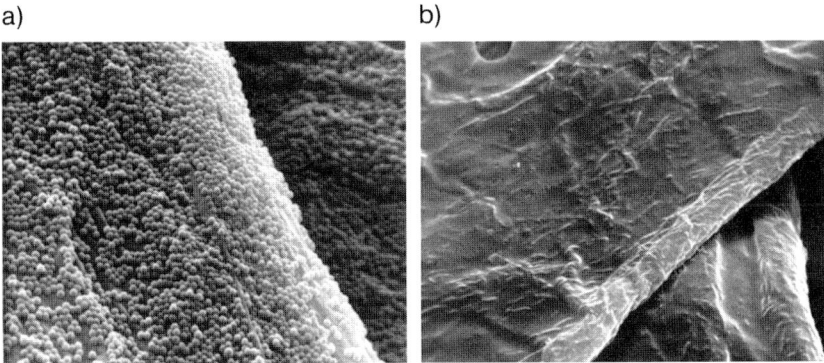

Figure 1.5 (a) Part of a fiber, fully coated by polystyrene latex particles; (b) after annealing.

In the last two examples, particle detachment was absent ($k_2 = 0$). This is because of the relatively low shear in laboratory experiments, which is not typical of papermaking, where usually a dynamic equilibrium is established between deposition and detachment. Often, even in the absence of retention aids, particle deposition on fibers occurs (e.g. when electrostatic repulsion is screened by the high ionic strength of process water or when positive particles such as precipitated $CaCO_3$ are used), but the bond strength between the colloid and fiber is weak, resulting in large detachment rate constants. The role of a retention aid is then mainly in strengthening the bond, thus reducing the detachment rate constant.

As we have seen, in most cases polymer adsorbs preferentially on fibers. This is obviously not true when the polymer does not adsorb on the fiber, in which case only adsorption on colloids occurs. This is the case for PEO, which does not adsorb on

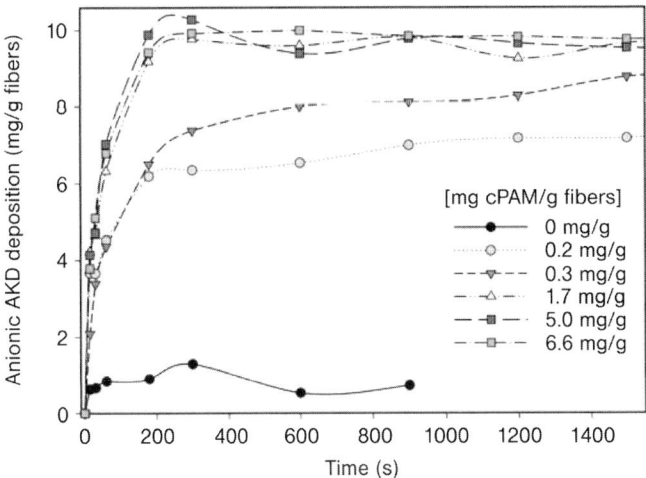

Figure 1.6 Deposition of AKD particles stabilized by CMC on pulp fibers induced by cPAM as a function of time, for various cPAM dosages. [AKD] = 10 mg g^{-1} fiber (after Ref. [9]).

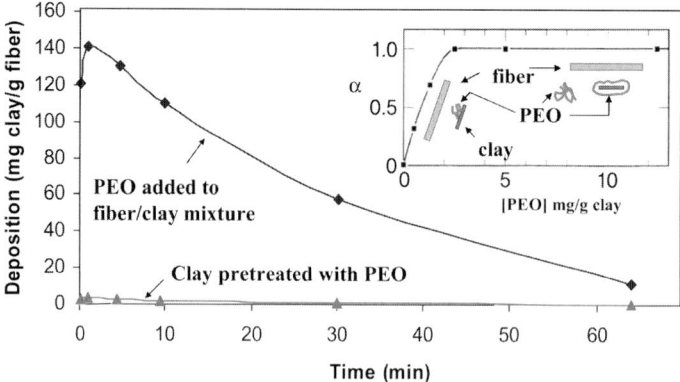

Figure 1.7 Deposition of clay on fibers, induced by PEO. In the absence of PEO or for clay pretreated with PEO for 30 min or longer, no deposition takes place. Because PEO does not adsorb on fibers, this is an example of asymmetric polymer bridging, in which PEO first adsorbs on clay, prior to clay deposition on fibers. Deposition efficiencies observed for various PEO dosages are shown in the inset and agree with theoretical predictions. With excess PEO, deposition is independent of PEO dosage, since excess PEO does not adsorb on fibers (adapted from Ref. [31]).

most types of fibers. Interestingly, after the PEO adsorbs on the colloids, the colloidal particles can deposit on the fibers, a process we call asymmetric polymer bridging [31]. Polymer adsorption is a competition between enthalpy, which favors adsorption, and entropy which tries to prevent it. When adsorbing on colloids, the polymer has attained a new conformation with lower entropy and the enthalpy gain of adsorption on fibers is now sufficient to overcome any additional entropy loss. As an example of asymmetric polymer bridging Figure 1.7 shows kaolin clay deposition on fibers by PEO. Without PEO, no deposition occurs. In asymmetric bridging the deposition efficiency equals: $\alpha = \theta_2$, θ_2, being the polymer coverage on the colloidal particles. One thus predicts that the deposition efficiency increases with polymer dosage and reaches a maximum when all the colloids are fully coated by PEO. This is indeed observed (inset of Figure 1.7).

An additional interesting observation is that the deposition is transient and the particles leave the fibers with time. This is due to a flattening of PEO, which increases the electrostatic repulsion between fibers and clay particles. Adding PEO to clay prior to adding it to fibers, and allowing time for the PEO to flatten onto the clay, leads to no deposition (Figure 1.7).

1.2.3
Polymer Transfer

As we have seen, when adding polymer to a mixture of fibers and colloids subjected to shear, most polymer ends up on the fibers, because the effective volume fraction of fibers is usually much larger than that of colloids. However, polymer can also be

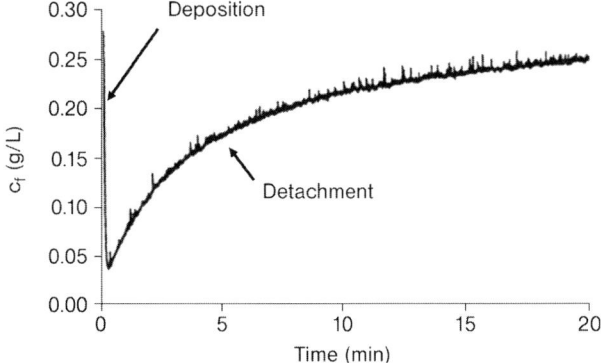

Figure 1.8 Deposition of papermaking fines on pulp fibers coated by cPAM, expressed as the variation in the concentration of fines in the supernatant. Owing to polymer transfer the deposition efficiency decreases from 1 to about 0.2, resulting in a net detachment of fines (adapted from Ref. [32]).

transferred when colloidal particles detach from fibers, taking polymer with them in the process. This process of polymer transfer is usually much faster than direct adsorption. Figure 1.8 gives an example of polymer transfer, showing the deposition and subsequent detachment of papermaking fines from fibers pretreated with cPAM [32]. The initial deposition (expressed as a reduction in the concentration of fines in the supernatant) is consistent with the predictions of Equation 1.6 with $\alpha = 1$. However, the theory predicts that the deposition should reach a plateau at long times. Instead it goes through a maximum, with fines leaving the surface of the fibers until, at long times, most of the fines are resuspended freely. This is because when fines detach from fibers they take cPAM with them, resulting in a redistribution of cPAM over the fibers and fines. This take-up was confirmed by fines electrophoresis measurements, which showed a reduction in negative charge at the end of the deposition experiment [32]. Because the surface area of fines is about 10 times that of fibers [33], about 90% of cPAM ends up on the fines and 10% on fibers; hence, after polymer transfer, $\theta_1 \approx \theta_2 \approx 0.1$ and $\alpha \approx 0.18$. Thus, the deposition efficiency decreases with time due to polymer transfer. Even when the detachment rate constant is not altered, this will lead to a reduction of fines on the fibers (Figure 1.8).

Because, to begin with, only the fibers were coated and the fines were not, initially transfer occurred from fibers to fines. With time, as more and more polymer ends up on the fines, the transfer of polymer from fines to fibers will take place as well. For fillers treated with polymer before the fibers, the initial transfer is from fillers to fibers. How the polymer is distributed over the fibers and fillers depends on their respective surface areas and the affinities of the polymer for the two types of surfaces [33].

Polymer transfer is important in papermaking. For instance, if one wants to operate a paper machine in such a way that filler aggregation is absent, it is not sufficient to treat the fiber with a retention aid. Fillers will indeed deposit on such fibers, but filler detachment from fibers will lead to fillers partially coated by polymer, due to transfer. Not all these fillers will be retained in the sheet during papermaking.

The non-retained fillers end up in the whitewater that is returned to the paper machine. In the short circulation loop, these fillers can flocculate and later can become incorporated as aggregates in the sheet. To avoid this, one must operate the paper machine at a very high first pass filler retention, so that few fillers end up in the whitewater. If one wants a high level of filler loading in the sheet, filler aggregation is desirable and one can operate the paper machine with a low first pass filler retention, giving fillers with transferred polymer time to flocculate in the short circulation loop.

1.2.4
Time Dependence of Deposition and Detachment Rate Coefficients

We have seen that many processes can lead to a time dependence of the deposition and detachment rate coefficients, such as polymer transfer and polymer rearrangement. In addition, in papermaking the shear is continuously changing as a fiber suspension passes through pipes, screens, pumps, headbox, slice and drainage section, which also leads to a time dependence of these rate coefficients. Some changes are reversible, most are not. An example of a reversible change is the deposition and flocculation of particles by cPAM, after enough time has past to reach a steady state distribution of polymer on all surfaces. By increasing the shear, particles detach from fibers and aggregates break up, because detachment and break-up rate constants increase with shear more strongly than deposition or flocculation rate constants. When the shear is reduced, the deposition and flocculation return to their previous level. The shear dependence of the extent of flocculation is of crucial importance in papermaking. Flocs usually break up according to their size, because the hydrodynamic forces exerted on them increase proportionally with surface area. Thus flocs of fibers are more easily broken up than flocs of fines or fillers, whereas the detachment of fillers from fibers is the most difficult, because of the small size of the filler. These differences allow the retention of fillers on fibers to occur with minimal fiber floc formation, making it possible to have at the same time both good retention and good paper uniformity. It follows that it is important to have the right time-dependence of deposition and detachment rate constants. If polymers are not sufficiently strong to keep the particles attached to fibers under high shear, one can make use of microparticulate retention aid systems, usually consisting of a cationic polymer (such as cPAM or cationic starch) and a nanocolloid (such as montmorillonite or silica). The nanocolloids form a bridge between two polymer patches on two particles, thus bridging them together, and in the process modifying both the detachment rate constant and the deposition rate constant. An example is shown in Figure 1.9, for the deposition of $CaCO_3$ particles on fibers by cPAM, added first, followed by bentonite (whose main ingredient is montmorillonite, consisting of 0.9 nm thin platelets) [8]. After cPAM addition the amount of fillers on the fibers reaches about 75 mg g^{-1} (after a small overshoot), which corresponds to a dynamic equilibrium between deposition and detachment. After 2 min the shear was increased to 5000 rpm, resulting in much reduced deposition, which tends toward a new equilibrium with a larger deposition and a much larger detachment rate constant. When subsequently bentonite is added, the detachment rate is much

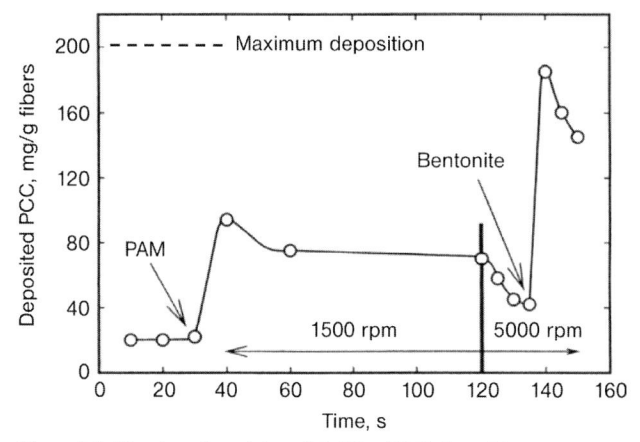

Figure 1.9 Kinetics of precipitated CaCO$_3$ (PCC) deposition on pulp fibers, with cPAM (added first) and bentonite, added after an increase in shear. The addition mimics that on a paper machine where bentonite is added last, near the headbox, under high shear conditions (adapted from Ref. [8]). PCC addition was 200 mg g^{-1} fiber.

reduced, leading to a large increase in deposition (close to 100%). With time the detachment rate constant decreases, possibly due to a poisoning of bentonite with cPAM. This example shows that with the right chemical additives the kinetics of deposition and detachment can be controlled and optimized for a particular mill application.

For some polymers, such as PEO, the detachment rate increases with time even at a fixed shear rate (cf. Figure 1.7). This also happens when PEO is used with a cofactor, although in this case the decrease is slowed down. This type of deposition is irreversible (because in a flattened conformation, PEO is unable to bridge particles). Reducing the shear will not lead to a redeposition. Increasing the shear will accelerate the detachment.

1.3
Heteroflocculation Among Colloids

A papermaking suspension can contain many types of colloids. Many additives are added in the form of colloids, such as fillers, internal sizing agents and color pigments. As already mentioned, other colloids can come from ingredients in the wood, such as pitch, or from residual particles left over from recycled paper or from reslushed broke, which might contain clay and latex particles (so-called white pitch). Clearly, all these particles will interact hydrodynamically and colloidally. Most colloidal particles in a papermaking suspension are negatively charged, but positively charged particles can be present as well, such as PCC (precipitated calcium carbonate) and solidified sizing droplets, coated by cationic starch. PCC loses it positive charge when exposed too long to process water [34], whereas starch-stabilized sizing agents appear to retain their positive charge [10]. Most other colloids are negative, such as ground calcium carbonate (GCC), pitch, ink particles, most latexes, and so on. Some particles can be positive or negative, depending on pH, such as clays or titanium dioxide. At papermaking pHs these particles are commonly negatively charged, although clays are amphoteric, consisting of plates with negative faces and positive edges. It is commonly believed that negatively charged particles heteroflocculate with positive particles but, surprisingly, this is not always the case. An example is the interaction between negatively charged calcium carbonate and solidified AKD droplets, stabilized by cationic starch. When these particles interact, cationic starch is transferred from AKD to $CaCO_3$, resulting in starch-induced homoflocculation of $CaCO_3$ and homoflocculation of AKD, because of the loss of its stabilizer. Little or no heteroflocculation occurs [10]. Transient doublets of $CaCO_3$ and AKD are formed, but these are broken up in shear, transferring starch from AKD to $CaCO_3$ in the process.

Heteroflocculation can lead to several types of aggregates, depending on the size ratio a_2/a_1 of the particles and the relative particle concentrations c_i. When $a_2 \ll a_1$, heteroflocculation is similar to the deposition of small particles on fibers (cf. Figure 1.10a), with the difference that the small particles can induce flocculation of the large ones at partial coverages. In principle this can occur for pulp fibers as well, but because the fibers are large and subjected to shear the hydrodynamic forces are large enough to break up doublets of fibers. For much smaller colloids, this might not happen.

Even when the size ratio is of order unity, one type of particle can stabilize the other by coating it completely, provided $c_2 \gg c_1$ (Figure 1.10b). When the sizes and concentrations are similar, large flocs can be formed in which particles of type 1 and type 2 alternate (Figure 1.10c).

When $n_2 \gg n_1$ aggregates of type a or b in Figure 1.10 will be formed and the heteroflocculation will stop when particles of type 1 are fully coated by particles of type 2. The kinetics of heteroflocculation can then be well described by Equation 1.5, with θ being the fractional coverage of the particles of type 1 by the other ones (type 2) and n_0 being the dimensionless concentration of the particles in excess (type 2), relative to

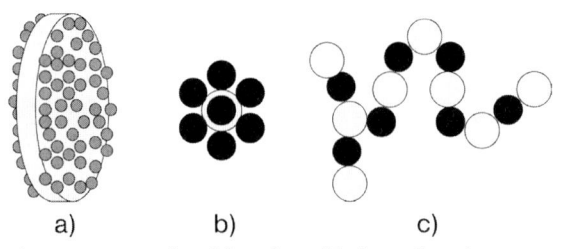

Figure 1.10 Examples of flocs formed by heteroflocculation. (a) Large disk-shaped particle coated by small colloids; (b) colloid of type 1 (gray) surrounded by colloids of type 2 (black); (c) floc consisting of alternating particles of types 1 and 2.

Figure 1.11 Kaolin clay, used as filler in papermaking. (a) Bare clay particles; (b) clay fully coated by cationic polystyrene latex (130 nm diameter). Scale bars: 10 and 1 μm, respectively.

monolayer coverage on the other particles (type 1). The deposition rate constant in this case is:

$$k_1 = \alpha \frac{4\gamma G}{3}(a_1 + a_2)^3 n_1 \tag{1.10}$$

When $a_2 \ll a_1$, Equation 1.10 reduces to Equation 1.6, with $\varphi = \varphi_1$. Alternatively we can describe the heteroflocculation as:

$$\frac{dn_1}{dt} = \frac{dn_2}{dt} = -\frac{4}{3}\alpha \gamma G(a_1 + a_2)^3 n_1 n_2 \tag{1.11}$$

An example of heteroflocculation between large and small particles is that between clay and cationic latex. Kaolin clay fully coated by polystyrene latex ($a_2 = 65$ nm) is shown in Figure 1.11. At lower latex dosages, less than required for full coverage, the latex acts as a cationic polyelectrolyte, flocculating the clay. This can be seen in Figure 1.12, where the stability ratio $W = 1/\alpha$, is shown as a function of latex addition, together with the electrophoretic mobility of the clay.

It can be seen that the latex particles reverse the charge of the clay. The deposition efficiency and the stability ratio are one around the point of zero charge of the clay, implying that the latex acts as a charge modifier, similarly to highly charged polyelectrolytes, such as PDADMAC and PEI [35]. At excess latex, the clay particles, covered by latex as seen in Figure 1.11b, become stable again.

Similar behavior was found for other fillers, such as PCC. PCC is positively charged, but reverses its charge in tap or process water [34]. Figure 1.13 shows results for the flocculation efficiency $\alpha \, (= 1/W)$ for both positive PCC (in distilled water) and negative PCC (in tap water). For positive PCC no heteroflocculation with cationic latex occurs, but latex destabilizes negative PCC in the same way as negative clay, shown above. Again the point of zero charge coincides with the fast flocculation regime.

An anomalous behavior was found with cationic PSB latex (Figure 1.14). No heteroflocculation between (positively charged) PCC and cationic PS latex occurs, as

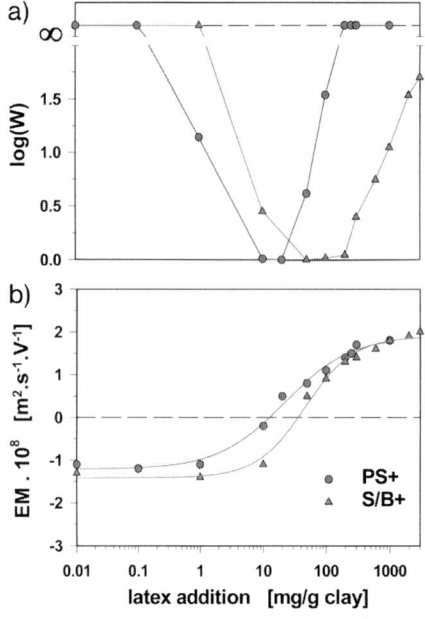

Figure 1.12 Stability ratio (a) and electrophoretic mobility (b) of kaolin clay as a function of latex addition for two types of latex (PS$^+$, cationic polystyrene, diameter 130 nm and S/B$^+$, cationic polystyrene butadiene, diameter 120 nm).

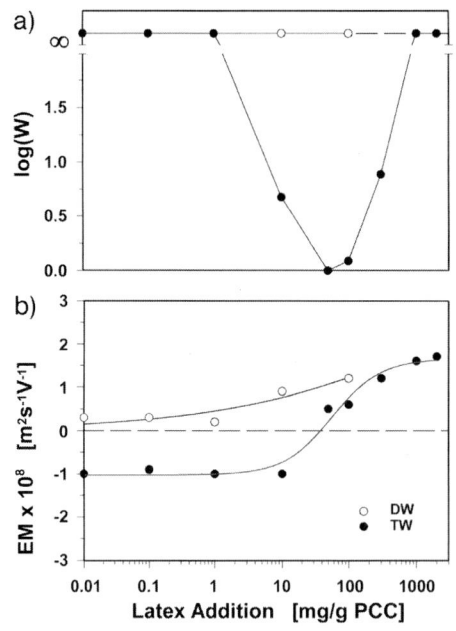

Figure 1.13 Stability ratio (a) and electrophoretic mobility (b) of precipitated calcium carbonate (PCC) as a function of polystyrene latex addition, in distilled water (DW) and in tap water (TW).

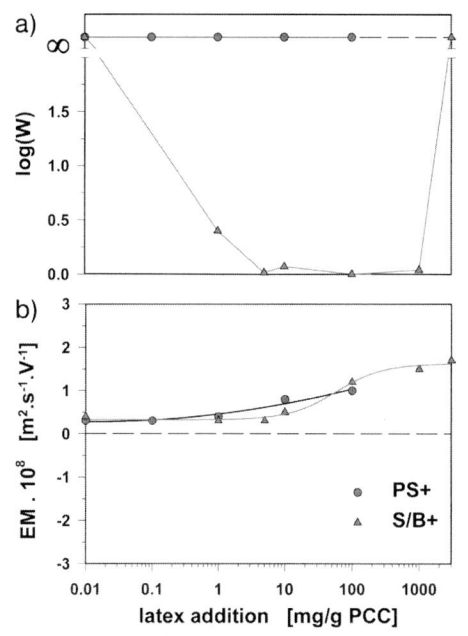

Figure 1.14 Stability ratio (a) and electrophoretic mobility (b) of precipitated calcium carbonate (PCC) as a function of polystyrene and PSB latex addition.

perhaps expected. Although both types of latex particles deposit on positive PCC (as can be concluded from the increase in electrophoretic mobility), PS does not bridge PCC particles together. However, cationic PSB latex does cause PCC to flocculate.

Both PS and PSB latex were made in our laboratory in the absence of emulsifiers, using the cationic comonomer diethylaminoethyl methacrylate (DEAEMA) and subjected to flash evaporation. As such there should be no "free charge," which could interfere with the deposition [28]. The electrophoretic mobility increases with latex addition, implying that PSB latex deposits on PCC, thus increasing its charge. So far the reason for the anomalous heteroflocculation is unknown. A possible explanation is that the soft PSB latex particles adhere more strongly to PCC, due to flattening and an increase in contact area, than to the hard PS latex particles. When two PCC particles with latex particles deposited on them collide in shear, they could form a doublet of PCC particles, bridged by a latex particle. For PS latex, the bridge is too weak to withstand the shear, leading to the break-up of the doublet, but the stronger PSB latex bridge could survive, leading to a permanent doublet.

1.4
Heteroflocculation of Fines and Colloids

So far we have only considered the interactions among colloids and between fibers and colloids, both in the absence and presence of polymers. Other important particles in papermaking suspensions are fines. These are fragments of fibers, or other types of cells in wood, liberated during pulping. In papers made from mechanical fibers, the fines content in the paper is typically 30%, whereas the fines content in the papermaking suspension can be as high as 60%. This corresponds to a 50% retention of fines on the paper machine, with the non-retained fines being recirculated back to the papermaking suspension. In papers made from chemical fibers, the fines content in the paper is typically 10–15%, with fines concentrations in the headbox of a paper machine about twice these values. Clearly, fines are an important ingredient in papermaking. Fines can be divided into two categories, long thin fibrillar fines and large "chunky" fines [36]. The fibrillar fines impart strength to the paper, due to their large surface area, whereas the chunky fines improve the optical properties of paper, because they scatter light efficiently.

Obviously, these fines can interact with other components in the papermaking suspension, such as colloids, polymers and polyelectrolytes. Despite their complicated shapes and polydispersity in size and shape, the homoflocculation of fines subjected to shear follows Smoluchowski kinetics, since flocculation times were found to follow adequately the predictions of Smoluchowski's theory [11]. Hence heteroflocculation is expected to follow these predictions as well. Heteroflocculation between papermaking fines and clay, treated with PEI, has been studied experimentally, as was that between MCC (microcrystalline cellulose) and clay [11]. The behavior of these systems is very similar to the heteroflocculation of clay with latex (discussed above). When the clay concentration was below monolayer coverage of clay on fines (or MCC), the fines flocculated, whereas when the clay concentration was above monolayer coverage, the system was stable, that is it consisted of fines, fully coated by clay.

1.5
Concluding Remarks

From the examples presented in this chapter papermaking suspensions are clearly fascinating complex systems that show a richness of interesting phenomena. Both colloidal and hydrodynamic phenomena play a crucial role. The colloidal interactions can be modified, and thus optimized and controlled, by polymers and poly electrolytes. The time scales of polymer adsorption, particle deposition on fibers, particle detachment polymer transfer, flocculation and break-up of colloidal aggregates determine how a papermaking suspension behaves on a paper machine. These time scales can be controlled by dosage and addition points. Some of the relevant time scales can be predicted by theory, as some of the examples given here show, whereas others require experimental determination, such as polymer transfer rates, particle detachment and floc break-up rates, which are difficult to predict from first principles. Therefore, expensive pilot and mill trials are usually required to optimize and fine-tune the use of additives on a paper machine. Nevertheless, laboratory experiments can provide useful trends and help to elucidate the mechanisms by which additives function.

References

1 Roberts, J.C. (ed.) (1996) *Paper Chemistry*, Blackie Academic & Professional, Chapman & Hall.

2 Au, C.O. and Thorn, I. (eds) (1995) *Applications of Wet-End Paper Chemistry*, Blackie Academic & Professional, Chapman & Hall.

3 van de Ven, T.G.M. (2000) Polymers and Polyelectrolytes in wet-end papermaking. *Japan Tappi Journal*, **54** (4), 516–522.

4 van de Ven, T.G.M. (2005) Filler and fines retention in papermaking, Proceedings of the 13th FRC Symposium, Cambridge, UK, September 11–16, Pira International, Leatherhead, UK.

5 Kamiti, M. and van de Ven, T.G.M. (1994) Kinetics of deposition of calcium carbonate particles onto pulp fibres. *Journal of Pulp and Paper Science*, **20** (7), 1199–1205.

6 Koper, G.J.M., Vanerek, A. and van de Ven, T.G.M. (1999) Poly(propylene imine) dendrimers as retention aid for the deposition of calcium carbonate on pulp fibers. *Journal of Pulp and Paper Science*, **25** (3), 81–83.

7 Vanerek, A., Alince, B. and van de Ven, T.G.M. (2000) Interaction of calcium carbonate fillers with pulp fibers: effect of surface charge and cationic polyelectrolytes. *Journal of Pulp and Paper Science*, **26** (9), 317–322.

8 Alince, B., Bednar, F. and van de Ven, T.G.M. (2001) Deposition of calcium carbonate particles on fiber surfaces induced by cationic polyelectrolyte and bentonite. *Colloids and Surfaces A: Physicochemical and Engineering Aspects*, **190** (1–2), 71–80.

9 Poraj-Kozminski, A., Hill, R.J. and van de Ven, T.G.M. (2007) Asymmetric polymer bridging between starch-coated colloidal particles and pulp fibers by cationic polyacrylamides, *Canadian Journal of Chemical Engineering*, **85**, 580–585.

10 Poraj-Kozminski, A., Hill, R.J. and van de Ven, T.G.M. (2007) Flocculation of starch-coated solidified emulsion droplets and calcium carbonate particles. *Journal of Colloid and Interface Science*, **309**, 99–105.

11 Porubská, J., Alince, B. and van de Ven, T.G.M. (2002) Homo- and heteroflocculation of papermaking fines and fillers. *Colloids and Surfaces A: Physicochemical and Engineering Aspects*, **210**, 223–230.

12 van de Ven, T.G.M. (2006) Interactions between fibers and colloidal particles subjected to flow. *Annual Transactions Nordic Rheology Society*, **14**, 9–18.

13 van de Ven, T.G.M. (1993) Particle deposition on pulp fibers: The influence of added chemicals. *Nordic Pulp & Paper Research Journal*, **1** (8), 130–134.

14 von Smoluchowski, M. (1917) Versuch einer mathematischen Theorie der Koagulationskinetik kolloider Losungen. *Zeitschrift für Physikalische Chemie*, **92** (2), 129–168.

15 Petlicki, J. and van de Ven, T.G.M. (1992) Shear-induced deposition of colloidal particles on spheroids. *Journal of Colloid and Interface Science*, **148** (1), 14–22.

16 Ödberg, L., Swerin, A. and Tanaka, H. (1993) Kinetic aspects of the adsorption of polymers on cellulose fibers. *Nordic Pulp & Paper Research Journal*, **8**, 16.

17 Petlicki, J. and van de Ven, T.G.M. (1994) Adsorption of polyethylenimine onto cellulose fibers. *Colloids & Surfaces A*, **83**, 9–23.

18 Alince, B., Vanerek, A. and van de Ven, T.G.M. (1996) Effects of surface topography, pH and salt on the adsorption of poly disperse polyethylenimine onto pulp fibers. *Berichte der Bunsen-Gesellschaft Physical Chemistry Chemical Physics*, **100** (6), 954–962.

19 Shirazi, M., van de Ven, T.G.M. and Garnier, G. (2003) Adsorption of modified starch on pulp fibers. *Langmuir*, **19**, 10835–10842.

20 van de Ven, T.G.M. and Alince, B. (1996) Association-induced polymer bridging:

New insights into the retention of fillers with PEO. *Journal of Pulp and Paper Science*, **22** (7), 1257–1263.
21 Adamczyk, Z. (2006) *Particles at Interfaces, Interactions, Deposition, Structure*, Elsevier, pp. 705–726.
22 LaMer, V.K. and Healy, T.W. (1963) Adsorption-flocculation reactions of macromolecules at the solid-liquid interface. *Reviews of Pure and Applied Chemistry*, **13**, 112.
23 Petlicki, J. and van de Ven, T.G.M. (1990) Particle trajectories near freely rotating spheroids in simple shear flow. *International Journal of Multiphase Flow*, **16** (4), 713–725.
24 Dabros, T. and van de Ven, T.G.M. (1994) Collision-induced dispersion of droplets attached to solid particles. *Journal of Colloid and Interface Science*, **163**, 28–36.
25 Alince, B., Petlicki, J. and van de Ven, T.G.M. (1991) Kinetics of colloidal particle deposition on pulp fibers. I. Deposition of clay on fibers of opposite charge. *Colloids & Surfaces, A*, **59**, 265–277.
26 Alince, B. and van de Ven, T.G.M. (1993) Kinetics of colloidal particle deposition on pulp fibers. 2. Deposition of clay on fibers in the presence of poly(ethylenimine). *Colloids & Surfaces A*, **71**, 105–114.
27 van de Ven, T.G.M., Abdallah Qasaimeh, M. and Paris, J. (2005) Fines deposition on fibers and fines flocculation in a turbulent flow loop. *Industrial & Engineering Chemistry Research Journal*, **44**, 1291–1295.
28 van de Ven, T.G.M. (1997) Mechanisms of fines and filler retention with PEO/Cofactor dual retention aid systems. *Journal of Pulp and Paper Science*, **23** (9), 1447–1451.
29 Alince, B. and van de Ven, T.G.M. (1997) Porosity of swollen pulp fibers evaluated by polymer adsorption, in *The Fundamentals of Papermaking Materials*, Vol. 2 (ed. C.F. Baker), Transactions of 11th Fundamental Research Symposium, Cambridge, UK, Sept. Pira International, Surrey, UK, pp. 771–788.
30 Alince, B., Kinkal, J., Bednar, F. and van de Ven, T.G.M. (2001) The role of "free charge" in the deposition of latex particles onto pulp fibers, in "Polymer colloids: Science and Technology of Latex Systems", ACS Symposium Series 801, Chapter 5, pp. 52–70.
31 van de Ven, T.G.M. and Alince, B. (1996) Heteroflocculation by asymmetric polymer bridging. *Journal of Colloid and Interface Science*, **181**, 73–78.
32 Asselman, T. and Garnier, G. (2000) Mechanism of poly electrolyte transfer during heteroflocculation. *Langmuir*, **16**, 4871.
33 Ödberg, L., Tanaka, H., Glad-Nordmark, G. and Swerin, A. (1994) Transfer of polymers from cellulosic fibers to filler particles. *Colloids and Surfaces A: Physicochemical and Engineering Aspects*, **86**, 201–205.
34 Vanerek, A., Alince, B. and van de Ven, T.G.M. (2000) Colloidal behavior of ground and precipitated calcium carbonate fillers: Effects of cationic polyelectrolytes and water quality. *Journal of Pulp and Paper Science*, **26** (4), 135–139.
35 Alince, B. and van de Ven, T.G.M. (1993) Stability of clay suspensions – effect of pH and polyethylenimine. *Journal of Colloid and Interface Science*, **155**, 465–470.
36 de Silveira, G., Zhang, X., Berry, R. and Wood, J.R. (1996) Location of fines in mechanical pulp handsheets using scanning electron microscopy. *Journal of Pulp and Paper Science*, **22** (9), J315.

2
Uptake and Release of Active Species into and from Microgel Particles

Melanie Bradley, Paul Davies, and Brian Vincent

2.1
Introduction to Microgel Particles

Microgel particles are crosslinked polymer particles that change their volume (i.e. swell or deswell) according to: (i) the solvency conditions of the medium in which they are dispersed; (ii) the density of the crosslinking moieties within the particles. For comprehensive reviews of these types of particles, including their preparation and characterization, one is referred to previous articles [1–3].

Since microgel particles are usually prepared in the deswollen state, by a dispersion or emulsion polymerization route, there are normally charged groups present at the periphery of the particles. These surface-charge groups arise from the initiator residues at the end of the constituent polymer chains within the particles, and help stabilize the microgel particles against coagulation in their deswollen state. In their swollen state their tendency to coagulate is greatly reduced, since the Hamaker constant of the particles then closely matches that of the continuous phase [4]. In addition to any *surface* charge, microgel particles may acquire a *bulk* charge from ionizable groups associated with the monomer(s) used in their preparation.

To discuss the swelling/de-swelling behavior of microgel particles in thermodynamic terms, one may write an osmotic balance between the two terms, (i) and (ii) above, as follows [5]:

$$\Pi_{osm} + \Pi_{el} = 0 \qquad (2.1)$$

The first term in Equation 2.1 is the osmotic mixing term, which is related to the polymer/solvent interaction parameter (χ) through the expression:

$$\Pi_{osm} = -\frac{N_A kT}{v_s}\left[\phi + \ln(1-\phi) + \chi\phi^2\right] \qquad (2.2)$$

where ϕ is the (average) polymer volume fraction in the microgel particles, N_A is the Avogadro constant, k is the Boltzmann constant, T is the temperature and v_s is the molecular volume of the solvent molecules.

Highlights in Colloid Science. Edited by Dimo Platikanov and Dotchi Exerowa
Copyright © 2009 WILEY-VCH Verlag GmbH & Co. KGaA, Weinheim
ISBN: 978-3-527-32037-0

The usual ways in which χ may be varied systematically are either by a change in the solvent (composition) or by a change in the temperature. Poly(N-isopropylacrylamide) [polyNIPAM] microgel particles are perhaps the most commonly studied systems of this sort, which show both a temperature response [1–3] and a response to solvency changes, for example, adding short-chain alcohols to the aqueous phase [6]. With respect to the temperature response, polyNIPAM microgel particles de-swell on heating (aqueous solutions of corresponding high MW, linear polyNIPAM chains have a lower critical solution temperature around 32 °C). Bouillot and Vincent were able to make microgel particles which demonstrated the *inverse* temperature effect, that is, they swelled on heating [7]. These particles were based on interpenetrating networks of poly(acrylamide) (PAM) and poly(acrylic acid) (PAA).

For microgel particles having a bulk charge there is an additional osmotic contribution from the charge-balancing counter-ions within the microgel particles [5]. In the simplest case, where the counter-ions may be treated ideally, one may write [4] this additional (ionic) contribution to the osmotic term as:

$$\Pi_{ion} = kT \frac{f}{V} \tag{2.3}$$

where f is the number of *mobile* (i.e. non-*condensed*) counter-ions per particle; f clearly relates to the bulk particle charge, Q_b.

The second term in Equation 2.1 refers to the elastic term, which restricts swelling, and is given by:

$$\Pi_{el} = \frac{2N_A X kT}{V_0} \left[\left(\frac{\phi}{2\phi_0} \right) - \left(\frac{\phi}{\phi_0} \right)^{\frac{1}{3}} \right] \tag{2.4}$$

where X is the number of crosslink sites per particle, and V_0 and ϕ_0 are, respectively, the particle volume and volume fraction of polymer in the "unswollen" microgel particles. Note that ϕ_0 is not actually equal to unity, since there is strong evidence that, once swollen with solvent, microgel particles always retain some solvent, even when de-swollen at high temperatures, provided they remain in dispersion. For example, Rasmusson et al. [8] estimated ϕ for polyNIPAM particles, containing 9% crosslinking monomer, as a function of temperature, in the absence of electrolyte. For this particular system, they showed that ϕ is ~0.1 at 25 °C, increasing to ~0.2 in the region of the LCST for polyNIPAM solutions (~32 °C) and to ~0.5 at 40 °C. Crowther and Vincent [6] showed that, by adding small-chain alcohols to aqueous dispersions of polyNIPAM particles, at low concentration of alcohol the microgel particles deswelled, even at 50 °C, that is, above the LCST for polyNIPAM. This is due to the alcohol molecules essentially "dehydrating" the microgel particles further.

By substituting Equations 2.2 and 2.4 into Equation 2.1 (and realizing that $\phi/\phi_0 = V_0/V$, where V is the particle volume in the swollen state) one may obtain, for neutral microgel particles, a relationship between V/V_0 (the volume swelling ratio, R, which is always >1) and χ, for a given crosslink density (X/V_0), although this relationship has to be solved numerically.

Because of the uncertainty in the value for V_0, discussed above, it is more usual to express experimental values for the volume swelling ratio as $S = V/V_{max}$ (always <1), where V_{max} is the *maximum* value of V, for a given crosslink density. One may also use (equivalently) the *diameter* swelling ratio, $S_d = d/d_{max}$, where d and d_{max} are the corresponding particle diameters (as obtained, for example, using dynamic light scattering); clearly, $S = S_d^3$.

For weakly-charged microgels, the solution pH may be used to vary Q_b. For example, poly(vinylpyridine) (polyVP) microgel particles have no bulk charge at neutral or higher pH but become progressively more positively charged as the pH is lowered below 4.5, through protonation of the N-atoms in the (weakly basic) pyridine groups [9]. Alternatively, one may alter the ionic balance, inside and outside the particles, by adding electrolyte to the system. One may think of this, equivalently, as a "screening effect" on the charge repulsion within the particles.

Microgel particles may swell or deswell in the presence of other species, in addition to small ions or added co-solvent molecules. This is of particular relevance in their potential application as controlled uptake/release systems, which is the main focus of this review. In the following sections we discuss this aspect of microgel particles, in particular in the context of "active" species used in biomedical and industrial applications.

2.2
Absorption of Small Molecules

Interest in the uptake and release of small molecules into and from microgel particles stems partly from their potential use as novel drug delivery systems, as well as other applications, such as the release of bactericides. The open network structure of microgel particles and their ability to undergo large swelling–deswelling transitions allows small drug molecules (as well as larger moieties such as proteins and peptides – see Section 2.4) to be incorporated and then released from their interior. Microgel particles are typically in the size range of a few 100 nm to several μm, so they are able to travel freely in the bloodstream and have the potential to be targeted to certain diseased tissues outside the bloodstream and even be taken up into the intracellular compartment of target cells [10]. Release systems based on specific environmental triggers can control the release of rapidly metabolized drugs and/or have the ability to protect sensitive drugs. Most research efforts are being dedicated to pH-sensitive microgels in the biomedical arena, exploring the fact, for example, that tumor tissues are slightly acidic (pH 6.8) and the endosomal and lysosomal compartments of cells have even lower pH (5–6) with respect to the physiological pH (7.4).

In the design of pH-responsive microgel particles, for the uptake and release of small molecules, it is possible to prepare a particle with permanent (covalent) or semi-permanent (hydrolyzable) crosslinks and introduce functional groups into the microgel network that are environmentally responsive. In the design of oral drug delivery systems, for example, microgels that contain weak acid groups within their network have the potential to transport drug through the stomach for colon-specific drug delivery. In the acidic environment of the stomach, the microgel particles are uncharged, and hence collapsed, and protect any encapsulated drug from degradation. Upon entering the neutral pH of the intestine, charge builds within the network, resulting in particle swelling and release of the trapped drug. The release profiles of small drug molecules from such microgels have been studied under both simulated gastric and intestinal pH conditions and the release was reported to be faster in pH 7.4 buffer than in pH 1.0 solution [11–14].

Microgels with acid-cleavable crosslinks have also been investigated for the release of drugs under mildly acidic conditions [15, 16]. The drug molecules were trapped within the microgel network during particle preparation and subsequently released when exposed to an acidic environment or an enzyme that degraded the polymer network and disintegrated the particles. Pioneering work in the preparation of degradable microgels, using bisacrylamide acetal crosslinker with a p-methoxy substituent as an acid degradable linkage [17], are described in Section 2.4.

Another novel approach in the drug delivery domain is to prepare particles with temporary physical crosslinks that can be broken in response to environmental triggers. In this case microgel particles are modified to contain target recognition molecules. In the first particles of this kind an antigen responsive microgel was synthesized where the reversible binding between an antigen and an antibody acted as a crosslinking mechanism in a semi-interpenetrating network microgel. The particles swelled in the presence of free antigen because the intra-chain antigen–antibody binding was dissociated by exchange of the grafted antigen for free antigen (Figure 2.1) [18].

Much research has focused on insulin release from microgel particles [19–22] and in one approach sugar-based microgel particles were doped with a functional group that interacted with the microgel network to form physical crosslinks [23]. When the insulin-containing microgels were exposed to an increase in glucose concentration the crosslinks within the particles were broken as the crosslinking agent partitioned between the microgel network and free glucose, thus releasing the insulin.

Attention has recently turned towards microgel particles that target specific sites and release small molecules within these sites. It has been demonstrated [24] that microgel particles can be functionalized to facilitate release and to provide surface-based targeting using specific surface ligands. The conjugation of ligands onto the surface of temperature-responsive microgel particles was effective in targeting the receptors on cancer cells and resulted in internalization of the particles by the cells [24]. The receptor-mediated internalization of pH-responsive biocompatible microgels, followed by pH-triggered delivery of the drug once in the acidic cell environment, has also been reported [25, 26]. Figure 2.2 [25] shows that microgel particles with targeting ligands are effective in delivering small molecules to receptor cells.

Microgel particles are of interest in biomedical applications because they closely resemble natural living tissues due to their high water content, their gel-like consistency and their potential to minimize irritation to surrounding tissues. A synthetic microgel has been shown to mimic the properties of the secretory granule in biological systems [27]. This was achieved through the preparation of carboxylated microgel particles that were loaded with small ionic molecules trapped by pH-induced collapse of the microgel network. The surface of the microgel particles was

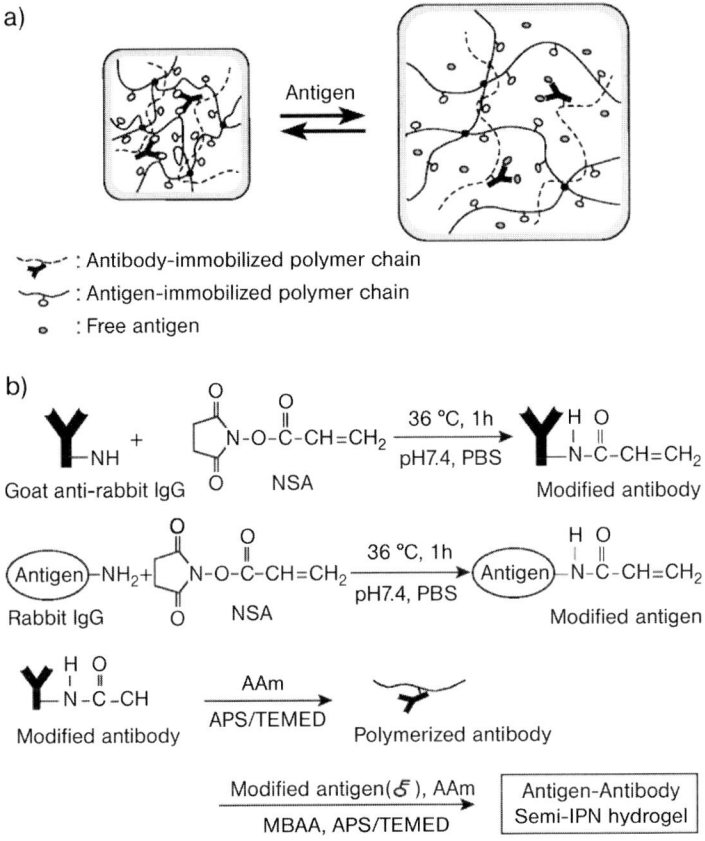

Figure 2.1 (a) Suggested mechanism for the swelling of an antigen–antibody semi-IPN hydrogel in response to a free antigen; (b) synthesis of the antigen–antibody semi-IPN hydrogel [18].

subsequently coated with an ion-impermeable lipid bilayer. Bilayer disruption by detergent solubilization or electroporation (by an externally applied electrical field) resulted in the release of the molecules by ion exchange. Figure 2.3 shows a schematic representation of this process.

2.3
Absorption of Surfactants

The absorption of surfactant molecules into microgel particles affects the swelling and, in many cases, also the electrophoretic properties of the microgel particles. The mechanism for this absorption process largely depends on the properties of the microgel particles, and may be associated with hydrophobic bonding, H-bonding or electrostatic attraction. The primary driving force for the uptake of the anionic surfactant sodium dodecyl sulfate (SDS) into polyNIPAM microgel particles is the

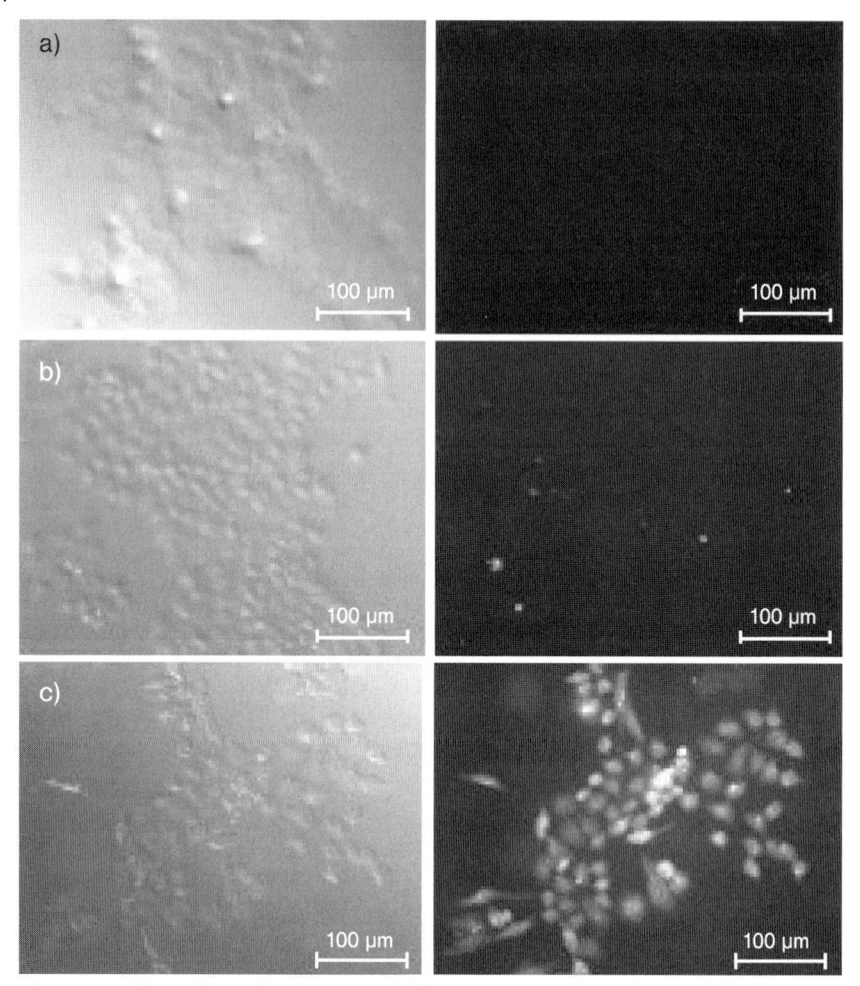

Figure 2.2 Differential interference contrast (left) and epifluorescent (right) images of HeLa cells after incubation with drug loaded microgels that were not conjugated to surface ligands (a), conjugated to albumin (b) and conjugated to transferrin (c) [25].

hydrophobic interaction between the hydrocarbon tails of the surfactant and the isopropyl moieties within the microgel particles [28]. At concentrations below the critical aggregation concentration (CAC) of the surfactant molecules with the polymer chains comprising the microgel particles, the absorbed amount of the surfactant into the microgel network was low and there was little effect on the microgel swelling. Above the CAC, surfactant aggregate formation within the microgel network caused a sharp increase in the absorbed amount and the internal electrostatic repulsion between these aggregates within the microgel particles

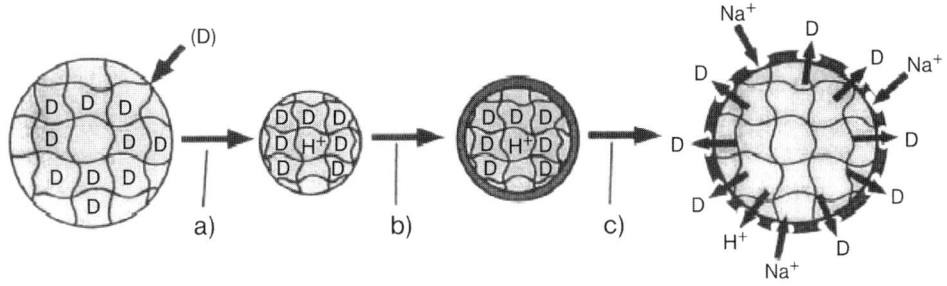

Figure 2.3 Mechano-chemical device for triggered release of drug. (a) Loading of microgel with drug and collapse of the polymer network, (b) lipid coating of microgel and (c) rupture of lipid layer followed by swelling of the microgel network and drug release [27].

resulted in their swelling with water. The increased hydrophilic character within the microgel network also resulted in an increase in the volume phase transition temperature of the particles [29–31]. The electrophoretic mobility of the particles increased with increasing SDS concentration and temperature. In the latter case this was due to greater SDS binding to the microgel at higher temperature.

The absorption of SDS and also the photodegradable anionic surfactant 4-hexyl-phenylazosulfonate (C_6PAS) into (cationic) poly(2-vinylpyridine) (poly2VP) microgel particles has also been investigated [32, 33]. Strong electrostatic attraction occurs between the head group of the anionic surfactant molecules and the protonated microgel network (at pH < 4), resulting in particle de-swelling. At surfactant concentrations above the CAC the particles showed a small degree of swelling due to aggregate formation within the microgel network (Figure 2.4a). The electrophoretic mobility of the particles, as a function of surfactant concentration, became increasingly less positive and eventually showed charge-reversal. The microgel particles containing the photodegradable surfactant were found to swell upon exposure to UV-irradiation (Figure 2.4b).

The Bristol group has investigated the uptake and release of surfactant molecules from copolymer microgel particles based on NIPAM plus a functionalized monomer containing acidic or basic groups. The interaction of non-ionic surfactant, containing ethylene oxide in the head group, with poly(NIPAM-co-AAc) (AAc = acrylic acid) microgel particles demonstrated that, when the AAc groups were protonated (pH < 4), H-bonding interactions were sufficient to cause uptake of surfactant into the microgel network [34].

The addition of the cationic surfactant cetylpyridinium chloride (CPC) to aqueous dispersions of polyNIPAM-co-AAc microgel particles leads to absorption of the CPC into the microgel particles [35]. Uptake of the surfactant depends on two distinct attractive interactions between the surfactant molecules and the microgel particles: (i) electrostatic (at high pH when the carboxylic acid groups are de-protonated) and (ii) hydrophobic. Adjusting either the temperature or pH was shown to cause release of the surfactant (Figure 2.5).

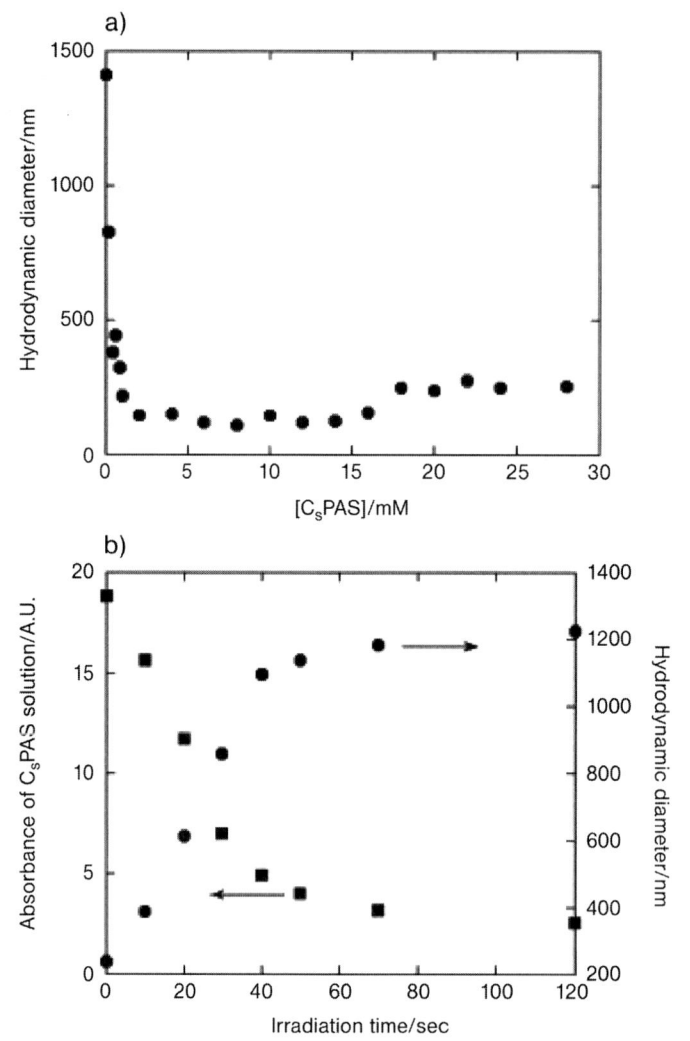

Figure 2.4 (a) Poly2VP hydrodynamic diameter as a function of C_6PAS concentration for dispersions at pH 3; (b) absorbance of 2 mM C_6PAS solution at 307 nm and poly2VP hydrodynamic diameter as a function of UV irradiation time [33].

Similar results were observed with the uptake and release of an anionic surfactant into and from poly(NIPAM-co-2VP) microgel particles [36]. A comparison has also been made between the effect of added CPC on the hydrodynamic diameter of the free microgel particles in dispersion and the thickness, as determined from ellipsometry measurements, of deposited monolayers of the same microgel particles on cationically-modified, oxidized silicon surfaces. The behavior of the microgel layers reflects the dispersion properties of the particles in the presence of CPC [35].

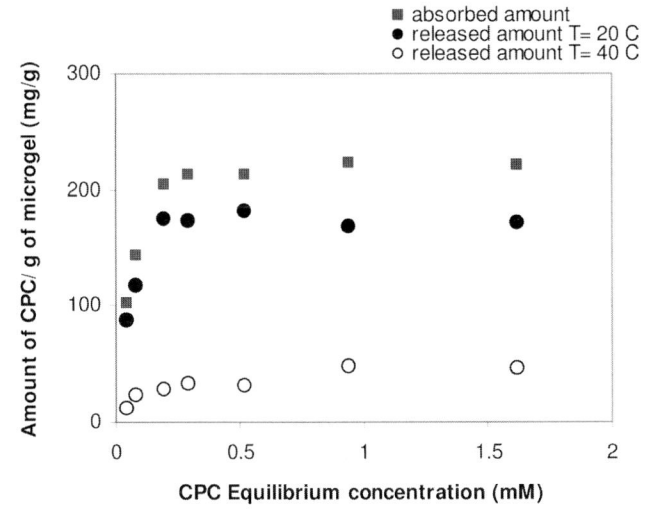

Figure 2.5 Absorbed and released amount of CPC as a function of CPC equilibrium concentration, for poly(NIPAM-co-AAc) microgel particles (with 6%AAc and 6%BA). (■) absorbed amount at pH 8 and 20 °C; (●) released amount at pH 3 and 20 °C; (○) released amount at pH 3 and 40 °C [35].

2.4
Absorption of Polymers and Proteins

An alternative way of using osmotic balancing, inside and outside microgel particles, to control the particle size is to add a *non-penetrating* polymer. Saunders and Vincent demonstrated this effect for polyNIPAM microgel particles in aqueous solution, to which poly(ethylene oxide) (PEO) chains were added [37], and also for poly(styrene) (PS) microgel particles in ethyl benzene, to which free PS chains were added [38]. If the polymer chains do *not* enter the microgel particles (because their solution dimensions are significantly greater than the average pore-size within the microgel particles) then the microgel particles deswell, even under good solvent conditions; this is due to the increase in osmotic pressure outside the particles tending to balance that inside.

Even for systems where the free polymer solution dimensions are actually smaller than the average pore-size within the microgel particles, in order for the PEO molecules to enter the microgel particles a "driving force" (i.e. an attractive interaction between the two species) is required. PEO is known to H-bond to carboxylic acid groups (in the *undissociated* form, that is, at pH < ~4). The swelling and de-swelling properties of poly(NIPAM-co-AAc) copolymer microgel particles have been investigated in aqueous poly(ethylene oxide) (PEO) solutions [39]. The uptake of PEO into the microgel particles at low pH is due to the H-bonding association between PEO and the –COOH of the acrylic acid (AAc) moieties, referred to above. The effects of PEO molecular weight (2000–300 000) and microgel crosslinker density (i.e. average pore size) on the swelling of the microgel particles and uptake of PEO were investigated (at pH 3).

The microgel particles were shown to *swell* upon addition of low concentrations of lower MW PEO. Upon increasing the concentration of PEO, beyond some critical value, then *deswelling* occurred, once the microgel particles had become saturated with PEO chains. The critical concentration of PEO required to cause de-swelling decreased with increasing MW. There was an upper PEO MW limit after which swelling of the particles was prevented (Figure 2.6a). The maximum absorbed amount decreased with increasing MW; this was rationalized in terms of the ease of penetration of lower MW species into the microgel particles (Figure 2.6b).

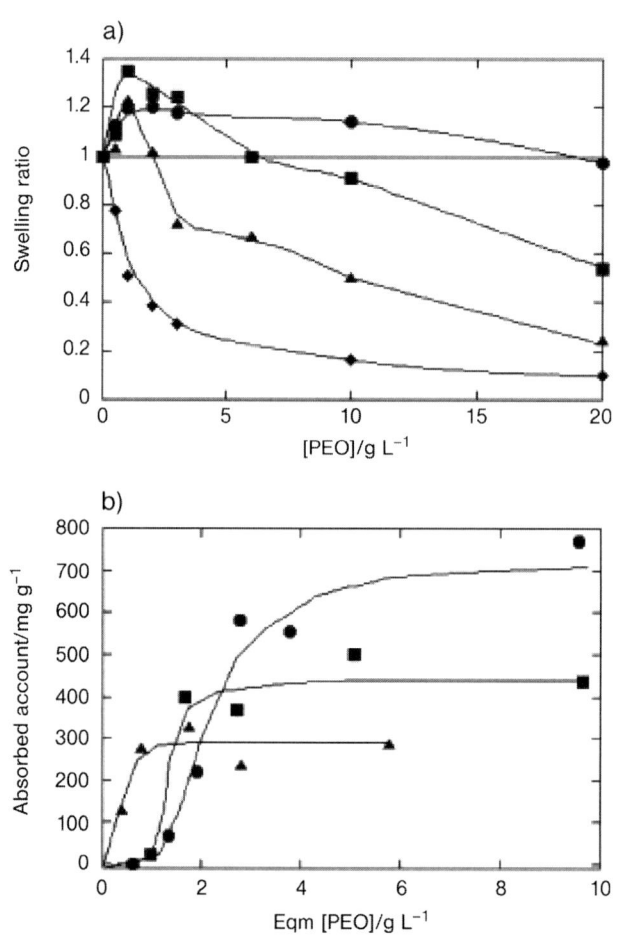

Figure 2.6 Effect of PEO MW on the swelling ratio (a) and absorbed amount (b) as a function of PEO concentration for 10 wt% crosslinked PNIPAM-co-AAc (10 wt% AAc) microgel particles at pH 3: (●) 2000; (■) 20 000; (▲) 100 000; (♦) 300 000 [39].

A greater amount of swelling was observed with decreasing crosslinker concentration at low PEO concentration, with de-swelling occurring at much lower PEO concentrations with decreasing crosslinker concentration (Figure 2.7a). The lowest crosslinked microgel has a high-affinity type isotherm while the higher crosslinked particles have low-affinity type isotherms (Figure 2.7b). However, the absorbed amount is much lower for the lower crosslinked particles. This trend is due to the greater ease of penetration of the PEO molecules into the microgel particles. The higher absorbed amount, at higher crosslinker concentration, is due to the particle being more swollen at these concentrations. Finally, the *ab*sorption and desorption of the PEO molecules into and out of the microgel particles were shown to be extremely

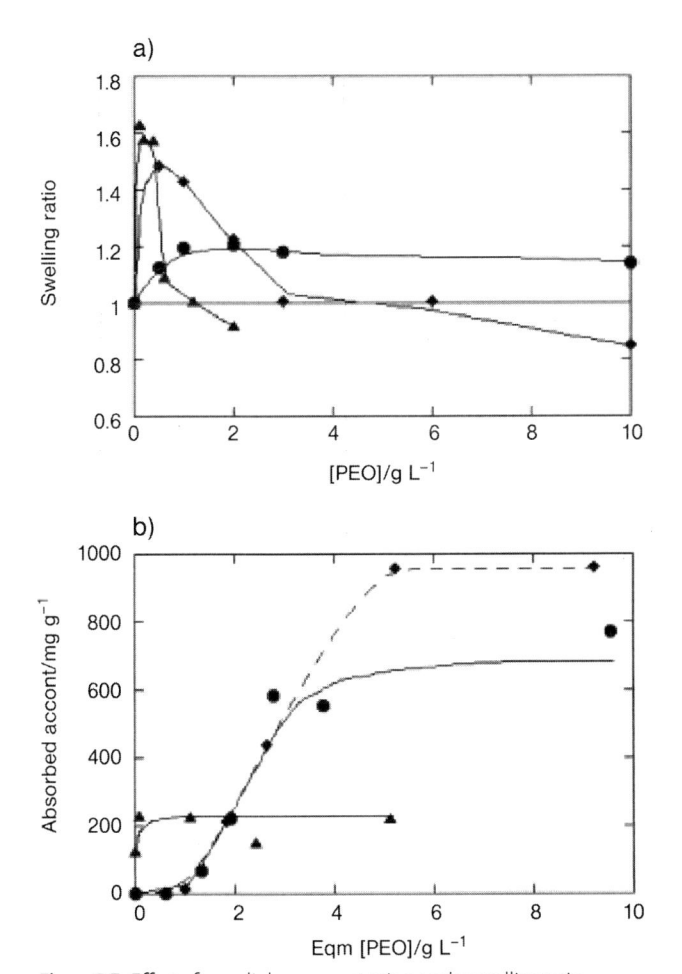

Figure 2.7 Effect of crosslinker concentration on the swelling ratio (a) and absorbed amount (b) as a function of PEO (2000) concentration at pH 3 for PNIPAM-co-AAc (10 wt% AAc) microgel particles: (▲) 6, (♦) 8 and (●) 10 wt%. [39].

slow compared to normal diffusion timescales for polymer adsorption onto rigid surfaces.

The swelling behavior of polyNIPAM microgel particles by PEO chains has been explored theoretically using thermodynamic considerations [40]. The particle expansion was the result of an increase in osmotic pressure resulting from penetration of PEO chains into the microgel network. The different contributions to the Gibbs free energy changes in the system and the chemical potential balance were used to predict the swelling and collapse behavior of the microgel particles.

As with small-molecule drugs, there has been a growing interest in the use of microgel particles as drug-delivery systems for proteins and peptides. The entrapment of proteins within a microgel network is advantageous to avoid protein-denaturing effects. As with the PEO example, discussed above, the loading efficiency and the quantity of the protein encapsulated within the gel are affected by the pore size of the microgel and, hence, the crosslink density. Since the pore size within a microgel network can be adjusted, the possibility of preserving spatial structure of proteins during encapsulation exists. Proteins have been trapped within a microgel network through immobilization during microgel synthesis [41], or absorbed into a post-fabricated microgel network using the stimulus responsive properties of the microgel particles to expand its network to allow diffusion of the protein within the interior [19]. In both cases the behavior of the native and microgel-coupled proteins was similar, and the protein–microgel derivatives had stabilities similar to [19] or superior to [41] those of the native protein.

The use of degradable microgels for protein release, based on a bisacrylamide acetal crosslinker with a p-methoxy substituent as an acid degradable linkage [17], is shown schematically in Figure 2.8.

BSA protein was encapsulated by including the protein during the polymerization process. The rate of hydrolysis of the microgel particles, and therefore the rate of release of the protein, was pH dependent, being faster under more acidic conditions. In a recent study the biocompatible polymer poly(hydroxyethylmethacrylate) (PHEMA) was crosslinked with a vinyl-functionalized acid-labile crosslinker [42]. The protein-release profile of the microgels with low crosslinker density was found to be slightly more sensitive to pH than the release profile of the microgels with high crosslink density. This was attributed to the fact that, within the microgel particles with higher crosslink density, the number of crosslinks that needs to be cleaved, to create pores larger than the diameter of protein, is higher than the number of crosslinks within microgel particles with lower crosslinker density.

Similar to the earlier discussion around Figure 2.2, core–shell microgel particles have been prepared [43], where the core has been modified with the ligand biotin and the shell possesses an acid-cleavable crosslinker. The protein avidin, for which biotin is a tightly binding ligand, was able to penetrate into the core upon cleavage of the microgel shell.

The interaction of proteins with microgel particles, aside from specific drug delivery applications, has been investigated in relation to their potential use in bio-separation and as biosensors. polyNIPAM microgel particles and their derivatives have been the focus of several studies on the adsorption of proteins onto the microgel

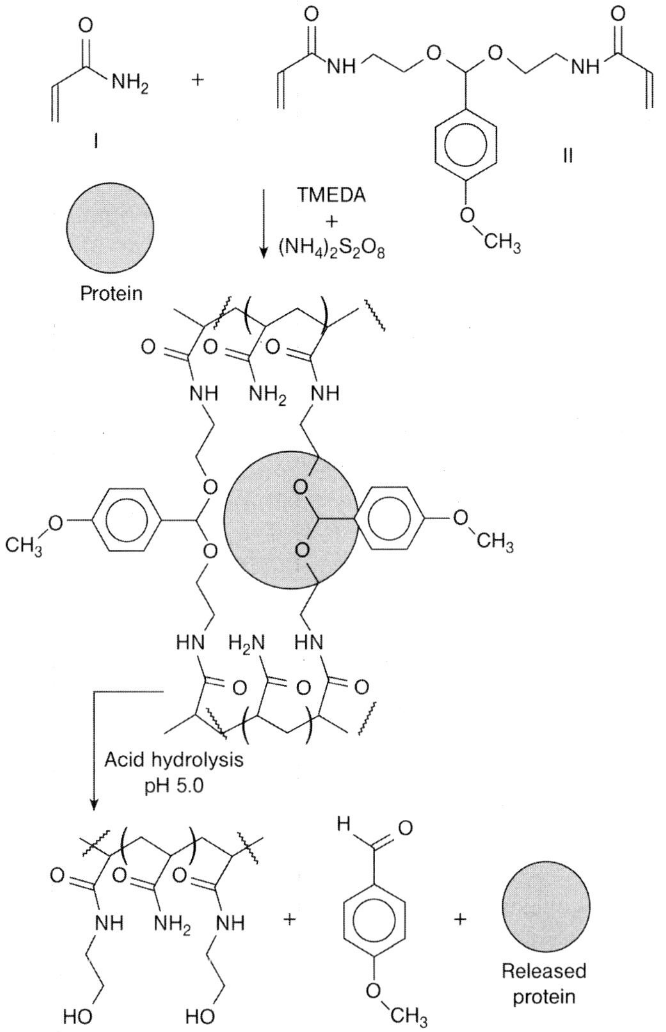

Figure 2.8 Schematic representation of acid-degradable protein-loaded microgel [17].

surface. polyNIPAM has attracted interest in the biomedical field because its volume phase transition (VPT) occurs just below body temperature. The most common protein studied is bovine serum albumin (BSA). Although the surface area of polyNIPAM microgel particles decreases with increasing temperature above the VPT, it has been shown that the absorbed amount of BSA increases due to the stronger hydrophobic interaction of the protein with the microgel particles [44–47]. The absorption of protein into microgel particles based on NIPAM, copolymerized with a monomer containing acidic functional groups, has also been shown to be pH dependent [47], with protein absorption occurring in the pH range when there is electrostatic attraction between the protein and microgel network.

It is often desirable to prevent the ad- or absorption of protein when using polymeric carriers as drug delivery systems. Surface modification of the particles using adsorbed or grafted PEO chains is a proven approach to protect particles from protein binding and, hence, achieve a longer circulation time in the blood. This idea has been extended to the design of polyNIPAM microgel particles. It has been shown that polyNIPAM microgel particles, modified with poly(ethylene glycol) (PEG) on their surface, exhibit a considerably reduced amount of protein absorption over a wide temperature range [44]. Even when the PEO chains were buried in the particle core surrounded by a polyNIPAM shell, protein adsorption was inhibited above the VPT temperature, suggesting that PEO somehow reduces the particle surface hydrophobicity even when it is not purposely localized in the shell. polyNIPAM microgel particles have also been crosslinked by copolymerization with low MW PEO-diacrylate [46]. Below the VPT there was a decrease in the amount of protein absorbed as the chain length and PEO content increased, due to an increase in the hydrophilicity of the network. Above the VPT there was only a decrease in the amount of protein absorption for the longest PEO chain systems, suggesting that the structure of these deswollen particles was one where the PEO chains were either mobile enough to protrude from the collapsed network or were enriched at the surface due to phase separation during polymerization.

2.5
Absorption of Nanoparticles

The uptake and release of nanoparticles by microgel particles has been an area of growing interest in recent years. Such systems have potential applications in purification, size separation of nanoparticles and catalyst support systems, among others. As with polymer and protein chains, discussed in the previous section, the incorporation of nanoparticles into a gel matrix can protect the solid particles, giving them enhanced stability and providing access to versatile surface functionalities. The composite materials produced often display material properties that combine the structural and responsive properties of the microgel particles with the unique, size-dependent, optical and electronic properties afforded by certain classes of colloidal nanoparticles. As well as absorption of the nanoparticles into preformed microgel particles, similar composite materials may be produced by the direct synthesis of the nanoparticles within the preformed microgel particles [48] or, alternatively, by synthesis of the microgel particles in the presence of the preformed nanoparticles [49]. However, these systems lie outside the scope of this chapter and will not be discussed in any detail here.

The optical properties of quantum dots (QDs), based on cadmium selenide (CdSe), have been exploited for the development of multiplexed optical coders for use in biological imaging by embedding of the QDs within polystyrene (PS) microgel particles [50]. QDs are semiconducting nanoparticles that fluoresce when stimulated by visible light, with the wavelength of maximum fluorescence being dependent on particle size. The uptake of (ZnS-capped) CdSe QDs was achieved by varying the solvent conditions to swell the PS microgel particles in the presence of pre-formed QDs, in common organic solvents. Single color-coded microgel particles, with emission wavelengths nearly identical to the original QDs, were produced by this method, with the porous structure of the swollen microgel acting as a matrix to spatially separate the embedded QDs. Multi-color QD-tagged microgels were also prepared by the sequential uptake of QD particles of different sizes, and hence differing fluorescence maxima. These multi-color QDs could be incorporated into the microgel at precisely controlled ratios. The use of these QD-tagged PS microgel particles for biological assays has been demonstrated by employing them in DNA hybridization studies.

An attempt to determine the distribution of CdSe QDs within larger microgel particles (diameter ~25 μm) has been made using confocal microscopy [51]. A similar method of uptake was used as described above, namely swelling the PS microgel particles in the presence of pre-formed QDs, then changing the solvent conditions such that the microgel particles were caused to deswell. Such experiments were shown to be useful in revealing the internal structure and porosity of the large microgel particles. The pore size was seen to decrease towards the center of the microgel particle, as indicated by the fact that in moderately poor solvent conditions for PS the QDs did not fully penetrate into the core of the microgel. By increasing the extent of swelling of the PS microgel particles, by choosing a much better solvent for the PS microgel particles, it then became possible for the CdSe QDs to penetrate into the central region of the microgel particles. This work shows the potential of using sets of nanoparticles, of a known size, to estimate the pore-size distribution within these materials, which is not easy to achieve by other methods.

The uptake of QDs into microgel particles in aqueous solution has also been reported. The absorption of water-soluble, thioglycolic acid (TGA)-capped, cadmium telluride (CdTe) QDs into poly(*N*-isopropylacrylamide-*co*-4-vinylpyridine) [poly(NI-PAM-co-4VP)] microgels particles is one such example [52]. These microgel particles exhibit both temperature- and pH-dependent swelling, due to the presence of the NIPAM and 4VP moieties respectively. At pH 3 the poly(NIPAM-co-4VP) microgel particles are swollen, since the 4VP groups become protonated. Incubation of the swollen microgel particles with a dispersion of CdTe QDs led to uptake of the QDs into the microgel. On subsequently increasing the pH, the microgel particles collapsed and the QDs became entrapped within the polymer network. The process led to a uniform spatial distribution of CdTe particles throughout each of the microgel particles. The amount of QDs per microgel particle could be tuned by varying the concentration of the QDs. As discussed earlier with regard to the non-aqueous systems, multicolor-encoded hybrid particles could be produced by incorporating different-sized QDs into one set of microgel particles. Note that the presence of the gel network effectively prevents aggregation of the embedded particles. Such incorporation of water-dispersible nanoparticles into microgel particles may provide promising fluorescent markers for bio-molecule and other chemical sensing. The subsequent release of CdTe QDs from the loaded poly(NIPAM-co-4VP) microgel particles was also studied. The QDs were released by increasing the system pH above 11, effectively "squeezing" the QDs out of the microgel particles as these became increasingly collapsed.

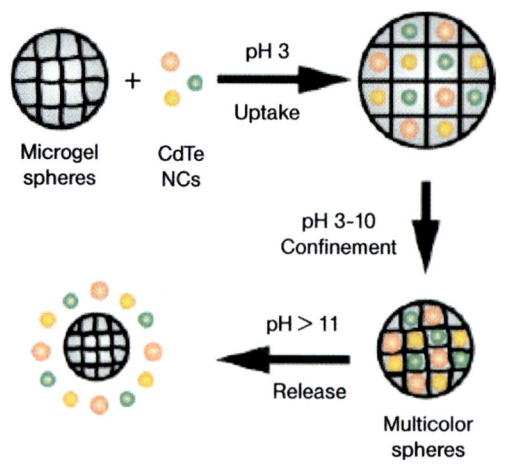

Figure 2.9 Schematic mechanism for the uptake and release of QDs by pH-responsive microgels [52].

Figure 2.9 shows a schematic representation of the uptake and release mechanisms employed in this pH-responsive system. By tuning the pH and electrolyte content of the system, a 98% release of loaded material could be achieved.

Similarly, CdTe QDs capped with TGA and/or thioglycerol have been entrapped in temperature-responsive microgel particles based on PNIPAM [53]. CdTe QDs stabilized by only TGA could be absorbed into NIPAM-based microgel particles at temperatures higher than their LCST. All the QDs were released when the temperature was reduced to ambient, that is, on re-swelling the microgel particles. CdTe QDs capped with both ligands allowed hydrogen-bonding between the thioglycerol and the amide groups on the polyNIPAM chains to occur. In this system the loading capacity of the microgel particles depended on incubation temperature, with the amount of QDs incorporated into the particles increasing with increasing temperature. In this case, the CdTe QDs remained associated with the microgel on cooling to below the LCST, due to H-bonding interactions, in contrast to the TGA-capped QDs. Release of the dual-capped CdTe QDs could be triggered, however, by the addition of urea, a known hydrogen-bond breaker.

The uptake of CdTe QDs into poly(NIPAM-co-AAc) microgel particles has used to prepare materials for controlled, two-dimensional self-assembly [54]. TGA-capped QDs were first coated with cetyltrimethylammonium bromide (CTAB) to impart a positive charge. These particles were then absorbed into the into poly(NIPAM-co-AAc) microgel particles at pH 6, where the AAc groups are de-protonated, and hence the microgel particles swollen. The electrostatic attraction between the CTAB-coated QDs and the negatively-charged AAc groups within the microgel network provided the driving force for uptake. After collapsing the microgel particles, by adjusting the system pH below the pK_a of AAc (~4.8), the composite materials were self-assembled on a glass surface, producing fractal and dendritic patterns on a large scale. The combination of the properties of the inorganic CdTe QDs with the responsive microgel particles affected the self-assembly process, with the alignment of the dipole moments of the CdTe QDs acting to induce the self-assembly of the hybrid particles.

As well as QDs, the uptake of metal nanoparticles has also been of interest. Metal nanoparticles exhibit a characteristic plasmon resonance, which is dependent on particle size and shape, and have reasonable biocompatibility. However, many types of metal nanoparticles are unstable to environmental changes. Hence, their uptake into microgel particles can provide a route for imparting stability to such systems. For example, the uptake of citrate-stabilized gold nanoparticles into thermo-responsive polyNIPAM and poly(NIPAM-co-MAc) (MAc = methacrylic acid) microgel particles has been reported [55]. Repeated incubation of the microgel particles with a dispersion of the gold nanoparticles allowed control over the absorbed amount of the nanoparticles. A high loading efficiency was achieved by this method, for both the charged and the uncharged microgel particles, indicating that electrostatic attraction makes little or no contribution to the nanoparticle uptake in these cases. Entrapment instead largely arises from physical entanglement of the gold nanoparticles within the polymer network. The composite materials produced retained the thermo-responsive nature of the parent microgel particles. However, the pH-responsiveness of the poly(NIPAM-co-MAc) microgel particles was suppressed, as the gold nanoparticles act to crosslink polymer chains within the microgel particles by binding to the MAc groups, displacing citrate groups from the gold particle surface. Gold-loaded poly(NIPAM-co-MAc) microgel particles could be transferred into various organic solvents without any deterioration the properties of the nanoparticles, thereby demonstrating their successful protection within the microgel particles.

In situ mineralization within poly(NIPAM-co-MAc) microgel particles, pre-loaded with gold nanoparticles, by the method just described, was used to create calcium carbonate ($CaCO_3$) shells around the nanoparticles [56]. In this manner, protection and stabilization of the gold NPs within the microgel particle gives access to reactions otherwise not possible with dispersions of gold NPs. The increased size of the $CaCO_3$-coated nanoparticles led to their expulsion from within the microgel network. Their gold cores could subsequently be etched out to give hollow $CaCO_3$ shells, showing the potential of this route to produce bio-mineral nanocapsules.

Citrate-stabilized gold NPs have also been incorporated into a film of polyNIPAM particles, via a "swelling-in" process [57]. A shrunken polymer film at 40 °C was allowed to swell in the presence of a dispersion of nanoparticles at 20 °C, before being re-immersed in an aqueous solution at 40 °C to re-collapse the gel layer. As with dispersions of microgel particles, repeated cycles allowed for a controlled amount of NPs to be introduced into the polyNIPAM film, and the incorporated particles were physically entrapped within the film. A reversible color change was observed in the composite film when switching between the collapsed and swollen states, as the separation distance of the nanoparticles within the polymer network was changed, with the NPs coming into closer contact in the shrunken film. The closeness of the gold NPs also imparted to the composite film an enhanced electrical conductivity compared to an un-doped polyNIPAM film. Such signal-triggered nanoparticle/

microgel particle hybrids can, therefore, provide unique materials having controllable optical and electronic properties.

While the vast majority of work in this area has focused on the uptake of spherical nanoparticles, it is not limited to such systems. Gold nanorods have been loaded into polyNIPAM microgel particles, producing hybrid microgels that are photo-responsive in the near-IR range [58]. The ability to use near-IR radiation as a trigger is important as such radiation can penetrate human skin, allowing the technique to be used for *in vivo* applications. Cationic gold nanorods, stabilized by adsorbed CTAB, were embedded in the anionic microgels at pH 6 and 25 °C, where the microgel particles are in their swollen state. Through varying of the aspect ratio of the nanorods, they could be tuned to absorb light in the near-IR range. Therefore, irradiation of the hybrid materials at pH 4, with laser-light of wavelength 810 nm, can be used to trigger a strong, reversible, photo-thermally induced decrease in microgel particle size due to local heating of the microgel network via the gold nanorods. This work was developed further to produce a hybrid system that could undergo such a photo-thermally induced microgel particle size under physiological conditions [59]. Loading of the same nanorods into poly(NIPAM-*co*-maleic acid) microgel particles achieved this aim, with potential application in drug delivery and photodynamic therapy. The method here would be to use the photo-thermally triggered shrinkage of the hybrid material to release a pre-loaded drug from within the microgel particle interior.

Various mechanisms have been demonstrated for the uptake and release of nanoparticles into and from responsive microgel particles and films, in both aqueous and organic media. Uptake of nanoparticles into microgels acts to protect the nanoparticles while, in general, retaining both the responsive nature of the microgel particles and the unique optical and electronic properties of the nanoparticles themselves. The degree of uptake of NPs into microgels can be readily controlled, as discussed.

As with the uptake of other types of active species discussed in this chapter, microgel particle/nanoparticle composite materials show a large range of potential applications, making this whole field of microgel particle uptake and release a rapidly developing area.

References

1 Murray, M.J. and Snowden, M.J. (1995) *Advances in Colloid and Interface Science*, **54**, 73.

2 Saunders, B.R. and Vincent, B. (1999) *Advances in Colloid and Interface Science*, **80**, 1.

3 Saunders, B.R. and Vincent, B. (2002) *Encyclopedia of Surface and Colloid Science*, Marcel Dekker, USA, p. 4544.

4 Routh, A. and Vincent, B. (2002) *Langmuir*, **18**, 5366.

5 Fernández-Nieves, A., Fernández-Barbero, A. and Vincent, B. (2003) *Journal of Chemical Physics*, **119**, 10383.

6 Crowther, H.M. and Vincent, B. (1998) *Colloid and Polymer Science*, **276**, 46.

7 Bouillot, P. and Vincent, B. (2000) *Colloid and Polymer Science*, **278**, 74.

8 Rasmusson, M., Routh, A. and Vincent, B. (2004) *Langmuir*, **20**, 3536.

9 Loxley, A. and Vincent, B. (1997) *Colloid and Polymer Science*, **38**, 6129.

10 Eichenbaum, G.M., Kiser, P.F., Dobrynin, A.V., Simon, S.A. and Needham, D. (1999) *Macromolecules*, **32**, 4867.

11 Lowman, A.M., Morishita, M., Kajita, M. et al. (1999) *Journal of Pharmacological Sciences*, **88**, 933.

12 Soppimath, K.S., Kulkarni, A.R. and Aminabhavi, T.M. (2001) *Journal of Controlled Release*, **75**, 331.

13 Vinogradov, S.V., Zeman, A.D., Batrakova, E.V. and Kabanov, A.V. (2005) *Journal of Controlled Release*, **107**, 143.

14 Lin, Y., Chen, Q. and Luo, H. (2007) *Carbohydrate Research*, **342**, 87.

15 Chan, Y., Bulmus, V., Zareie, M.H. et al. (2006) *Journal of Controlled Release*, **115**, 197.

16 Kim, I.S., Jeong, Y.I. and Kim, S.H. (2000) *International Journal of Pharmaceutics*, **205**, 109.

17 Murthy, N., Thng, Y.X., Schuck, S. et al. (2002) *Journal of the American Chemical Society*, **124**, 12398.

18 Miyata, T., Asami, N. and Uragami, T. (1999) *Nature*, **39**, 766.

19 Zhang, Y., Zhu, W., Wang, B. and Ding, J. (2005) *Journal of Controlled Release*, **105**, 260.

20 Wang, L.Y., Gu, Y.H., Zhou, Q.Z. et al. (2006) *Colloids and Surfaces B: Biointerfaces*, **50**, 126.

21 Nolan, C.M., Serpe, M.J. and Lyon, L.A. (2004) *Biomacromolecules*, **5**, 1940.

22 Nolan, C.M., Gelbaun, L.T. and Lyon, L.A. (2006) *Biomacromolecules*, **7**, 2918.

23 Kim, J.J. and Park, K.J. (2001) *Journal of Controlled Release*, **77**, 39.

24 Nayak, S., Lee, H., Chmielewski, J. and Lyon, L.A. (2004) *Journal of the American Chemical Society*, **126**, 10258.

25 Das, M., Mardyani, S., Chan, W.C.W. and Kumacheva, E. (2006) *Advanced Materials*, **18**, 80.

26 Zhang, H., Mardyani, S., Chan, W.C.W. and Kumacheva, E. (2006) *Biomacromolecules*, **7**, 1568.

27 Kiser, P.F., Wilson, G. and Needham, D. (2000) *Journal of Controlled Release*, **68**, 9.

28 Mears, S.J., Deng, Y., Cosgrove, T. and Pelton, R.H. (1997) *Langmuir*, **13**, 1901.

29 Tam, K.C., Ragaram, S. and Pelton, R.H. (1994) *Langmuir*, **10**, 418.

30 Wu, C. and Zhou, S. (1996) *Journal of Polymer Science Part B: Polymer Physics*, **34**, 1597.

31 Woodward, N.C., Chowdhry, B.Z., Leharne, S.A. and Snowden, M.J. (2000) *European Polymer Journal*, **36**, 1355.

32 Crowther, H.M., Morris, G.E., Vincent, B. and Wright, N.G. (2003) in *Role of Interfaces in Environmental Protection* (ed. S. Barany), Kluwer Academic Publishers, Netherlands, p. 169.

33 Bradley, M., Vincent, B., Warren, N., Eastoe, J. and Vesperinas, A. (2006) *Langmuir*, **22**, 101.

34 Bradley, M. and Vincent, B. (2005) *Langmuir*, **21**, 8630.

35 Nerapusri, V., Keddie, J.L., Vincent, B. and Bushnak, I.A. (2007) Langmuir, in press.

36 Bradley, M., Vincent, B. and Burnett, G. (2007) Langmuir, in press.

37 Saunders, B.R. and Vincent, B. (1996) *Journal of the Chemical Society-Faraday Transactions*, **92**, 3385.

38 Saunders, B.R. and Vincent, B. (1997) *Colloid and Polymer Science*, **275**, 9.

39 Bradley, M., Ramos, J. and Vincent, B. (2005) *Langmuir*, **21**, 1209.

40 Routh, A.F., Fernandez-Nieves, A., Bradley, M. and Vincent, B. (2006) *The Journal of Physical Chemistry B*, **110**, 12721.

41 Otero, C., Robledo, L. and Alcantara, A.R. (1995) *Journal of Molecular Catalysis B-Enzymatic*, **1**, 23.

42 Bulmus, V., Chan, Y., Nguyen, Q. and Tran, H.L. (2007) *Macromolecular Bioscience*, **7**, 446.

43 Nayak, S. and Lyon, L.A. (2004) *Angewandte Chemie-International Edition*, **43**, 6706.

44 Gan, D. and Lyon, L.A. (2002) *Macromolecules*, **35**, 9634.

45 Grabstain, V. and Bianco-Peled, H. (2003) *Biotechnology Progress*, **19**, 1728.

46 Nolan, C.M., Reyes, C.D., Debord, J.D. et al. (2005) *Biomacromolecules*, **6**, 2032.

47 Silva, C.S.O., Baptista, R.P., Santos, A.M. et al. (2006) *Biotechnology Letters*, **28**, 2019.

48 Zhang, J.G., Xu, S.Q. and Kumacheva, E. (2004) *Journal of the American Chemical Society*, **126**, 7908.
49 Zhu, M.Q., Wang, L.Q., Exarhos, G.J. and Li, A.D.Q. (2004) *Journal of the American Chemical Society*, **126**, 2656.
50 Han, M.Y., Gao, X.H., Su, J.Z. and Nie, S. (2001) *Nature Biotechnology*, **19**, 631.
51 Bradley, M., Bruno, N. and Vincent, B. (2005) *Langmuir*, **21**, 2750.
52 Kuang, M., Wang, D.Y., Bao, H.B. *et al.* (2005) *Advanced Materials*, **17**, 267.
53 Gong, Y.J., Gao, M.Y., Wang, D.Y. and Moehwald, H. (2005) *Chemistry of Materials*, **17**, 2648.
54 Li, J., Liu, B. and Li, J.H. (2006) *Langmuir*, **22**, 528.
55 Kuang, M., Wang, D.Y. and Moehwald, H. (2005) *Advanced Functional Materials*, **15**, 1611.
56 Kuang, M., Wang, D.Y. and Moehwald, H. (2006) *Chemistry of Materials*, **18**, 1073.
57 Sheeney-Haj-Ichia, L., Sharabi, G. and Willner, I. (2002) *Advanced Functional Materials*, **12**, 27.
58 Gorelikov, I., Field, L.M. and Kumacheva, E. (2004) *Journal of the American Chemical Society*, **126**, 15938.
59 Das, M., Sanson, N., Fava, D. and Kumacheva, E. (2007) *Langmuir*, **23**, 196.

3
Stability of Fluorinated Systems: Structure-Mechanical Barrier as a Factor of Strong Stabilization

Eugene D. Shchukin, Elena A. Amelina†, and Aksana M. Parfenova

3.1
Introduction

The development of methods for guaranteeing the high stability and non-toxicity of finely dispersed oil-in-water (o/w) emulsions of fluorinated organic liquid phases (FL), which have become widely used in various areas, including medicine and biology, calls for a detailed investigation of the properties of the boundary layers between FL and aqueous solutions of surfactants [1–8]. Good emulsifying properties with respect to o/w emulsions of FL (compounds that are peculiar in many respects in comparison to hydrocarbons) are displayed by nonionic surfactants, particularly block-copolymers of ethylene oxide and propylene oxide, Pluronics, and Proxanols with sufficiently high molecular weights ($\sim$8000 or more). At the same time, the emulsifying ability of these surfactants depends on the nature of the phase being emulsified and is different for different FL [2, 9–11]. One could believe that "related" fluorinated surfactants may work even better. However, our first experiments showed that fluorinated surfactants were often poorer stabilizers for fluorocarbon emulsions with respect to coalescence than the usual organic amphiphiles, such as Pluronics [12, 13]. Both practical aspects, the use of fluorocarbon aqueous emulsions as artificial blood substitutes, and interest in colloid-chemical peculiarities of interfacial layers in such systems initiated our comparative investigations of the properties and structure of adsorption layers of various surfactants – "ordinary" (hydrocarbon) and fluorinated ones at various interfacial boundaries between their aqueous solutions and nonpolar phases – either hydrocarbon or perfluorocarbon liquids, with a special attention to variety brought about by their character. An important role in the analysis and explanation of the obtained data is played by our work, following on from Rehbinder, on the lyophilic structure-mechanical barrier that is formed by a surfactant at a given interface, serving as a strong factor of stabilization due to the mechanical strength of such a layer [14–16]. This chapter presents a short survey of some results from the cycle of studies in this field made at the Colloid Chemical Department of the Chemical Faculty of Moscow State University in collaboration with

Highlights in Colloid Science. Edited by Dimo Platikanov and Dotchi Exerowa
Copyright © 2009 WILEY-VCH Verlag GmbH & Co. KGaA, Weinheim
ISBN: 978-3-527-32037-0

the Institute of Physical Chemistry and Electrochemistry and Institute of Elemento-Organic Compounds of the Russian Academy of Sciences [17–21].

As hydrocarbon surfactants (HS), the common water-soluble nonionic amphiphiles (block-copolymers of ethylene and propylene oxides) were used: Pluronic F-68 (mol. wt 8700) and several proxanols, particularly Proxanol 168 (mol. wt 8000) and Proxanol 268 (mol. wt 13 000). As a water-soluble nonionic fluorinated surfactant (FS), perfluorodiisononenyl-poly(ethylene glycol) C_9F_{17}–O–$(EO)_{20}$–C_9F_{17} (φ-PEG) was selected in studies described here. Normal alkanes C_6–C_8 were used as hydrocarbon liquids (HL). Various fluoroorganic compounds served as fluorocarbon liquids (FL), including those with normal and cyclic structures, as well as tertiary amines, piperidines, ethers, and so on [20]. The data presented here are mostly for three typical examples: perfluorodecalin $C_{10}F_{18}$ (PFD), which is moderately soluble in water, perfluoromethylcyclohexylpiperidine, $CF_3(C_6F_{10})(NC_5F_{10})$ (PFMCHP), which is less soluble than PFD, and perfluorotributylamine, $(C_4F_9)_3N$ (PFTBA), which is practically insoluble.

Several independent experimental methods were applied that allowed comparison of the properties of these systems [18–20]. We present only the principal results of the three following approaches: (i) rheological studies of interfacial adsorption layers (IAL) by the rotating suspension method; (ii) observation of the compression of two nonpolar droplets in the surfactant aqueous solution, with measurement of the force f_{coal} needed for their coalescence; and (iii) evaluation of the free energy of interaction between nonpolar groups of IAL and various nonpolar liquids by measuring the contact rupture force between two methylated (or fluorinated) smooth solid particles in a given liquid.

3.2
Rheological Studies of Interfacial Adsorption Layers in Fluorinated Systems

The rheological behavior of the IAL formed between a nonpolar phase (HL, FL) and an aqueous solution of a surfactant (HS, FS) was studied by the rotation suspension ("torque pendulum") method [13–15]. Its principle is illustrated in Figure 3.1. As a rule, the conditions of constant rate of revolution Ω of the cylindrical vessel were used, that is, the constant shear strain rate. Both the revolution rate and the surfactant concentrations were varied over a broad range.

As a typical example, Figure 3.2 presents the $\tau = \tau(t)$ curves obtained with Pluronic F-68 (HS) adsorption layers at the interfaces with heptane (HL) and with PFD and PFTBA (FL), where τ is the shear stress related to the whole thickness of the layer (in the $N\,m^{-1}$) and t is the time to which the shear strain (deformation) is proportional, with $\Omega = 0.084\,rad\,s^{-1}$ [13, 20, 22].

The rheological behavior of IAL at the interface between Pluronic aqueous solution and heptane (HS/HL system) is substantially different from that of the same Pluronic aqueous solution in contact with PFD and PFTBA (HS/FL systems). In the first case (HS/HL), the development of deformation is similar to liquid flow. In the second case (HS/FL), nonlinear solid-like behavior is observed. After some quasi-elastic

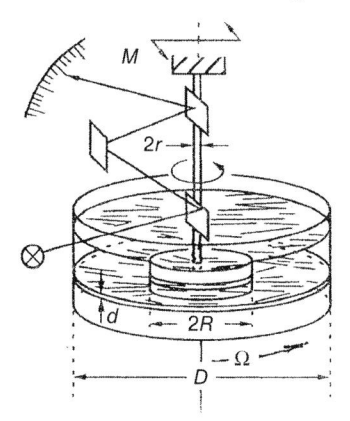

Figure 3.1 Scheme illustrating the use of rotating suspension for studying the rheological characteristics of the interfacial adsorption layer between polar and nonpolar liquids [15].

(reversible) deformation, a typical sharp maximum appears, indicating the existence of the critical shear stress, that is, specifically, the *mechanical strength* of the IAL. Alongside the principal differences between HS/HL and HS/FL systems, one can also see some differences between two HS/FL couples. For PFTBA the strength is higher than for PFD. We mention here that in the case of PFTBA the free energy of interaction (F) is larger, and the work of adhesion (W) is smaller, than for PFD (Section 3.4). Thus, the nature of the nonpolar liquid phase affects the mechanical properties of the IAL in both systems.

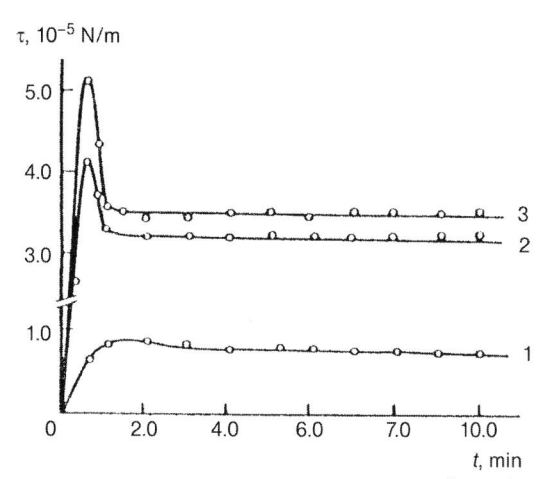

Figure 3.2 Dependence of the shear stress (τ, 10^{-5} N m^{-1}) on the deformation time t (min) of the adsorption layer of Pluronic F-68 at the boundaries between its 10^{-3} M solution and nonpolar liquids: heptane (1), PFD (2) and PFTBA (3); $\Omega = 0.084$ rad s^{-1} [20].

Stress–strain curves analogues to those shown in Figure 3.2 are also observed for the IAL formed at the boundary between the same PFD and solutions of different Proxanols [20, 22]. For all such systems nonlinear rheological behavior takes place, with typical maxima in $\tau = \tau(t)$ curves, which correspond to the critical shear stress (the strength of IAL). The levels of these maxima differ: the highest strength corresponds to the most developed hydrophilic (ethylene oxide) part of surfactant molecules.

3.3
Studies of the Rupture and Coalescence of Individual Droplets

The method proposed by us was used, based on measurements of the forces providing coalescence of two identical droplets and rupture of the column so-formed [13, 20, 22, 23]. These forces are measured using an electromagnetic system (a galvanometer) as the highly sensitive dynamometer (Figure 3.3), described in detail in Refs [15, 21, 24], and applied also in measurements of cohesion forces between solid particles (Section 3.4). Two drops of nonpolar liquid, HL or FL, in the form of spherical segments are applied to the tip surfaces of the holders – glass rods of identical diameter of $2r = 0.5$–0.7 mm immersed in the surfactant aqueous solution. These surfaces are perfectly smooth and strictly perpendicular to the axes of the rods. The shape of the holders guarantees the movement of samples co-axially along the

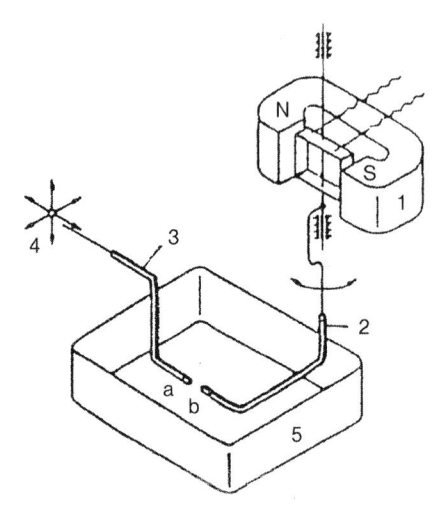

Figure 3.3 Scheme of the device for compressing and coalescence of two nonpolar droplets and for rupturing the resulting one in aqueous surfactant solution. Droplets a and b are fixed within a container (5) at the end faces of cylindrical holders (3 and 2) that are connected, respectively, to a manipulator (4) and the coil of a magneto-electric galvanometer (1). The compressive (p) and tensile (f) forces are determined by the current passing through the coil of galvanometer; the approach and break of droplets are observed through a microscope. Solid particles can be used also as samples a and b [15, 21].

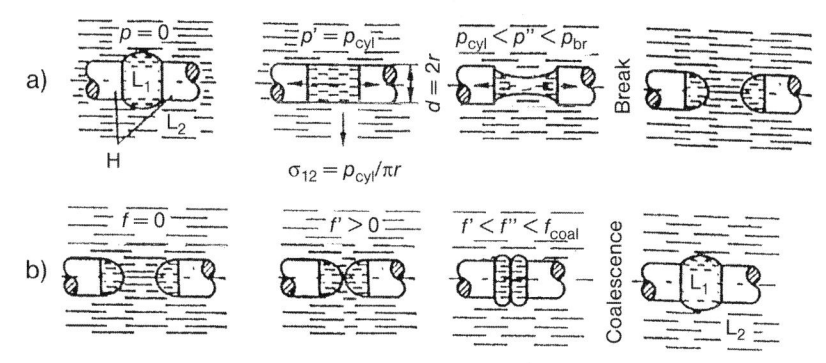

Figure 3.4 Scheme of experiments: Extension and breaking of a liquid (L_1) droplet, placed between cylindrical holders H with the diameter $2r$ in the dispersion medium L_2, by the tensile force p (a); and coalescence of the two resultant droplets due to the compressive force f (b). The p_{cyl} force at which the droplet assumes the form of the cylindrical column gives the interfacial tension $\sigma_{12} = p_{cyl}/\pi r$. An increase in the tensile force results in a break of the column at p_{br}. The compressive force f_{coal} determines the strength of the (symmetrical) film between droplets of L_1 [20, 23].

normal to the surfaces. The drops are fixed by hydrophobization of the tip surfaces of the holders.

During measurement, two drops applied to the holders are first brought together until they merge and form one spherical segment. Then a tensile force that gradually extends the column is applied (Figure 3.4). The force p_{cyl} needed to maintain the cylindrical shape of the column is determined with good accuracy. This measured force directly relates to the value of the interfacial tension: $\sigma_{12} = p_{cyl}/\pi r$. The time during which the cylindrical shape of the column is maintained can be varied from several seconds to several hours; this allows study of the adsorption kinetics. Subsequently, the tensile force is increased. The column breaks into two drops: an elementary event of dispersion (rupture of an asymmetric adsorption monolayer) takes place, when p_{br} of the tensile force is reached. Observations showed that the column always separated to form two equal spherical segments.

The drops formed by the split of the column are then brought again into contact. Their flattening and subsequent coalescence are observed. In the absence of any surfactants (in pure water!), the drops of pure HL (or FL) merge spontaneously. In a surfactant solution, it is necessary to apply a definite compressive force f_{coal} for the coalescence of HL or FL drops (rupture of a symmetric adsorption bilayer). Hydrostatic effects were avoided by carrying out the experiments with small volumes of nonpolar liquid; the ratio l/r of the length of the cylindrical column to its radius is 1.0–1.2.

The general feature of the results obtained in the described experiments is the essential dependence of both p_{br} and, especially, f_{coal} on the nature of a nonpolar liquid. Figure 3.5 presents three plots of f_{coal} versus the Pluronic F-68 concentrations (C): for PFTBA, PFD and heptane. In all cases, the initial resistance f_{coal} in pure water is zero. The f_{coal} reaches its maximum at 10^{-7}–10^{-6} M, and drops moderately at higher concentrations. This behavior is in accordance with known data on emulsions

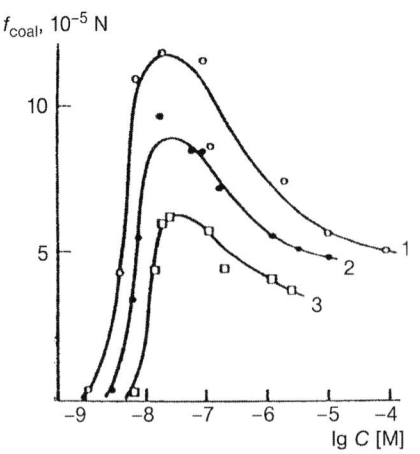

Figure 3.5 Dependence of compressive force (f_{coal}, 10^{-5} N) needed to cause the coalescence of droplets on the concentration C of Pluronic F-68: (1) PFTBA, (2) PFD and (3) heptane [22, 23].

stability [25–28]. However, the levels of the maxima are significantly different. These data become more illustrative and informative when f_{coal} values are compared at the same, reasonably low, Pluronic concentrations (Table 3.1). At F-68 concentration of 5×10^{-9} M, the resistance towards the coalescence of droplets for the HS/FL systems is up to two orders of magnitude higher than that for the HS/HL systems. Among the HS/FL systems, the resistance for PFTBA is more than that for PFD.

Such behavior is ultimately related to various interactions between the Pluronic molecules in the adsorption layer and between the same molecules, especially their hydrophobic parts and the nonpolar phase. It is reasonable to assume that for surfactant molecules with the same hydrophilic groups the perfect similarity (high *affinity*) between their hydrophobic groups and the nonpolar phase, that is, perfect adhesion and mutual solubility, result in deep immersion of such groups into this phase, without forming a dense layer. Conversely, lesser physical-chemical similarity (poorer affinity, lesser adhesion) leads to their "pushing out," to the aqueous phase,

Table 3.1 Coalescence forces f_{coal}, 10^{-5} N of the droplets of various nonpolar liquids: in the Pluronic F-68 aqueous solution (5×10^{-9} M) and in the φ-PEG aqueous solution (5×10^{-6} wt%), and decrease in the interfacial tension $\Delta\sigma$, mJ m^{-2} caused by the surfactant adsorption at corresponding boundaries [18, 20].

Nonpolar drop	In F-68 solution				In φ-PEG solution	
	Heptane	PFD	PFMCHP	PFTBA	Heptane	PFD
f_{coal} (10^{-5} N)	<0.01	2.5	9	>10	1.2	0.1
$\Delta\sigma$ (mJ m^{-2})	15	17	12	10	4	14

where attraction between the hydrophobic groups is stronger; this results in increasing density, viscosity and strength of IAL.

An inverse ("mirror-like") behavior can be expected in the case of surfactants containing a perfluorinated radical as hydrophobic part. Specifically, the tests performed on heptane and PFD drops in aqueous solutions of φ-PEG show that the stability is higher for heptane (FS/HL system) than for PFD (FS/FL). For example, at $C \approx 5 \times 10^{-6}$ wt%, f_{coal} is approximately 1.2 dyn and only about 0.1 dyn for heptane and PFD, respectively.

Table 3.1 also gives values for $\Delta\sigma$ [$= \sigma_0 - \sigma(C)$], the decrease in interfacial tension at the corresponding boundaries between nonpolar drops and aqueous medium. Differences among these values are not significant; however, some general trend may be noted: the higher the resistance to coalescence, the lesser is the corresponding decrease in interfacial tension. This, again, is in agreement with the notion that resistance to coalescence is related to some deficiency in adhesion between hydrophobic groups of surfactant molecules and nonpolar liquid, that is, in Rehbinder's terms, to the "nonperfect compensation" of intermolecular bonds at the boundary [14, 15].

3.4
Studies of the Interaction of Hydrophobized Solid Surfaces in Nonpolar Liquids

The adhesion between the IAL (its hydrophobic part) and an adjacent nonpolar phase can be modeled by the adhesion, in the same liquids, of the modified solid surfaces simulating the hydrophobic parts of corresponding surfactants. In this method, experimental complications connected with the mutual solubility of components are prevented. For the solid/liquid interface, the principal quantitative characteristic of interaction, the free energy of interaction F (mJ m^{-2}, in the plain-parallel gap or film) can be established experimentally from Derjaguin's equation: $p = \pi R F$, where p is the cohesive force in the immediate contact between two spherical particles immersed in the corresponding liquid medium [29, 30].

For such measurements, we model the considered systems as two spherical, molecularly smooth, glass particles with radius R of 1–1.5 mm, with hydrophobized surfaces: methylated surfaces, simulating hydrophobic parts of HS, and fluorinated surfaces, simulating hydrophobic part of FS, in various hydrocarbon and fluorocarbon media (HL, FL). Practically, only dispersion interactions take place in these systems, making their quantitative consideration particularly definite – uncomplicated by combination with various polar components [16, 31, 32]. The cohesion force p between these two particles is measured by a device based on the magneto-electric dynamometer, similar to the apparatus used in experiments with droplets [15, 23, 33].

The data obtained for model HS/FL and HS/HL systems are reported in Table 3.2 (traditionally, as $F/2$ values, by modulus). Differences in the F values for these two systems reach orders of magnitude. For two related phases, HS/HL, very low F values characterize the full lyophilicity. For all studied HS/FL systems, using many fluoroorganic compounds mentioned in Section 3.1, the values of F are in the range

Table 3.2 Free energies of interaction $F/2$ (mJ m^{-2}) of adhesion W (mJ m^{-2}), and interfacial σ_{SL} (mJ m^{-2}) for solid hydrocarbon surfaces in various nonpolar liquids; symbols without subscript refer to the authors' experimental data, those with subscripts (A) and (F) are to approximations following Antonov and Fowkes, respectively (see subsequent text for details) [20, 31].

Liquid phase	$F/2$	W	W^*	$W_{(A)}$	$W_{(F)}$	σ^*_{SL}	$\sigma_{SL(A)}$	$\sigma_{SL(F)}$
PFTBA	7.0	33	33	34	39	6.5	5	0.3
PFMCHP	6.8	35	38	39	41.5	3.5	2.5	<0.1
PFD	4.7	37	38	39	42	4	2	<0.1
Heptane	<0.01	42	41	41	42	<0.01	<1.0	0

Data based on measurements of the angle of wetting are marked with*.

$F/2 \gtrsim 5$ mJ m^{-2}, which is evidence of the essentially lower affinity between phases [20]. The F values decrease from the highest for PFTBA (insoluble in water and showing the best stabilization with respect to coalescence) to lower values for the more soluble and less stable PFMCHP and PFD. Correspondingly, similar behavior takes place for the inverse ("mirror-like") systems FS/HL and FS/FL; for FS/HL, $F/2$ values also reach several mJ m^{-2}. Indeed, the perfect stabilization with respect to coalescence is connected with some excess in the free energy of interaction, that is, the deficiency in affinity between nonpolar parts of the surfactant molecules and the adjacent nonpolar phase (and low mutual solubility of components).

It is expedient to relate these "F-measurements" with the estimations, both experimental and theoretical, of the energy (work) of adhesion and interfacial energy between two neighboring nonpolar phases.

After Gibbs, we relate the free energy values of the liquid (subscript L) film between two identical solid (S) surfaces σ_{film}, interfacial σ_{SL}, and of the interaction $F_{(SLS)} = F$ (represented by modulus): $\sigma_{film} = 2\sigma_{SL} - F$. The Dupre rule states for the energy of adhesion $W_{(SL)} = W$: $\sigma_{SL} + W = \sigma_S + \sigma_L$ (no adsorption takes place). With model HS/FL and HS/HL systems, σ_S is the surface energy of the methylated surfaces in air, and σ_L represents the surface tension of the various nonpolar liquids studied.

We have shown that for methylated surfaces σ_{film} is small with respect to σ, and Derjaguin's equation gives: $\sigma_{SL} \approx F/2 \approx p/2\pi R$ [15, 16, 21]. For methylated surfaces in air $F/2 = p/2\pi R \approx 22$ mJ m^{-2}; this is a reasonable estimation of the surface free energy of a hydrocarbon. The Dupre rule can then be expressed in a form allowing experimental estimation: $W = \sigma_S + \sigma_L - F/2$, where all three terms are measured independently.

After Fowkes [34, 35], the dispersion (d) component of the work of adhesion equals: $W_{(F)} = 2(\sigma_S^d \sigma_L^d)^{1/2}$, and $\sigma_{SL(F)} = [(\sigma_S^d)^{1/2} - (\sigma_L^d)^{1/2}]^2 = F_{(F)}/2$. Similarly, the geometric mean approximation is used to describe molecular interactions in the Hildebrand theory of solubility parameters [36], and also for calculating the combined Hamaker const. [16, 27, 30]).

A different estimation gives Antonov's empirical rule, known for two liquids, when applied also to S/L interface: $\sigma_{12} \approx \sigma_1 - \sigma_2$, that is, $W_{(A)} = 2\sigma_L$, and $F_{(A)}/2 \approx \sigma_S - \sigma_L$

This means that the "rupture plane" between phases is located not exactly at their boundary but in the weakest one.

In all these approximations, the value W is characteristic of the strength of adhesion between molecules of two phases, while σ_{SL} and $F/2$ all characterize the deficiency in this adhesion energy (i.e. they are antibatic to the previous ones).

In Table 3.2, these approaches are applied to our systems: methylated solid surfaces and four characteristic nonpolar liquids. The adhesion energy W values are estimated as mentioned above: $W = \sigma_S + \sigma_L - F/2$. W^* values are independently obtained data on the work of adhesion for considered *nonpolar phases* on the basis of contact angle of wetting measurements [20, 34] as $W^* = \sigma_L(1 + \cos\vartheta)$; σ_{SL}^* are evaluations of interfacial energy in these systems as $\sigma_{SL}^* = \sigma_S - \sigma_L \cos\vartheta$. $W_{(A)}$ and $W_{(F)}$, $\sigma_{SL(A)}$ and $\sigma_{SL(F)}$ are the estimations calculated for the same systems following Antonov and Fowkes, respectively, with the same σ_S and σ_L.

A principal tendency can be seen in these data, despite their relatively coarse estimations (assumptions include small σ_{film}, scatter in the σ_S values for different samples up to 3%, etc.). For heptane, that is, for the HS/HL system, all these values are in complete agreement. However, for all HS/FL systems, the adhesion energy $W_{(F)}$ is obviously overestimated. Correspondingly, the data in Table 3.2 show significant underestimation of the $\sigma_{SL(F)}$ values for the same HS/FL couples. This means that the concept of the "geometrical mean," applicable to hydrocarbon systems, is not valid in the case of the HS/FL contact, although the dispersion forces only take place in both cases [16, 31, 34]. The real adhesive contact HS/FL is weaker (compensation of molecular forces is not perfect), and the traditional "$(\sqrt{\ } - \sqrt{\ })^2$," which gives very low estimations for σ_{12} and $F/2$, is not applicable.

Thus, it should be stressed that interactions between hydrophobic parts of surfactant molecules and nonpolar liquid phase play a critical role in controlling the emulsion stability (Davis [2]). Of course, each particular system needs an individual approach in the quantitative evaluation of such interactions. However, we see that in the case of fluorocarbon emulsions stabilization (with respect to coalescence), the high stability relates to some deficiency in the HS/FL adhesion, when some kind of "squeezing out" of hydrophobic radicals from the nonpolar liquid phase can take place.

3.5
Discussion

What has been described up to this point allows us to propose a scheme that generalizes the results obtained in this work.

In the earlier studies of Peter A. Rehbinder and the present authors, the idea and role of a *lyophilic structure-rheological*, or *structure-mechanical barrier* as a factor of strong stabilization in colloid systems was formulated and developed [14, 15, 37]. The *mechanical strength* of the IAL is considered as the principal characteristic of the layer. This mechanical strength in combination with the increased viscosity and elasticity (complex, nonlinear behavior) gives rise to the resistance of thin interfacial films to rupture, preventing coalescence of the drops (bubbles) in real, concentrated disperse systems at any salinity, temperature, and so on (properties known for protective colloids). Such a *strong* factor of stabilization, including the obligatory peculiar mechanical parameter ("non-thermodynamic" one [15]), the strength, may be opposed to relatively weak factors of stabilization: of thermodynamic nature (DLVO electrostatic barrier, elasticity of thin films after Gibbs), and also those predicted by kinetics (Marangoni–Gibbs effect), or of hydrodynamic origin (Reynolds flow).

The authors believe that the idea of the structure-mechanical barrier can serve as a basis for explaining the observations encountered in this study. The mechanical strength of IAL found, particularly in the FS/HL and HS/FL systems, can arise only due to the presence of some special, sufficiently dense structure in these layers. For water-soluble surfactants, such strength can not be provided by the hydrophilic ethylene oxide tails and loops. Being highly hydrated they are spread in the aqueous phase like tentacles. These tentacles take part in stabilization towards coagulation, but they play only a limited role in the resistance of the interfacial film to thinning, mainly at relatively large distances ("the steric factor"). However, their role can be found in "pulling out" hydrophobic parts of the surfactant molecules from the nonpolar phase close to the geometrical interface. Such an effect is more significant the more developed are the hydrophilic parts of the surfactant molecules (as was observed for different Pluronics) – and the less is their affinity for the nonpolar phase (i.e. in the case of PFTBA).

In this aspect, the most important factor is the interaction between hydrophobic parts of surfactant molecules and nonpolar liquid. In the case of high affinity between them, when HS/HL and FS/FL systems are considered (good adhesion, very low values of the free energy of interaction F), the hydrophobic tails and loops can also behave like tentacles, being unable to form a mechanically strong structure. Conversely, if the adhesion of such hydrophobic parts to the nonpolar phase is relatively poor (deficiency in affinity, high F values), the hydrophobic loops and tails are "pushed out" from the bulk of nonpolar liquid to the boundary with water, where their mutual attraction is stronger, allowing them to aggregate into a compact structure – the real carrier of mechanical strength. These are the HS/FL and FS/HL systems.

In our view, Rehbinder's doctrine on the structure-mechanical barrier had been proposed much earlier than the idea of steric stabilization. The latter, related (at least, initially) to the conformational statistics of hydrophilic tails and loops, presents only the entropic part of elasticity and is not responsible for the mechanical strength.

The authors stress that the structure-rheological properties of the IAL (mechanical strength), obviously important in preventing *coalescence*, may be not sufficient in themselves for the full stabilization. Preventing *coagulation* requires a high lyophilicity (hydrophilicity in our case) of this structure-rheological barrier with respect to the ambient polar medium, often through some classical ionic additives (such as SDS). The mechanical strength of IAL can not withstand the Ostwald ripening effect; however, the rate of such ripening does depend upon the composition of the IAL [18, 38].

3.6
Conclusion

The stabilizing effect of interfacial adsorption layers (IAL) formed by ordinary hydrocarbon surfactants (HS) and by the fluorinated (FS) ones has been studied at the boundaries of their aqueous solutions and hydrocarbon (HL) or fluorocarbon (FL) nonpolar liquid phase.

The mechanical (rheological) studies of the considered IALs by the rotating suspension method show an essential difference between the "liquid-like" behavior of such layers for HS/HL and FS/FL systems and the "solid-like" one (manifestation of the critical shear stress, the strength) in both FS/HL and HS/FL.

Direct observations of the individual droplets of nonpolar liquids coalescence in aqueous solutions of regular and fluorinated surfactants reveal that the compressive forces (f_{coal}) resulting in coalescence of two droplets, that is, the strength of IAL, can be much more in the case of FS/HL and HS/FL systems than for HS/HL and FS/FL ones.

Measurements of free energy of interaction F of either methylated or fluorinated solid surfaces in hydrocarbon and in fluorocarbon liquids give F values as low as $0.01\,\text{mJ}\,\text{m}^{-2}$ for two methylated surfaces in hydrocarbon liquids and for both fluorinated phases (the full lyophilicity). However, for methylated surfaces in fluorinated liquids (and for fluorinated surfaces in hydrocarbon liquids), the F reaches several $\text{mJ}\,\text{m}^{-2}$. These F-measurements and corresponding evaluations of the work of adhesion and of the free interfacial energy between two nonpolar phases show that the interaction (adhesion) between hydrocarbon and fluorocarbon phases is significantly weaker, and that the interfacial energy is much higher than for both hydrocarbon, or both fluorocarbon, neighbors (this means that the traditional approach in estimating mutual dispersion interactions as a geometrical mean does not work here).

The studies presented show undoubtedly that the nature of the nonpolar liquid phase (after Davis) and its interaction with the hydrophobic parts of surfactant molecules play important roles in the stabilization of emulsions. All these data may be explained as a manifestation of the structure-mechanical barrier formed by the IAL serving as a factor of the strong stabilization with respect to coalescence. In particular, in fluorinated systems such a barrier can arise in the case of relatively low similarity (affinity) between nonpolar liquid phase and the nonpolar groups of surfactant molecules, causing the "expulsion" of the latter from a liquid and the formation of dense IAL with stronger molecular interaction and, consequently, high mechanical strength. Further elucidation of this factor may assist in selecting optimal surfactants and conditions for controlling the stability of emulsions.

References

1 Patrick, F.M. (1982) in *Preparations, Properties and Industrial Applications of Organofluorine Compounds*, (Banks, R. E., ed.), Ellis Horwood Series in Chem. Sci., Ellis Horwood, p. 323.

2 Davis, S.S., Perwall, T.S., Buscal, R., Smith, A.L. et al. (1987) *Proceedings of the International Conference on Colloid and Interface Science*, Vol. 2, Academic Press, New York, p. 265.
3 Napoli, M., Fraccaro, C., Scipioni, A. and Alessi, P. (1991) *Journal of Fluorine Chemistry*, **51**, 103.
4 Krafft, M.P. and Riess, J.D. (1998) *Biochimie*, **80**, 489.
5 Pabon, M. and Corpart, J.M. (2002) *Journal of Fluorine Chemistry*, **114**, 149.
6 Mohamed, M.Z., Ghazy, E.A., Badawi, A.M. and Kandil, N.G. (2005) *Journal of Surfactants and Detergents*, **8**, 181.
7 Polidori, A., Presset, M., Lebaupain, F., Ameduri, B. et al. (2006) *Bioorganic & Medicinal Chemistry*, **16**, 5827.
8 Soloshonok, V.A., Mikami, K., Yamazaki, T. et al. (eds) (2007) *Current Fluoroorganic Chemistry*, Oxford University Press.
9 Binks, B.P., Fletcher, P.D.I., Kotsev, S.N. and Thompson, R.L. (1997) *Langmuir*, **13**, 6669.
10 Kiss, E. (ed.) (2001) *Fluorinated Surfactants and Repellants*, Surfactants Science Series, Vol. 97, CRC Press.
11 Kirsch, P. (2004) *Modern Fluoroorganic Chemistry*, Wiley-VCH.
12 Amelina, E.A., Parfenova, A.M., Safronova, N.A., Shchukin, E.D. et al. (1984) *Colloid Journal*, **46**, 561, 1045, 1047.
13 Shchukin, E.D., Amelina, E.A., Parfenova, A.M. et al. (1988) *Colloid Journal*, **50**, 677.
14 Rehbinder, P.A. (1978) *Izbrannye Trudy: Kolloidnaya Khimiya (Selected Works: Colloid Chemistry)*, Nauka, Moscow.
15 Shchukin, E.D. (1997) *Colloid Journal*, **59**, 248.
16 Shchukin, E.D., Pertsov, A.V., Amelina, E.A. and Zelenev, A.S. (2001) *Colloid and Surface Chemistry*, Elsevier.
17 Amelina, E.A., Makarov, K.N., Safronova, N.A. and Gervits, L.L. (1984) *Zhurnal Vsesojuznogo Khimicheskogo Obshchestva imeni D.I. Mendeleeva (Journal of the D.I. Mendeleev All-Union Chemical Society)*, **29**, 468, 588.
18 Amelina, E.A., Kumacheva, E.E., Pertsov, A.V. and Shchukin, E.D. (1990) *Colloid Journal*, **52**, 185.
19 Amelina, E.A., Kumacheva, E.E., Chalykh, A.E. and Shchukin, E.D. (1996) *Colloid Journal*, **58**, 415, 420.
20 Shchukin, E.D., Amelina, E.A. and Parfenova, A.M. (2001) *Colloids and Surfaces A: Physicochemical and Engineering Aspects*, **176**, 35.
21 Shchukin, E.D. and Amelina, E.A. (2003) *Journal of Dispersion Science and Technology*, **24**, 377.
22 Parfenova, A.M., Amelina, E.A., Vitvitskii, V.M. et al. (1990) *Colloid Journal*, **52**, 700.
23 Shchukin, E.D. (2002) *Journal of Colloid Interface Science*, **256**, 159.
24 Somasundaran, P., Lee, H.K., Shchukin, E.D. and Wang, J. (2005) *Colloids and Surfaces*, **266**, 2.
25 Shchukin, E.D. and Pertsov, A.V. (2007) *Colloid and Interface Science Series*, Vol. 1, Part 1, Wiley-VCH Verlag, Gmbh & Co. KGaA, p. 23.
26 Sjoblom F J. (ed.) (2001) *Encyclopedic Handbook of Emulsion Technology*, Marcel Decker, New York.
27 Holmberg K., Shah D.O., Schwuger F M.J. (eds) (2002) *Handbook of Applied Surface and Colloid Chemistry*, Vol. 2, John Wiley & Sons, Chichester.
28 Mittal, K.H., and Shah, D.O. (eds) (2003) *Adsorption and Aggregation of Surfactants in Solutions*, Decker Inc., New York.
29 Derjaguin, B.V. (1934) *Kolloid Zeitschrift*, **69**, 155.
30 Derjaguin, B.V., Churaev, N.V. and Muller, V.M. (1985) *Poverkhnostnye Sily (Surface Forces)*, Nauka, Moscow.
31 Lifschitz, E.M. (1954) *Doklady Akademii Nauk SSSR*, **97**, 643; (1955) *Zhurnal Eksperimentalnoi I Teoreticheskoi Fiziki*, **29**, 94.
32 van Oss, C.J. (1994) *Interfacial Forces in Aqueous Media*, Marcel Dekker, New York.
33 Shchukin, E.D. (2006) *Advances in Colloid and Interface Science*, **123–126**, 33.

34 Good, R.J. and Chaudhary, M.K. (1991) in *Fundamentals of Adhesion* (ed. L.H. Lee), Plenum Press, New York, p. 137.

35 Fowkes, F.M. (1972) *Journal of Adhesion*, **4**, 155.

36 Hildebrand, J.H. and Scott, R.L. (1950) *Solubility of Nonelectrolytes*, Reinhold Publishing Corp., New York.

37 Shchukin, E.D., Amelina, E.A. and Izmailova, V.N. (2003) *Role of Interfaces in Environmental Protection* (ed. S. Barany), Kluver, Netherlands, p. 81.

38 Kabalnov, A.S. and Shchukin, E.D. (1992) *Advances in Colloid and Interface Science*, **38**, 69.

4
Particle Characterization Using Electro-Acoustic Spectroscopy
Richard W. O'Brien, James K. Beattie, and Robert J. Hunter

4.1
Introduction

The Electrokinetic Sonic Amplitude (ESA) effect in this context refers to the generation of ultrasound by the application of an alternating electric field to a colloid. Previous reviews on the ESA have mainly focused on the determination of particle size and zeta potential from the ESA. While this is certainly a very important application of the ESA phenomenon, there is more information in the ESA spectrum than just particle size and zeta. It can be used, for instance, to determine the thickness of adsorbed polymer layers or the surface conductance under the shear plane. It is these other applications that will be our main interest here. To begin we will give an alternative explanation for the ESA phenomenon, one that allows a deeper understanding of the underlying physics.

4.2
Understanding the ESA Effect

The ESA effect occurs because the particles are electrically charged. The applied field causes the particles to move back and forth, and this backwards and forwards motion generates the ultrasound. This is the usual explanation for the ESA effect. Sound is a backwards and forwards motion, so it seems reasonable that a backwards and forwards motion of the particles will generate sound, and indeed this is usually the case. But this simple explanation does not account for two important aspects of the ESA effect: (i) the ESA effect does not occur in suspensions of neutrally buoyant particles, even though they move back and forth in the electric field, and (ii) the ESA sound waves are generated by the particles, and yet they appear to come from the electrodes.

To simplify things we focus on the ESA generated by a colloid that lies between two parallel plate electrodes. To begin with, the suspension is at rest and the electrodes are

Highlights in Colloid Science. Edited by Dimo Platikanov and Dotchi Exerowa
Copyright © 2009 WILEY-VCH Verlag GmbH & Co. KGaA, Weinheim
ISBN: 978-3-527-32037-0

uncharged. Then a sinusoidal voltage pulse is applied across the plates. We will assume that the distance between the plates is small compared to their height and width, and thus the electric field in the colloid is spatially uniform. Furthermore, the electrodes are taken to be rigid and immobile.

To understand what happens when the voltage is applied, we need to evoke the two rules that govern the motion of a suspension:

1. The rate of change of the momentum in a volume of suspension is equal to the net pressure force acting on that volume: This is more than just a statement of Newton's Second Law, for we are asserting that the only force on the suspension is the pressure force. Readers may be wondering why there is no mention of the electric force in this rule. Won't the electric force on the suspension alter its momentum? The answer is no, because the net electric force on a volume of suspension is zero. We are talking here about a volume that is large enough to enclose many particles. The particles are charged, so they feel an electric force, but this is balanced by an equal and opposite force on the ions in the liquid, because the suspension as a whole is electrically neutral. Thus there is no electric force on the suspension as a whole.[1] It is important to appreciate this point because if this were not the case the ESA effect would not appear to come from the electrodes as we have asserted above.

    Fluid mechanicians might also be puzzled by the fact that the viscous forces (i.e. internal friction of the liquid) do not appear in our statement of Newton's Second Law. They are present, but it can be shown [14] that these forces are much smaller than the pressure forces, and they can be ignored.

2. The pressure at a point in the suspension only changes if the material at that point is being squeezed or expanded: Thus, if there is a net volumetric flow of material into a region of the suspension, the pressure in that volume will rise, and conversely the pressure will fall if there is a flow out.

Immediately after the field is applied the particles will be set into motion. Since the field is spatially uniform, the particles will initially move with a spatially uniform velocity everywhere except within a few particle radii of the electrodes, where they are slowed down by the interactions with the electrode. Since the motion elsewhere is spatially uniform at this stage, there will not be any build up of material, and thus by Rule 2 the pressure will remain at its equilibrium value everywhere except at the electrodes.

Since the pressure is spatially uniform, there will not be any pressure gradients in the suspension, except possibly at the electrodes during this initial phase. Now consider Rule 1 applied to the slab shaped volume indicated by the broken line in Figure 4.1. The momentum in that volume can only change as a result of a pressure difference on the opposing faces of the slab, but at this stage there is no pressure

---

[1] This is only strictly true in the case of weak electric fields. There may be electric forces proportional to E^2, but these are assumed to be negligible in this analysis.

4.2 Understanding the ESA Effect

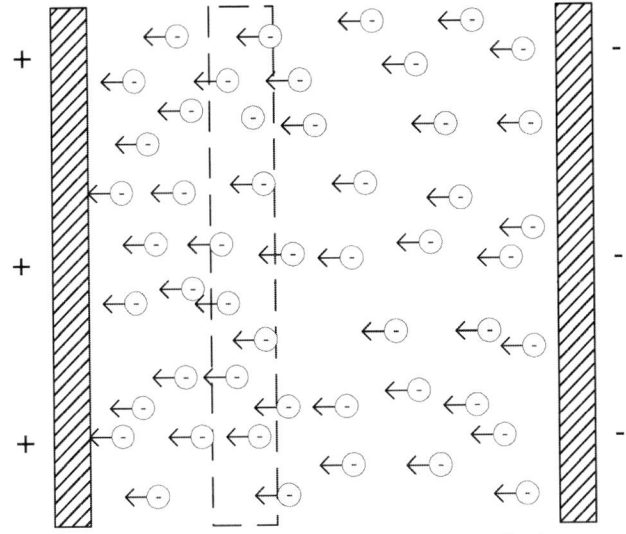

Figure 4.1 Motion of particles due to an applied voltage, for the case of negatively charged particles. The slab-shaped volume referred to in the text is indicated by the dashed lines.

difference, so the momentum is the same as it was before the switch was turned on, when the suspension was at rest.

The electric field is driving the particles towards the electrode of opposite charge, so the particles have momentum, but this momentum must be cancelled by that of the liquid. Thus there must be a flow of liquid towards the opposite electrode, and the magnitude of that flow is determined by the requirement that the total momentum must be zero.

To help simplify the description of what happens next, we focus on the case of negatively charged particles that are denser than the suspension medium. The applied field will cause these particles to move towards the anode. The momentum per unit volume of suspension can be written as the sum:

$$\rho_f v_f + \rho_p v_p \tag{4.1}$$

where ρ is the density and v is the volume flux.[2] The subscripts f and p refer to the fluid and the particles, respectively. As stated above, the momentum in the suspension is zero immediately after the field is applied. Since the particle density is greater than that of the fluid, the volume flux of fluid toward the cathode must be greater than that of the particles towards the anode if the sum (Equation 4.1) is to add up to zero. Thus the net volume flux:

$$v_f + v_p \tag{4.2}$$

---

2) The volume flux is the rate at which material volume sweeps across the left or right face of the slab, per unit area.

will be non-zero, and will be in the direction of the cathode. This must lead to a compression at the cathode and an expansion at the anode, and by Rule 2 the pressure must rise at the cathode and drop at the anode. So immediately after the voltage is turned on, a positive pressure develops at the cathode and a negative pressure at the anode. This is the start of the ESA effect.

Now consider Rule 1 applied to a slab shaped volume like the one in Figure 4.1, but displaced to the right so that one face is at the cathode. The pressure at that face is higher than it is on the other face, so there will be a net pressure force to the left acting on the volume. By Rule 1, this will cause the momentum of the suspension in that volume to increase in the direction of the anode. Thus the liquid and particles will accelerate in this direction, and this will reduce the net volume flux to the right. Since the magnitude of the volume flux in this slab is now smaller than in the region a little further out from the electrode, there will be a compression of the material in this slab. Thus the pressure in the slab will initially increase. This will then lead to a pressure gradient across the neighboring slab and so the process will be repeated, and a pressure disturbance will propagate out from the electrode. The resulting pressure waves are sketched in Figure 4.2, which shows a full cycle of the pressure wave moving away from the electrodes.

From this description it can be seen that the pressure waves do indeed emanate from the electrodes, and the reason they do is that the applied field drives the suspension towards or away from the electrodes, causing expansions and compressions that drive pressure waves out from the electrodes. If the particle density is the same as that of the liquid, then the requirement that the momentum (Equation 4.1) is zero also implies a zero volume flux (Equation 4.2). In this particular case, the volume flux of particles and liquid just cancel out and so there is no compression and

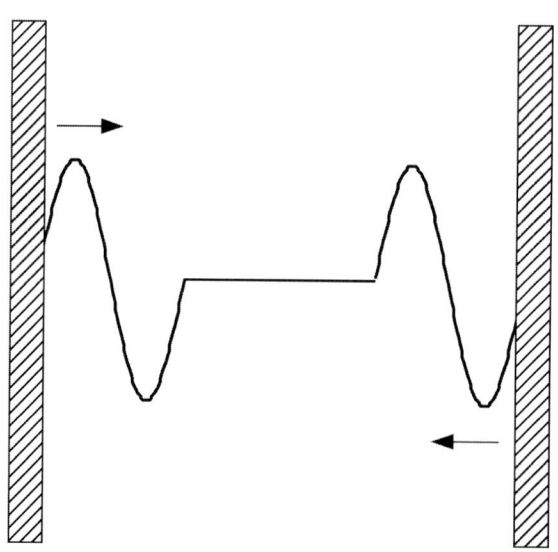

Figure 4.2 Pressure disturbance caused by a sinusoidal voltage pulse. Note that the two waves are 180° out of phase.

expansion at the electrodes. This is why there is no ESA effect in a suspension of neutrally buoyant particles.

Although we have now accounted for the two aspects of the ESA effect we set out to cover, there is still some tidying up to do.

First, if the particles are positively charged or if they are negative but they are *less* dense than the suspension medium, then the sign of the pressures in the above discussion will be reversed. The initial positive pressure will occur at the anode, rather than the cathode. Thus if the sign of the particle zeta potential is reversed, perhaps by changing the pH, the polarity of the ESA effect will change. This fact makes it easy to accurately determine the isoelectric point from the ESA measurement.

Second, if the electrodes are not rigid, the pressure changes will set them into motion, and sound waves will propagate into the material behind the electrodes as well as into the suspension. Consequently, waves travel out in front and behind the electrodes when the voltage is applied. In commercial ESA devices, pressure transducers are placed behind the electrodes to measure the ESA.

Note that after the voltage pulse is turned off there will still be ESA sound wave pulses bouncing back and forth between the electrodes. If the pulse duration is long enough, the waves from both electrodes will superpose. In the Colloidal Dynamics AcoustoSizer II the pulses are short and do not overlap. Only the first pulse generated at the electrode nearest to the sensing transducer is used. This is done because this first pulse is the easiest to interpret theoretically, because it has not been attenuated by passing through the suspension. In the Matec ESA 8000 device, a long pulse is applied and the frequency is adjusted so that the overlapping waves reinforce one another to set up a resonance. This has the advantage of giving a stronger ESA signal but it is restricted to discrete resonant frequencies, and the theoretical interpretation of the ESA in terms of particle properties is more complicated.

4.3
The Dynamic Mobility

In the previous section we showed how a sinusoidal voltage pulse generates sinusoidal pulses of sound that appear to come from the electrodes. These ultrasound signals are measured in practice by attaching a pressure transducer to the material behind one of the electrodes. Two parameters can be obtained from this measurement: the *amplitude* of the sound wave and the *phase* of the wave relative to the applied sinusoidal voltage.

To determine the link between these measurements and the particle properties it is necessary to solve the mathematical equations that represent Rules 1 and 2 in the previous section. For the planar geometry described in that section, O'Brien *et al.* [17][3]

---

3) This result has appeared in many earlier publications, but the derivation was not published until 2003.

have shown that the pressure amplitude behind the anode in Figure 4.2 is given by:

$$P = \frac{z_s z_b}{z_s + z_b} \phi \frac{\Delta \rho}{\rho} \mu_D E \tag{4.3}$$

where ϕ is the particle volume fraction, ρ is the solvent density, $\rho + \Delta \rho$ is the particle density and E is the amplitude of the electric field in the suspension. The quantities z_b and z_s are the "acoustic impedances" of the backing material behind the electrode and the suspension, respectively. The acoustic impedance of a material is defined as follows: if a plane solid boundary in contact with the material is moved back and forth with velocity amplitude V, then the pressure required to generate that motion is zV. So z is a kind of stiffness measure; the greater the impedance, the greater the force required to move the boundary at a given velocity. In the case of the backing material, the acoustic impedance is equal to its density times the speed of sound in that medium and is a known quantity.[4] For the suspension, the acoustic impedance formula is more complicated; in the AcoustoSizer II, z_s is determined by a separate ultrasonic measurement.

The final symbol in Equation 4.3 to be defined is μ_D. This is the "dynamic mobility" of the particles. It is the analogue for a sinusoidal electric field of the (DC) electrophoretic mobility. The dynamic mobility is a complex quantity and so it requires two real numbers to specify its value. Those two numbers can be represented as a point (x, y) on an Argand diagram, where the x axis represents the real part and the y axis measures the imaginary part of this complex number. This point can also be represented in polar coordinates as a *magnitude* – which is the distance from the origin to the point – and an *argument* (or phase angle), which is the angle between the x axis and the line from the origin to the point. The parameter μ_D is defined in terms of these polar coordinates as follows; an applied field:

$$E_0 \cos \omega t \tag{4.4}$$

will cause the particles to move with a sinusoidal velocity. This velocity can be written in the form:

$$V_0 \cos(\omega t - \theta) \tag{4.5}$$

where ω is the angular frequency. The dynamic mobility is defined in this polar form by the formulae:

$$\text{mag } \mu_D = \frac{V_0}{E_0}, \quad \text{and} \quad \arg \mu_D = -\theta \tag{4.6}$$

The argument, or phase, of the dynamic mobility is a measure of how much the particle velocity lags behind the applied field. As we shall see, there are several reasons why the particle velocity may lag, but most often this lag is caused by the particle inertia at the MHz frequencies that are used for ESA measurements.

---

4) This statement applies to thin-film electrodes. If the electrode is thick it is necessary to include the extra mass in the acoustic impedance formula.

The dynamic mobility is not the only complex quantity in Equation 4.3. In writing this formula we are using the standard convention of writing the sinusoidal pressure and electric field as the real part of $P\exp(i\omega t)$ and $E\exp(i\omega t)$ respectively.[5] The quantities P and E are called the complex amplitudes of the pressure and electric field, respectively.

The ESA pressure wave travels through the backing material behind the electrode where it generates an alternating voltage across a pressure transducer. This transducer voltage is the quantity that we use for our analysis, and we represent it by the symbol ESA. This ESA voltage is proportional to the pressure P but it also depends on the transducer characteristics and on effects such as attenuation and the spread of the beam as it travels through the backing material, and on the gain of the amplifiers used in the ESA measurement. We lump these instrument-dependent factors into a single quantity $A(\omega)$, and write [16].

$$\text{ESA} = A(\omega) \frac{z_s z_b}{z_s + z_b} \phi \frac{\Delta \rho}{\rho} \mu_D E \tag{4.7}$$

This instrument factor A is determined by calibration. This involves the measurement of the signal from a material that has a known ESA. In the AcoustoSizer II an electrolyte is used for this calibration. The reason for using an electrolyte, rather than a colloid, is that colloids are notoriously variable in their zeta potential over time, whereas the ESA signal from the electrolyte is reproducible and can be calculated from its thermodynamic and transport properties [5].

Once the instrument has been calibrated, it can be used for the determination of the dynamic mobility from the ESA measurements. The dynamic mobility clearly depends on the zeta potential, and from the above discussion about particle inertia it can be seen that the dynamic mobility also depends on particle size. This is illustrated by the measured mobility spectra for four different silica sols (Figure 4.3) The 1-micron silica particles have large phase lags and a pronounced drop in magnitude with increasing frequency, while the nanoparticles show almost no frequency dependence over this frequency range.

Clearly, the dynamic mobility can be used to measure the zeta potential, and also the particle size for the two larger particles in the figure, since these show significant inertia. Note that this particle size determination takes place without the need for any external colloid size standard as a reference. For spherical particles of known size the theoretical and experimental curves for both magnitude and phase angle of the mobility are in excellent agreement and the estimated zeta potential from those curves agrees with values obtained by more traditional methods.

The initial motivation for developing ESA instruments was the desire to be able to measure particle size and zeta in concentrated systems. But as the range of ESA applications has increased over time, it has become evident that the dynamic mobility

---

[5] The function $\exp(i\omega t) = \cos(\omega t) + i \sin(\omega t)$.
Thus if P is real, the real part of $P\exp(i\omega t)$ is $P\cos(\omega t)$. If the pressure is out of phase with the electric field, P will be complex.

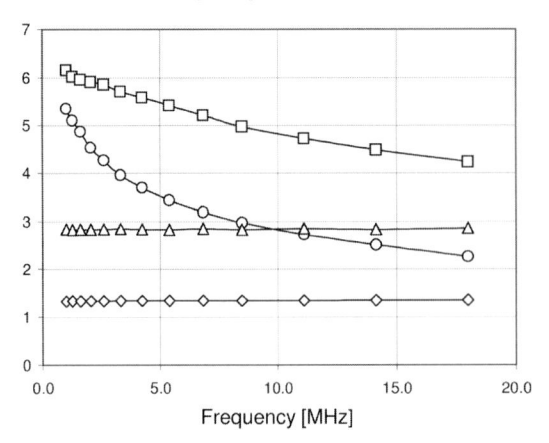

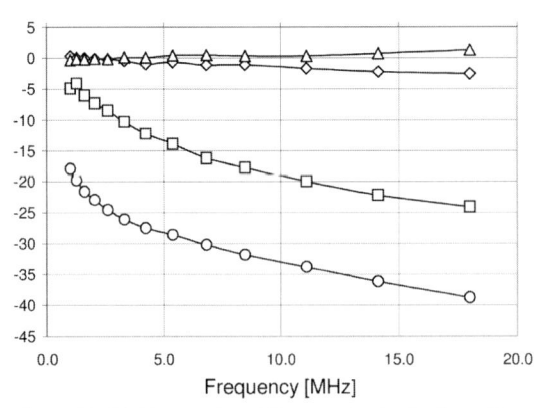

Figure 4.3 (a) Measured mobility magnitudes for four silica sols as a function of frequency. The units on the y-axis are $10^{-8} m^2 V^{-1} s^{-1}$. Diameter of silica particles: (○) 1 micron, (□) 300 nm, (open triangles) 70 nm, (open diamonds) 50 nm; (b) mobility arguments of the four silica suspensions, in degrees of angle.

also depends on other particle factors, and it is these other situations that we will be mainly concerned with here, since the determination of size and zeta from ESA has already been well covered in earlier reviews [6–8].

Before moving on to the link between dynamic mobility and particle parameters, we note that Equation 4.7 only applies to the AcoustoSizer II, which measures the ESA from the nearest electrode. It does not apply to the Matec ESA 9800, which uses a resonance type measurement.

4.4
The Dynamic Mobility for Thin Double Layer Systems

The double layer thickness around a charged particle depends on the electrolyte concentration (c). For a 1:1 symmetric electrolyte in water, the double layer thickness is approximately given by:

$$\kappa^{-1} = \frac{10}{\sqrt{c}} \text{nm} \tag{4.8}$$

where c is in mM. For a 1 mM electrolyte the double layer is 10 nm thick and for a 100 mM electrolyte it is 1 nm thick. This is a typical range of double layer thicknesses for aqueous systems. Colloidal particles are often more than 100 nm in diameter, and for those the double layer is often thin compared to the particle radius, a.

In this chapter we focus on the dynamic mobility of particles with thin double layers,[6] although we will discuss the important case of nanoparticles, which do not have thin double layers. Owing to lack of space we will not discuss colloids in nonpolar fluids.

The particle motion is driven by the action of the electric field on the double layer. Although the total electric force on the particle and double layer is zero – because the particle charge is balanced by the diffuse layer charge – a motion is still generated, for the field pushes the particle in one direction and it pulls the double layer ions, and the surrounding fluid, in the opposite direction. As described above, this motion is spatially uniform, except close to the electrodes, and so there is no ESA except from the electrodes.

In a frame of reference fixed to the particle, the double layer liquid moves with a tangential velocity that increases from zero at the shear surface[7] to a limiting value Δu beyond the double layer. This motion is called an "electro-osmotic flow." The relative velocity Δu is proportional to E_t, the tangential component of the electric field just outside the double layer:

$$\Delta u = S E_t \tag{4.9}$$

where the constant of proportionality S is here called the "slip coefficient," since it determines how fast the liquid beyond the double layer slips past the surface. For a particle with a smooth surface this relative velocity is given by the well-known Smoluchowski formula [19]:

$$S = -\frac{\varepsilon \zeta}{\eta} \tag{4.10}$$

---

6) Accurate results can be obtained from the theory discussed here for values of $\kappa a > 20$ and useful results for κa as low as 12.

7) This is the surface, some little distance from the particle surface, at which the fluid first begins to move past the particle. Various evidence suggests that it is some little distance into the diffuse layer.

in this case, where ε and η are the permittivity and viscosity, respectively, of the liquid and ζ is the particle zeta potential.

For an isolated spherical particle with a thin double layer, the dynamic mobility is given by [15]:

$$\mu_D = -\frac{8}{3\pi} S \frac{\langle E_t \rangle}{E} G\left(\frac{\omega a^2}{\nu}\right) \qquad (4.11)$$

where $\langle E_t \rangle$ is the average of the tangential field over the particle surface, E is the electric field far from the particle, a is the particle radius and ν is the kinematic viscosity of the liquid (equal to η/ρ). The function G is called the "inertia factor" and is defined by the formula:

$$G(x) = \frac{1 + (1+i)\sqrt{x/2}}{1 + (1+i)\sqrt{x/2} + i(x/9)(3 + 2\Delta\rho/\rho)} \qquad (4.12)$$

The non-dimensional quantity $\omega a^2/\nu$ is a measure of the ratio of the inertia to the viscous forces. If this ratio is small, the inertia is unimportant and the particle is able to keep up with the applied electric field, but as the ratio becomes of order unity the particle will show a significant phase lag due to this inertia term.

For many colloids, the particle permittivity is much less than that of water, and in this case the surface average tangential electric field is given by (see Equation 4.18 below for small λ):

$$\langle E_t \rangle = \frac{3\pi}{8} E \qquad (4.13)$$

If the surface is also smooth, Equation 4.11 for the dynamic mobility reduces to:

$$\mu_D = \frac{\varepsilon \zeta}{\eta} G\left(\frac{\omega a^2}{\nu}\right) \qquad (4.14)$$

a result obtained by combining Equations 4.10, 4.11 and 4.13. This is the formula that is most frequently quoted in papers on electroacoustics. Loewenberg and O'Brien [11] showed that Equation 4.14 also applies approximately to disk and rod shaped particles, provided that an effective radius, which depends on particle shape, is used for a.

The above formulae were derived for an isolated particle and so, strictly speaking, they can only be used for dilute suspensions, that is for suspensions with volume fractions of less than about 5%. The extension of the Equation 4.14 to suspensions of arbitrary concentrations is given in O'Brien et al. [17]. The extension was tested successfully with a 0.37 µm silica sample up to 43 wt% (25 vol.%) concentration and a 0.5 µm alumina sample up to 56 wt% (25 vol.%) concentration. An earlier approach, valid for particles with densities near to that of the suspension medium, had been used to measure oil-in-water emulsions up to 50 vol.% concentration [9]. Equation 4.14, and its extension to concentrated colloids, can be applied to many colloids, including simple oxide systems, but there are some important systems for which these formulae do *not* apply. It is to these systems that we now turn.

4.5
Particles with Adsorbed Polymer Layers

Polyelectrolytes and non-ionic polymers are often added to stabilize a colloid. The polymer molecules adsorb on the particle surface. In the case of the non-ionic polymers, the adsorbed layers provide a steric repulsion between the particles. For polyelectrolytes, the steric force is supplemented by electrostatic repulsion between the charged groups.

In their 1997 study of the effect of neutral polymer on the ESA signal from silica, Carasso et al. [1] used a "gel layer" model to derive an expression for the slip coefficient which has the form:[8]

$$S = -\frac{\varepsilon\zeta}{\eta} F(\kappa d, \alpha/\kappa^2\eta, \omega\eta/\gamma) \quad (4.15)$$

where d is the thickness of the polymer layer, α is the polymer drag coefficient and γ is the elastic shear modulus of the polymer layer.

In Figure 4.4a and b, the magnitude and argument of F for a uniform polymer layer are plotted as a function of the non-dimensional frequency:

$$\omega' = \frac{\omega\eta}{\gamma} \quad (4.16)$$

for $\alpha/(\kappa^2\eta) = 1$ and for κd values of 0.5, 1.0 and 2.0.

At high frequencies $F \to 1$ and the slip is unaffected by the polymer. This is because the amplitudes of the back and forth displacements are so small at high frequencies that there is almost no deformation of the polymer and so there are no elastic forces to hold back the liquid. The polymer slows down the motion most at low frequencies. At intermediate frequencies there is a phase difference between the tangential field and the slip velocity, and the fluid motion *leads* the applied field.

These effects become more pronounced as the polymer drag coefficient α is increased. For very large values of polymer drag coefficient, the low frequency limit of F is equal to $\exp(-\kappa d)$, which is equivalent to a displacement of the slipping plane by a distance d from the particle surface. In this limit the polymer is, in effect, immobile.

The shape of the slip coefficient curves does not depend on the elastic modulus (γ) of the polymer. That is, γ does not affect the high and low frequency limits of the magnitude or the maximum value in the argument of F, but it does determine the absolute frequency scale. As the elastic modulus is increased (as the polymer becomes stiffer) the curves are shifted to the right along the frequency axis, but the shapes do not change.

From these figures it is clear that the curves of slip coefficient are very sensitive to the thickness of the polymer layer relative to the double layer. Carasso et al. [1] used

---

8) There is a sign error in the formula in this reference. See Section 7 of Ref. [15] for the correct result. This was only a transcription error in the paper. Carasso et al. used the correct formula in their calculations.

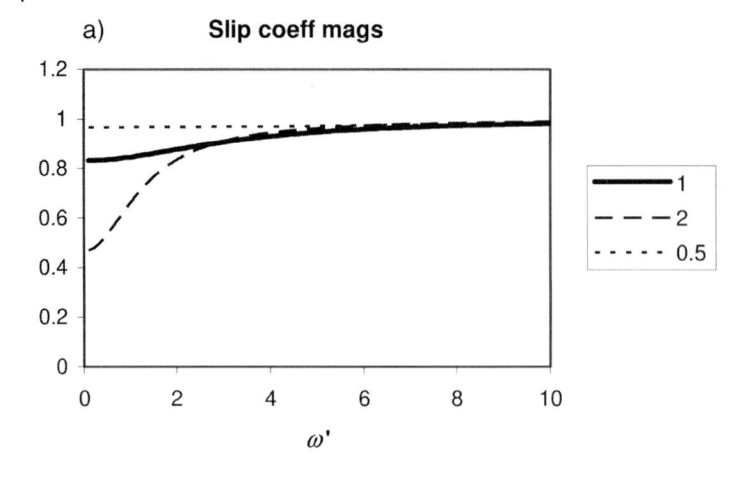

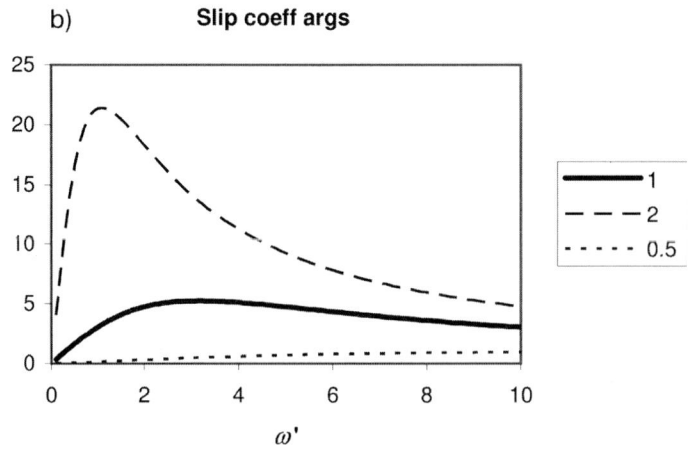

Figure 4.4 (a) Magnitude of F versus frequency for three different values of polymer layer thicknesses (κd); (b) arguments of F for the three polymer layer thicknesses, κd.

this sensitivity to determine the thickness of the adsorbed layers of poly(vinyl alcohol) and nonyl phenol ethoxylate from their AcoustoSizer II measurements. They determined the slip coefficient by first measuring the dynamic mobility of an electrostatically stabilized silica without added polymer. They then added polymer and determined the ratio of the dynamic mobility of the particles with and without polymer. From Equation 4.11 it can be seen that the mobility ratio is equal to the ratio of the slip coefficients with and without polymer. By fitting their measured spectra for this ratio with the theoretical curve using an α value calculated from the polymer geometry, they determined the polymer layer thickness and elastic modulus.

The same approach was used by Kong et al. in their study [10] of the adsorption of non-ionic surfactant on emulsion drops stabilized with SDS. They studied the same

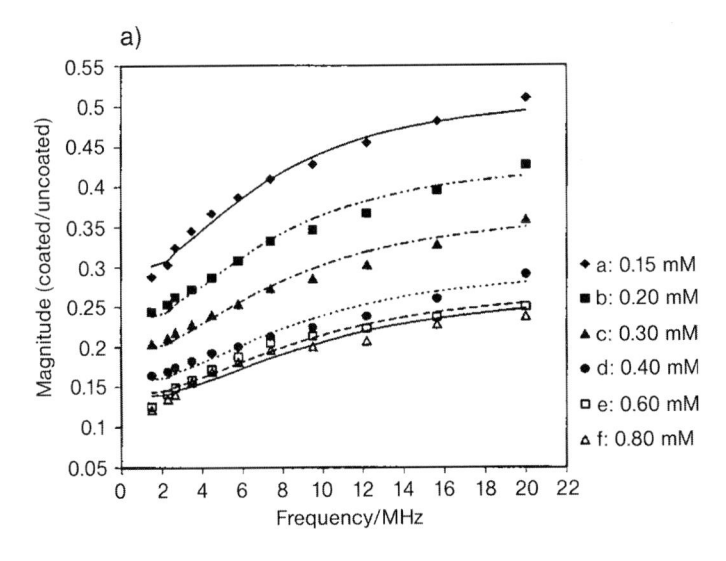

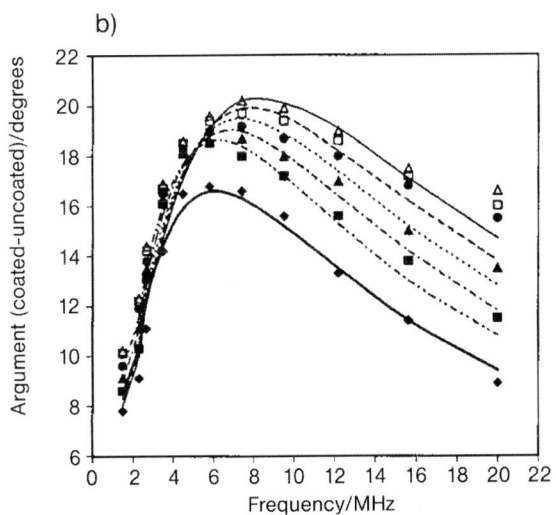

Figure 4.5 Ratio of the magnitudes (a) and the difference (b) in the arguments of the dynamic mobility of a 6 vol.% hexadecane emulsion stabilized with 3 mM SDS as a nonyl phenol ethoxylate surfactant is added at the concentrations shown.

surfactant as Carasso et al. [1] and obtained very similar adsorbed layer thicknesses. Figure 4.5 shows representative data for the effects of adding a nonyl phenol ethoxylate with ~100 ethylene oxide units to a 6 vol.% n-hexadecane emulsion stabilized with 3 mM SDS. As polymer is adsorbed on the emulsion drops the dynamic mobility magnitudes fall to only 10–20% of that of the uncoated drops and the phase lead is seen in the arguments, which increase with increased polymer concentration. The lines are theoretical fits with the gel-layer model. At the highest polymer concentration they give a polymer layer thickness of 8.7 nm, in addition to a displacement of the shear plane due to a dense polymer film of 4.4 nm.

Instead of trying to determine ratios of mobilities with and without surfactant as the above authors did, it is possible to obtain the polymer parameters directly from the mobility of the colloid with the surfactant coating. There is a limit to how much information can be obtained from the dynamic mobility spectrum due to its smooth nature, but in principle it should be possible to get up to four parameters with reasonable precision. Thus, if the particle size distribution is known by some other method, such as ultrasonic attenuation, then it should be possible to extract zeta and the three polymer layer parameters from the measured dynamic mobility spectrum. Unfortunately, no one has attempted this to date.

In 2002, O'Brien extended the non-ionic polymer theory to allow for the effect of weakly-charged polymers on the dynamic mobility [15]. In this case the slip velocity is given by:

$$S = -\frac{\sigma_s}{\eta\kappa}F(\kappa d,\ \alpha/\kappa^2\eta,\ \omega\eta/\gamma) - \frac{\rho_p d}{\eta\kappa}J(\kappa d,\ \alpha/\kappa^2\eta,\ \omega\eta/\gamma) \qquad (4.17)$$

The first term is the contribution from the particle surface charge; this is simply the Carasso gel layer result, but written in terms of the electrokinetic charge density rather than the zeta potential. The second term is the contribution from the polyelectrolyte charge, for the case when the charge density ρ_p is uniform across the polymer layer.

The function F has already been plotted in Figure 4.4. The new term J is plotted in the Figure 4.6a and b.

From these figures it can be seen that the effect of the polymer charge is opposite to that of adsorption of a neutral polymer: instead of a phase lead, it introduces a phase lag, and instead of the magnitudes increasing with frequency, they decrease. It should be possible to observe these competing trends by measuring the dynamic mobility of a polyelectrolyte that dissociates as a function of pH. As the polymer becomes more dissociated the mobility arguments should become more negative and the magnitudes should drop more steeply with frequency.

Equation 4.17 adds an extra parameter, namely the particle charge density, to the list of unknowns, but if the degree of dissociation of the charged group is known then it should be possible to relate this charge density to the thickness of the polymer layer, so that the number of parameters stays the same. Unfortunately, there have been no attempts made so far to apply this theory to the characterization of adsorbed polyelectrolyte layers. One possible application would be to explore whether this relationship can explain the occasional occurrence of DC mobilities that are greater than the maximum anticipated by the computer calculations of O'Brien and White [18].

To date, most ESA studies of polymer adsorption have *not* been concerned with determining the polymer layer properties, but have instead focused on the important problem of determining the optimum amount of polymer dispersant to add to a colloid. The method used to determine this optimum is straightforward: from the above discussion it is clear that any polymer, be it charged or uncharged, will alter the slip coefficient. Thus, if the standard mobility analysis (based on the mobility

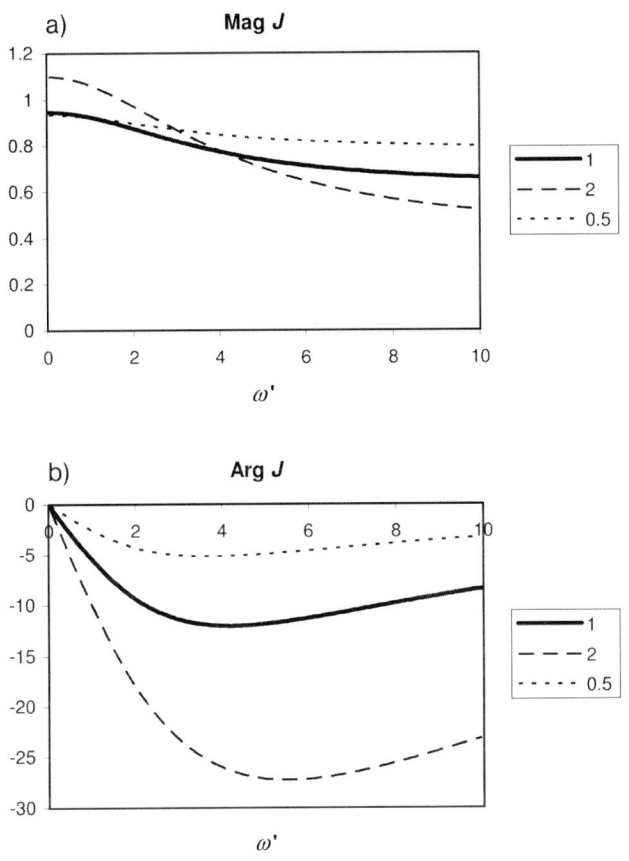

Figure 4.6 (a) Magnitude of the slip function J due to the polymer charge for various κd; (b) argument spectrum for J. The polymer charge introduces a phase lag.

Equation 4.14) is used for determining zeta from the mobility spectrum, the "effective zeta" that is obtained can still provide a useful parameter for measuring surface coverage. In this approach the effective zeta is plotted as a function of added polymer. The optimum polymer dose is determined from the point on the curve at which zeta levels out as a function of added polymer. Many examples of this method are described in Greenwood's review on this subject [3].

One of the most widely used polymers for stabilization of suspensions is sodium polyacrylate (PAA). On many surfaces (e.g. oxides and clay minerals) the negatively charged PAA tends to lie more or less flat on the surface, probably because the surface has both positive and negative sites at most pHs. We know this from our ESA measurements, for the PAA does not alter the dynamic mobility arguments and the shape of the magnitude spectrum does not change. Instead the magnitudes indicate an increasingly negative charge on the addition of the PAA. The effective zeta potential is then probably close to the actual zeta potential. The same applies for many other polyelectrolytes, both anionic and cationic. By making ESA measurements it is easy to see if a polymer lies flat or if it protrudes a significant distance into the double layer.

4.6
Surface Conductance

The tangential electric field drives electric currents around the particle surface. In most electrokinetic theories it is assumed that this current is dominated by the contribution from the diffuse layer ions, for the material inside the shear plane is taken to be immobile and non-conducting. While this assumption seems to be valid for many colloids, including oxide systems, for some systems there is significant electrical conduction in the region between the shear plane and the particle surface. Dielectric dispersion and ESA studies have shown this to be the case in kaolinite [21], bentonite [20], lattices [22, 23] and in emulsions [2]. We refer to this phenomenon as *stagnant layer conduction*.

The surface current leads to a build up of charge over one portion of the double layer and a depletion over the remainder. This in turn leads to a "back field," and this alters the average tangential field around the particle.

From Equation 6.10 in O'Brien's dynamic mobility paper [13], it can be shown that the average tangential field around a spherical dielectric particle is given by:

$$\frac{\langle E_t \rangle}{E} = \frac{3\pi}{4} \frac{1+i\tilde{\omega}}{2(1+i\tilde{\omega})+2\lambda+i\tilde{\omega}\varepsilon_p/\varepsilon} \tag{4.18}$$

where $\tilde{\omega}$ is a non-dimensional frequency defined by:

$$\tilde{\omega} = \frac{\omega\varepsilon}{K} \tag{4.19}$$

ε_p is the permittivity of the particle and:

$$\lambda = \frac{K_s}{Ka} \tag{4.20}$$

Here K is the conductivity of the electrolyte, a is the particle radius and K_s is the surface conductance of the particle, a quantity that includes contributions from the electrolyte both inside and beyond the surface of shear.

Figure 4.7 shows the effect that this surface conductance has on the frequency dependence of the tangential field. From these curves it can be seen that the surface conductance reduces the average tangential field at low frequencies and it introduces a phase lead at intermediate frequencies. These effects are superposed, through Equation 4.11, on the inertial drop off and increasing phase lag with frequencies. Thus, if a particle exhibits surface conductance, its magnitude spectrum will not drop off as rapidly with frequency as that of a non-conducting particle and its arguments will be less negative. If this effect is not taken into account the particle will appear to be smaller than it is.

a) tang field mags

Legend: 0, 0.5, 1

x-axis: non-dimensional frequency

b) tang field args

Legend: 0, 0.5, 1

x-axis: non-dimensional frequency

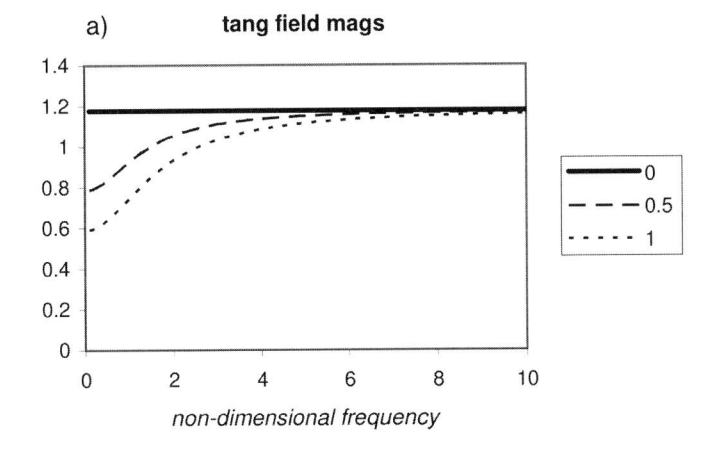

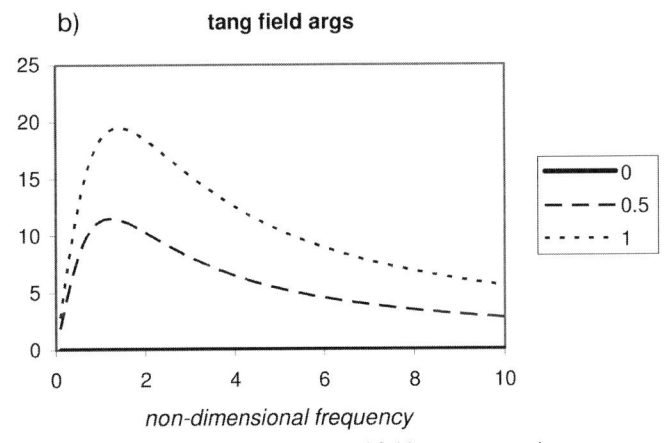

Figure 4.7 (a) Magnitude of the tangential field spectrum on the particle for different values of the surface conductance parameter λ; (b) effect of surface conductance on the tangential field argument (°).

The effects of surface conductance are most pronounced for small particles and at low electrolyte concentrations, for then the denominator of Equation 4.20 is small and so the λ factor is increased. For particles that do not exhibit stagnant layer conduction, the surface conductance is due to the motion of ions in the diffuse layer outside the shear surface. In this case K_s is determined by the zeta potential. This quantity is also larger at low electrolyte concentrations.

The surface conductance can dramatically alter the shape of the dynamic mobility spectrum. This is illustrated by Figure 4.8, which shows the dynamic mobility measurements on a 5 vol.% hexadecane emulsion stabilized with 1 mM SDS over a range of electrolyte concentrations [2]. At the lowest salt concentration (1 mM NaCl) the magnitudes show an increase with frequency and the argument displays a phase

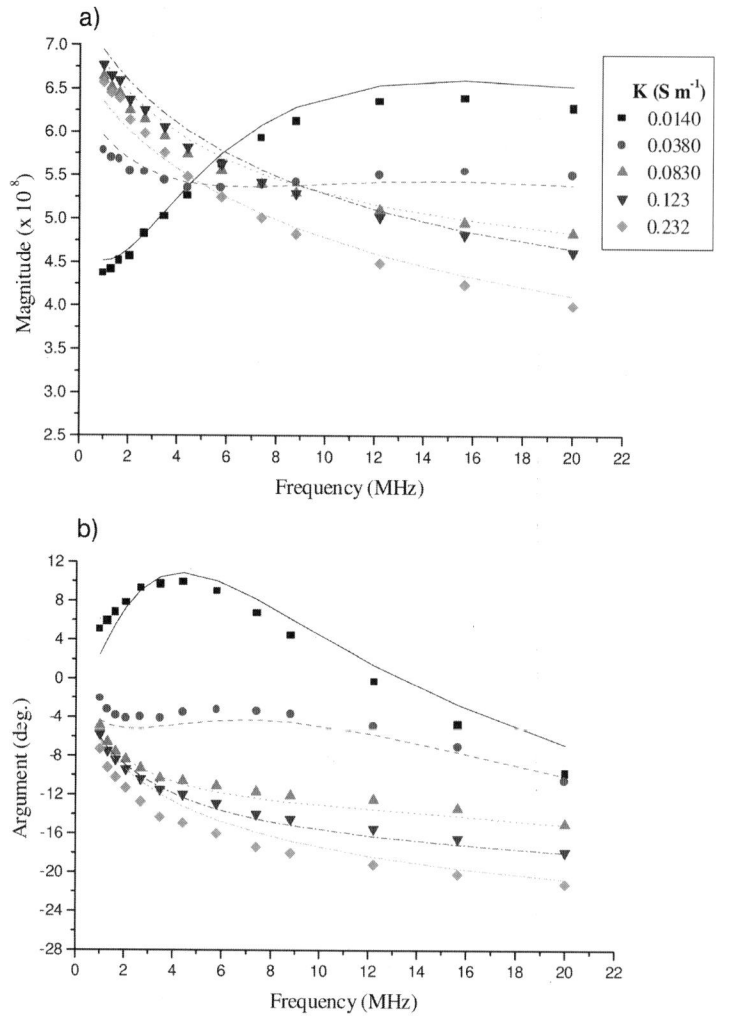

Figure 4.8 Dynamic mobility of a hexadecane in water emulsion at various electrolyte concentrations [magnitude (a) and argument (b)].

lead with a maximum at around 5 MHz. The continuous curves on Figure 4.8 are obtained from the above theoretical formulae with a λ value of 1.7 and a ζ of -141 mV. The diffuse layer contribution to λ at this ζ value is only 0.7 – so most of the surface conductance comes from stagnant layer conduction. As salt is added to increase the conductivity of the solvent, the relative effect of the surface conduction diminishes and the dynamic mobility spectra tend to their normal aspects, with a decreasing magnitude and increasing phase lag with frequency.

For systems that show stagnant layer conductance, there are three unknowns in the dynamic mobility formula: the zeta potential, the particle radius and λ. The smooth nature of the dynamic mobility curves makes it very difficult to extract all of these unknowns from the dynamic mobility measurements. One way around this problem is to use the attenuation measurements in the AcoustoSizer to obtain the particle size and then use a computer algorithm to adjust ζ and λ to fit the measured dynamic mobility spectrum. In the emulsion study referred to above, the authors had a prototype AcoustoSizer that did not include an attenuation sizing option. Instead they used the dynamic mobility measurement at the highest salt levels – where the effects of surface conductance can be neglected – to determine the droplet size distribution, and then on the assumption that the droplet size did not change with salt addition they used a manual fitting procedure to obtain ζ and λ from the dynamic mobility spectra at the lower electrolyte concentrations. They provided a cross check on their results by directly measuring λ with a dielectric response device.

Surface conductance also has pronounced effects on the usual electrophoretic mobility; it can lead to a large drop in the mobility and therefore to a large error in the reported zeta potential obtained using standard electrophoresis formulae. To determine the true zeta potential in these systems it is necessary to determine the surface conductance. The electroacoustic technique provides the most convenient method for doing this.

Before leaving the subject of thin double-layer systems we mention that Equation 4.18 for the tangential field involves the particle dielectric constant. In the studies referred to above, the particle dielectric constant (ε_p) was much smaller than that of the solvent, which was water, and so this term could be neglected; for such particles the field lines bend around the particle. However, for particles with high dielectric constants, such as titania, barium titanate and other semiconducting particles, ε_p is comparable to and sometimes much greater than that of the water. In this case the field lines cut into the particle and as a result the dynamic mobility spectrum has a different form. These particles show large phase lags that peak at around $\tilde{\omega} = 1$. See the paper by Guffond *et al.* [4] for examples of such mobility spectra. Since the mobility depends on ε_p in this case, the particle permittivity can be determined from the mobility spectrum if it is not already known. It is necessary to know the particle size distribution to do this, but that can be determined by another sizing technique such as the ultrasonic attenuation in the AcoustoSizer II.

4.7
Nanoparticles

Here we briefly discuss the dynamic mobility of particles that do *not* have thin double layers, that is systems in which the double layer thickness is more than about 1/20th of the particle radius. In Section 4.4 we pointed out that double layers are usually thin in aqueous based systems, for the salt concentration is often higher than 1 mM and so double layers are typically less then 10 nm thick. However, such double layers are *not* thin compared to nanoparticle radii and it is these systems that we are concerned with here.

Mean Mobility Spectrum

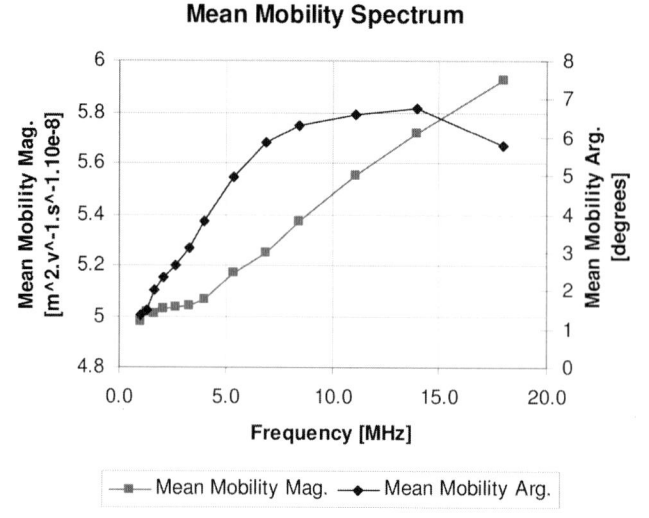

Figure 4.9 Magnitude and argument of 100 nm diameter particles surrounded by a 5 nm diffuse layer.

Figure 4.9 shows AcoustoSizer measurements of the dynamic mobility of a 100 nm diameter silica slurry. The ratio of particle radius to double layer thickness, κa, is around ten in this case. From these measurements it can be seen that the mobility magnitude and arguments have a similar form to the systems with surface conductance in the previous section, that is, the mobility has a phase lead and the magnitudes increase with frequency. In this case there are no inertia forces to slow the particles down, for the ratio of inertia to viscous forces:

$$\frac{\omega a^2}{\nu} \tag{4.21}$$

is only around 0.1 at the highest frequency.

The mobility curves have this shape because of *double layer distortion*. The applied field sweeps the double-layer ions back and forth and this leads to a change in the charge distribution around the particle. For highly charged particles, this has a significant effect on the flow field and the electric forces that act on the particle.

To understand why the double layer distortion affects the mobility in this way, it is useful to consider the representation of the sinusoidal electrical forces on the particle as "rotating vectors" in Figure 4.10. The force on the particle due to the applied electric field is denoted by F_a in the figure. The actual force is the projection of this vector on the horizontal axis. The vectors rotate with a constant angular velocity (ω), so their horizontal projection will be sinusoidal. The distortion of the double layer leads to the "back field" F_b. The phase of this force relative to the applied force is determined by the angle between these two vectors. The net electric force is the horizontal projection of F_r, the vector sum of F_a and F_b.

The double layer takes time to distort. If the applied field frequency is much greater than the electrolyte relaxation frequency, that is if $\tilde{\omega}$ is large, there is not enough time

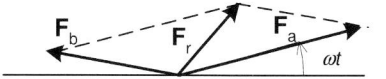

Figure 4.10 Electric forces on a colloidal particle, represented as the horizontal projection of rotating vectors.

during each cycle of the applied field for the double layer to distort. Thus, there is no backfield at these high frequencies. At the other extreme, when $\tilde{\omega}$ is small, the double layer is able to keep up with the instantaneous applied field, as if it were steady. The back force is a maximum at the low frequencies, and it acts in the opposite direction to the applied force, so it will be at an angle of 180° to the applied force in that case. In the diagram we have attempted to represent the back force at intermediate frequencies, when $\tilde{\omega}$ is around unity. At these frequencies, the double layer distortion will *lag* behind the applied field, so the vector F_b will be rotated clockwise away from its direction at low frequencies. From this figure it can be seen that the net force in this case *leads* the applied field. This is why we see a phase lead in the mobility, which peaks around $\tilde{\omega} = O(1)$, and why the mobility magnitude increases with frequency. Note that a phase lead does not imply that the particle moves *before* the field is applied. The discussion above refers to the situation after all the initial transients, caused by the switching on of the applied field, have died down.

The above discussion also applies to the thin double layer systems with surface conductance considered in the previous section. The surface conductance alters the double layer charge distribution, and so it gives rise to double layer distortion.

The double layer distortion is only significant for particles with moderate to high zeta potentials. The effect is most pronounced for κa values between 1 and 10. In this range it is significant if zeta is more than 50 mV in magnitude. For κa values outside this range the effect can still be important, but only at higher zeta potentials.

From the above discussion it can be seen that the double layer distortion will have its greatest effect on the electrophoretic mobility since it is measured at zero frequency. If the theoretical mobility for a fixed κa is plotted against zeta potential, the graph is initially proportional to zeta, then as this distortion becomes significant the graph bends below the linear form. Eventually, a peak is reached and after that the mobility decreases with zeta. Thus the zeta for a given mobility is not unique; the same mobility can yield two very different zetas, one below the maximum and the other above. It is not possible with electrophoresis to determine which zeta is correct, but the ESA method provides a unique zeta for these nanoparticle systems, for the higher zeta values lead to significant double layer distortion, while the low zeta values do not. In the latter case the mobility magnitudes are almost independent of frequency and the phase angles are close to zero.

Manglesdorf and White [12] have produced a computer program for calculating the dynamic mobility of a particle with an arbitrary double layer thickness, and this can be used for determining zeta from the measured mobility spectrum, provided κa is known. Thus, an independent measure of particle size is required to obtain the zeta in nanoparticle systems. Once again, the attenuation sizing in the AcoustoSizer can be used for this determination.

References

1 Carasso, M.L., Rowlands, W.N. and O'Brien, R.W. (1997) The effect of neutral polymer and nonionic surfactant adsorption on the electroacoustic signals of colloidal silica. *Journal of Colloid and Interface Science*, **193**, 200–214.

2 Djerdjev, A.M., Beattie, J.K. and Hunter, R.J. (2003) An electoacoustic and high frequency dielectric response study of stagnant layer conduction in emulsion systems. *Journal of Colloid and Interface Science*, **265**, 56–64.

3 Greenwood, R. (2003) Review of the measurement of zeta potentials in concentrated aqueous suspensions using electroacoustics. *Advances in Colloid and Interface Science*, **106**, 55–81.

4 Guffond, M.C., Hunter, R.J., O'Brien, R.W. and Beattie, J.K. (2001) The dynamic mobility of colloidal semiconducting antimony doped tin(IV) oxide particles. *Journal of Colloid and Interface Science*, **238**, 371–379.

5 Hodne, H. and Beattie, J.K. (2001) Verification of the electroacoustic calibration standards: comparison of the dynamic mobility of silicodecamolybdate and silicodecatungstate acids and salts. *Langmuir*, **17**, 3044–3046.

6 Hunter, R.J. (1998) Recent developments in the electroacoustic characterisation of colloidal suspensions and emulsions. *Colloids and Surfaces*, **141**, 37–65.

7 Hunter, R.J. (2004) More reliable zeta potentials using electroacoustics. *Progress in Colloid and Polymer Science*, **128**, 1–10.

8 Hunter, R.J. and O'Brien, R.W. (2001) Electroacoustics, Chapter in *Encyclopedia of Surfaces and Colloid Science* (ed. Arthur Adamson) Marcel Dekker, New York.

9 Kong, L., Beattie, J.K. and Hunter, R.J. (2001) Electroacoustic study of concentrated oil-in-water emulsions. *Journal of Colloid and Interface Science*, **238**, 70–79.

10 Kong, L., Beattie, J.K. and Hunter, R.J. (2001) Electroacoustic estimates of non-ionic surfactant layer thickness on emulsion drops. *Physical Chemistry Chemical Physics*, **3**, 87–93.

11 Loewenberg, M. and O'Brien, R.W. (1992) The dynamic mobility of nonspherical particles. *Journal of Colloid and Interface Science*, **150**, 158–168.

12 Manglesdorf, C.S. and White, L.R. (1992) Electrophoretic mobility of a spherical colloidal particle in an oscillating electric field. *Journal of the Chemical Society-Faraday Transactions 2*, **88**, 3567–3581.

13 O'Brien, R.W. (1988) Electroacoustic effects in a dilute suspension of spherical particles. *Journal of Fluid Mechanics*, **190**, 71–86.

14 O'Brien, R.W. (1990) The electroacoustic equations for a colloidal suspension. *Journal of Fluid Mechanics*, **212**, 81–93.

15 O'Brien, R.W. (2002) The effect of an absorbed polyelectrolyte layer on the dynamic mobility of a colloidal particle. *Particle & Particle Systems Characterization*, **19**, 186–194.

16 O'Brien, R.W., Cannon, D.W. and Rowlands, W.N. (1995) Electroacoustic determination of particle size and zeta potential. *Journal of Colloid and Interface Science*, **173**, 406–418.

17 O'Brien, R.W., Jones, A. and Rowlands, W.N. (2003) A new formula for the dynamic mobility in a concentrated colloid. *Colloids and Surfaces A-Physicochemical and Engineering, Aspects*, **218**, 89–101.

18 O'Brien, R.W. and White, L.R. (1978) Electrophoretic mobility of a spherical colloidal particle. *Journal of the Chemical Society-Faraday Transactions*, **2** (74), 1607–1626.

19 Overbeek, J.Th.G. (1952) in *Colloid Science*, Vol. 1 (ed. H.R. Kruyt), Elsevier, Amsterdam, p. 202.

20 Rasmusson, M., Rowlands, W.N., O'Brien, R.W. and Hunter, R.J. (1997) The dynamic mobility and dielectric response of sodium

bentonite. *Journal of Colloid and Interface Science*, **189**, 92–100.
21 Rowlands, W.N. and O'Brien, R.W. (1995) The dynamic mobility and dielectric response of kaolinite particles. *Journal of Colloid and Interface Science*, **175**, 190–200.
22 Russell, A.S., Scales, P.J., Mangelsdorf, C.S. and White, L.R. (1995) High-frequency dielectric response of highly charged sulfonate lattices. *Langmuir*, **11**, 1553–1558.
23 Shubin, V.E., Hunter, R.J. and O'Brien, R.W. (1993) Electroacoustic and dielectric study of surface conduction. *Journal of Colloid and Interface Science*, **159**, 174–183.

5
Modeling the Structure and Stability of Charged Hemi-Micelles at the Air–Water Interface

Johannes Lyklema, Ana B. Jódar-Reyes, and Frans A.M. Leermakers

5.1
Introduction

In this chapter we demonstrate the power of molecular modeling by addressing the formation of surfactant hemi-micelles at the air–water interface. The choice of this topic is fortuitous, because it allows us to handle two problems simultaneously.

The first is that the selection of the system calls for a modeling approach, because interfacial hemi-micelles (which, according to general consideration should exist) are barely visible. In particular, optical techniques to detect them are not abundant and not unambiguous, let alone that physical characteristics such as size shape and thickness can be established [1, 2].

The second is that it is a system where modeling is expected to reap rich awards, basically because we are looking for interfacial excesses with respect to surfactant solutions for which the modeling has been successful. It is known which types of approaches and which parameter values satisfactory describe reality. Moreover, as it is known how to model interfaces one can with confidence deal with the more complex systems in which interfaces and surfactants are both present.

The present chapter rests on many papers that have been reviewed in the recent literature. We mention in particular two review chapters on monolayers [3, 4], one on modeling of surfactants in solution [5], one on modeling fluid–fluid interfaces [6] and a recent book chapter [7] in which the modeling of the influence of the tail length and ion head group specificity on the adsorption and phase formation of surfactants in solution [8] was studied. The last mentioned issues recur in the present text.

We present here first results from the statistical thermodynamical modeling of the self-association of ionic surfactants that are adsorbed at the air–vapor interface, using advanced self-consistent field theory.

Self-consistent field (SCF) modeling gives accurate, close to quantitative, predictions for equilibrium surfactant self-assembly in solution. Both structural and thermodynamic properties of micelles follow experimental data and are consistent with simulation data. Recently, an in-depth SCF study has been presented, aimed at

Highlights in Colloid Science. Edited by Dimo Platikanov and Dotchi Exerowa
Copyright © 2009 WILEY-VCH Verlag GmbH & Co. KGaA, Weinheim
ISBN: 978-3-527-32037-0

describing sodium dodecyl sulfate self-assembly in an aqueous electrolyte solution [5]. The molecular architecture of the surfactant molecules were reasonably closely accounted for using a lattice model. The first and second critical micelle concentrations (CMCs) (i.e., the first appearance of spherical micelles and of wormlike micelles in the solution, respectively) were accurately reproduced. Here, we apply the same model focusing to ionic surfactants at the air–water interface.

The aim is to investigate when and how the adsorbed surfactants aggregate with each other, forming inhomogeneous interfacial layers. The effects of the surfactant tail length, the ionic strength and the size of the counterion on the hemi-micelle formation are also analyzed.

Even though the SCF theory and all details of the present molecular model are available in the literature, it is appropriate to recall the main approximations and discuss the choice of parameters. As only equilibrium situations are considered, we deal with those structures that are thermodynamically stable. In the following we start with some thermodynamic considerations. Then we will present some aspects of SCF theory and the molecular model. In the results section (Section 5.4) we consider the structure and stability of hemi-micellar aggregates of ionic surfactants at the air–water interface, focusing on the very first micelles that become stable in the low pressure (high area per molecule) region of the pressure–area isotherms. Where appropriate we will discuss how the aggregation of surfactants at the air–water interface is related to the corresponding phenomenon in the bulk solution.

5.2
Thermodynamics

The molecular state of a micellar solution is completely determined by the macroscopic parameters T, V and $\{n_i\}$. Hence, for an infinitesimal change in any of these variables:

$$dF = -SdT - pdV + \sum_i \mu_i dn_i \quad (5.1)$$

describes the corresponding variation of the Helmholtz energy. Self-association into micelles of surfactants is conveniently described (analyzed) using the thermodynamics of small systems [9, 10]. According to this theory the number of micelles $\mathcal{N}$ is introduced with the conjugated intensive parameter $\mathcal{E}$; the latter quantity is the partial excess Helmholtz energy, or, for that matter, the chemical potential of the micelle. The term $\mathcal{E}d\mathcal{N}$ does not occur in Equation 5.1 because $\mathcal{N}$ is not an externally applied state variable. Nevertheless, working with $\mathcal{N}$ can be used to optimize F: at equilibrium $\partial F/\partial \mathcal{N} = \mathcal{E} = 0$. As $\mathcal{E}$ has the nature of a chemical potential we can write:

$$\mathcal{E} = kT\ln\varphi_m + \mathcal{E}_m \quad (5.2)$$

where $kT\ln\varphi_m$ accounts for the entropical (translational) contribution (φ_m is the volume fraction of micelles) and $\mathcal{E}_m$ is the standard contribution, corresponding to the μ^0 of a dissolved low molecular weight solute. The term φ_m is also known as the translationally restricted grand potential per micelle. It is accessible from SCF theory. As will be discussed below, in SCF computations we can only evaluate properties of a most likely micelle fixed with its center to a specified point in the coordinate system. Here we stress that the cooperative degrees of freedom are the only ones that are constrained. On the molecular level we do not impose any limitations. Even though there is a rather sharp interface between the hydrophobic core and the aqueous solution, there are conformational fluctuations that give rise to a small overlap of heads and tails. Tails mainly reside in the core and protrude with an exponentially low probability into the water. The ions can also freely distribute. While doing so they can and will to a significant amount locally compensate for the head group charge and further produce a diffuse (Gouy–Chapman) layer, and of course to a much lesser extent enter the core. For a micelle with a given aggregation number we can compute $\mathcal{E}_m$ and Equation 5.2 is then used to evaluate the volume fraction of micelles φ_m, or, equivalently, the (true) volume available per micelle in a micellar solution. Besides focusing on most probable micelles we can evaluate the full micelle size distribution [5].

Let us next consider the two-phase air–water system with an interfacial area A and interfacial tension γ. For this system Equation 5.1 is extended with the term γdA. The interfacial tension is a decreasing function of the surfactant concentration or for that matter, the surface pressure $\pi = \gamma_0 - \gamma$, where γ_0 is the interfacial tension of the pure water–vapor interface, is a decreasing function of the area per adsorbed surfactant molecule. It will be clear from the above that macroscopic thermodynamics cannot provide detailed information on whether there are local micellar-like inhomogeneities at the interface. The reason is the same as why Equation 5.1 does not give information on the very existence of micelles. Analogously to micellization in the bulk we again invoke the approach of the thermodynamics of small systems [9, 10]. Extending the analogy, we postulate the existence of $\mathcal{N}_s$ surface micelles and introduce a conjugated quantity Ω that measures the excess Helmholtz energy associated with the presence of these surface entities. Optimization of the Helmholtz energy for given area A, using the number of surface structures proves that $\Omega = 0$ at equilibrium, as expected. Again, the corresponding SCF modeling contains the mixing entropy of the interfacial micelles associated to the translational degrees of freedom of these micelles and the translationally-restricted excess Helmholtz energy associated with the formation of the interfacial micelles Ω_m:

$$\Omega = kT\ln\phi_m + \Omega_m = 0 \tag{5.3}$$

At this stage we mention that Ω_m is a double excess quantity. It is the excess free energy that is at the interface with respect to the interfacial tension γ, which itself is also an excess free energy with respect to the two pure bulk phases joining at the Gibbs dividing plane. Again, Ω_m is the primary thermodynamic quantity in the SCF analysis and we will specify below how it is evaluated. We will also have to go into some details regarding the computation of aggregation numbers, which involves the definition of the so-called Gibbs dividing plane.

5.3
Fundamentals of SCF Theory and the Molecular Model

Here, we use the method of Scheutjens and Fleer, who introduced a convenient discretization scheme to solve the SCF equations [11, 12]. To outline the approach we first need to give details about the coordinate system. Using this coordinate system allows us to specify (on the segment-type level) discrete sets of two conjugated quantities, namely the volume fractions and the segment potentials. Physically, the segment potential is the Gibbs energy per segment, required to transport a segment from the bulk (reference state) to the specified coordinate. We note that the electrostatic potential is one of the contributions to this segment potential (i.e., for the segments that carry a charge) and that other contributions exist (see below). The volume fractions and segment potentials depend on each other. How to compute one from the other is the second main issue explained below in two consecutive subsections. In these paragraphs we will integrate the introduction of the molecular model and the parameters used in the modeling to save space. When all volume fractions and all potentials are mutually consistent it is possible to evaluate the Helmholtz energy, and from that all other relevant thermodynamic quantities can be derived. More specifically, it is possible to accurately evaluate $\mathcal{E}_m$ and Ω_m, cf. Equations 5.2 and 5.3. Thus, we will be able to judge the thermodynamic stability of a particular structure.

5.3.1
The Lattice

We anticipate that a typical micelle that occurs at the air–water interface is small and radially symmetric. This will be true sufficiently close to the first appearance of micelles at the surface, that is, just above the critical surface aggregation concentration (CSAC). A cylindrical coordinate system, capturing such interfacial micelles is shown in Figure 5.1 (see also, for example, Refs [14–16]). In this coordinate system the z-coordinate runs normal to the interface, low values are in the vapor, large values in the water-phase. In each z-plane we have concentric rings referred to by the shell ranking number r and with a number of lattice sites given by $L(z, r) = L(r) = \pi(2r-1)$. A mean field approximation is applied to all properties inside each ring at (z, r), which means that within each ring all volume fraction fluctuations and positional correlations are ignored. An interfacial micelle will be positioned at some position z (where the interface is) and with the center of mass at $r = 0$; lateral mobility is assumed absent. Indeed, below we will report on the center of mass of the adsorbed hemi-micelle and its position with respect to the Gibbs plane.

Within the lattice so-called transition probabilities $\lambda(z, r|z', r')$ are defined, which are normalized such that $1 = \sum_{z' = z-1, z, z+1} \sum_{r' = r-1, r, r+1} \lambda(z, r|z', r')$. These transition probabilities obey the internal balance equation $L(z, r)\lambda(z, r|z', r') = L(z', r') \lambda(z', r'|z, r)$. For very large values of r the curvature of the lattice rings vanishes and the transition probabilities converge to constant values λ.

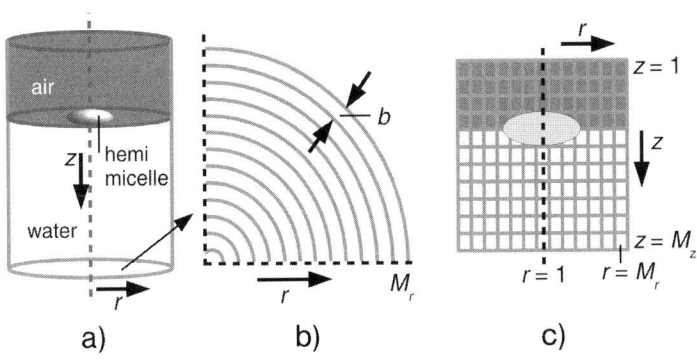

Figure 5.1 Schematic presentation of the system. (a) A 3D presentation. In gray the vapor phase exists at low values of z. The water phase is white. At the interface a hemi-micellar object is positioned. (b) A quarter of the concentric rings of lattice sites at each layer z. The last lattice layer in the r-direction, $r = M_r$, is indicated. The size of a lattice site is given by b and the volume of a lattice site is b^3. (c) A cross-section in the z-r plane, again with the vapor phase given in dark gray and the interfacial micelle in light gray. The first $z = 1$ and last $z = M_z$ lattice layer in the z-directions are indicated.

These transition probabilities are used in the site fraction of any property X that depends on the coordinates (z, r) as:

$$\langle X(z,r) \rangle = \sum_{r'} \sum_{z'} \lambda(z,r|z',r') X(z',r') \tag{5.4}$$

The site averages, used systematically in the computation of the interactions as well as the chain statistics, account for the local curvature in the coordinate system.

To convert volume fractions into molar concentrations, a choice is needed for the size of a lattice site b. Here we have used $b = 0.5$ nm, which is a compromise between the size of water molecules and surfactant segments. For a given molecule this conversion depends on the molar volume (v), which we will express in units b^3, and using the quantity N as the number of segments in a molecule (Table 5.1). For example, to convert the volume fraction for the ions one needs to multiply by roughly 10. For the surfactant the conversion factor is close to unity.

With this coordinate system in place, we can now focus on the computation of the volume fractions and the segment potentials.

5.3.2
From Volume Fractions to Potentials

At this stage we assume that at each coordinate $\mathbf{r} = (z, r)$ we know the volume fractions of all molecular species, which we generically will denote as $\varphi_A(\mathbf{r}) = n_A(\mathbf{r})/L(\mathbf{r})$, where n_A is the number of segments of type A. These distributions involve all allowed conformations of the molecular species in the system (see next section). Table 5.1 lists the properties of these molecules. Water is the first component, $i = 1$, and each molecule occupies five lattice sites (molecular volume $N = v/b^3 = 5$). In the case shown in the table, the positive ($i = 3$) and the negative ($i = 4$) ions have the same

Table 5.1 Molecular properties (number, name, molar volume $N = v/b^3$, architecture) and their segments (united atoms, A) (name, valence v), and relative dielectric permittivity ϵ used in the SCF calculations. Here, t is the number of segments in the tail of the surfactant and Y is the number of W segments around a central ion; these parameters are varied.

		Molecule			Segment		
i	Type	N	Architecture		A	v	ϵ
1	Water	5			W	0	80
2	Surfactant	$t+5$			CH_3	0	2
					CH_2	0	2
					X	0.2	80
3	P_Y	$1+Y$	for $Y=4$:		P	1	80
					W	0	80
4	N_Y	$1+Y$	for $Y=4$:		N	−1	80
					W	0	80
5	Vacancy	1			V	0	1

structure as water, albeit that the central "segment" has a unit positive or negatively charge, respectively. The number of W units around the central charge is indicated by Y and in general the ions have a molar volume of $(Y + 1)b^3$. The surfactant (component $i = 2$) occupies $N = v/b^3 = t + 5$ lattice sites, where t is the number of tail segments. The tail is composed of a terminal CH_3 (united) groups and $(t − 1)$ CH_2 segments. The charged head groups mimicking, for example, a sulfate group, occupy five sites (as indicated). The total charge on the surfactant is $v = −1e$, where the elementary charge e is distributed homogeneously over the head group segments (X). Finally, we allow for free volume (empty lattice site) in the system. These vacancies are denoted by $i = 5$ (V). As a result, the set of segment types is $A \in \{W, CH_3, CH_2, X, P, N, V\}$. The volume fractions are coupled through the incompressibility constraint:

$$\sum_A \varphi_A(\mathbf{r}) = 1 \tag{5.5}$$

To each segment type A a segment potential $u_A(\mathbf{r})$ can be assigned. Again, the segment potential is the reversible work needed to bring a segment of type A from the bulk (i.e., far from the interface in the water-phase, where all volume fractions assume the constant value of φ_A^b) to the position $\mathbf{r}$. We distinguish four contributions to this segment potential:

$$\frac{u_A(\mathbf{r})}{kT} = (u'(\mathbf{r}) - u'^b) + \sum_B \chi_{AB}(\langle \varphi_B(\mathbf{r}) \rangle - \varphi_B^b) + \frac{v_A e \psi(\mathbf{r})}{kT} - \frac{1}{kT}\frac{\epsilon_0}{2}(\epsilon_A - 1)E^2(\mathbf{r}) \tag{5.6}$$

Table 5.2 List of all Flory–Huggins interaction parameters. The segment types are defined in Table 5.1. The table is symmetric, that is, $\chi_{AB} = \chi_{BA}$ and has zero values on the diagonal $\chi_{AA} = 0$. All parameters are taken from Ref. [5].

χ	W	CH$_3$	CH$_2$	X	P	N	V
W	0	1.5	1.1	0.5	0	0	2.5
CH$_3$	1.5	0	0.5	2	2	2	1.5
CH$_2$	1.1	0.5	0	2	2	2	2
X	0.5	2	2	0	0	0	2.5
P	0	2	2	0	0	0	2.5
N	0	2	2	0	0	0	2.5
V	2.5	1.5	2	2.5	2.5	2.5	0

The first contribution can be interpreted as the work of transporting a vacant lattice site from the bulk to the interface. The numerical value is chosen such that locally Equation 5.5 is fulfilled. The second contribution accounts for the short-range interactions, as they occur when two segments occupy neighboring lattice sites (Bragg–Williams approximation). In a lattice model it is convenient to use for this the Flory–Huggins interaction parameters $\chi_{AB} = \chi_{BA}$, which only assume non-zero values when segments A and B are different. As there are seven types of segments, there are 21 parameters collected in Table 5.2.

All parameters have been used before and for a full justification we refer the reader to Ref. [5]. In short, the parameters were found in the following way. We first considered the water–V two-phase system and insist on a strong repulsion between W and V such that the interfacial tension is high. The binodal has a bulk volume fraction of the V unit in the water-rich phase of $\varphi_V^b = 0.04945$ which was subsequently fixed in all calculations. The interaction of V with the other components is also repulsive, but chosen such that the surfactants adsorb preferentially at the water–V interface with the tails pointing to the V-rich phase. The value of $\chi_{CH_2,W}$ was selected such that the CMC as a function of the tail length closely follows the experimental data. Repulsion between the charged units and the apolar segments prevents overcrowding of these segments in the hydrophobic core of the micelles. The CH$_3$ group is taken to be more hydrophobic than the CH$_2$ units and a small repulsion between these two segment types is assumed, similarly as done in the analysis of the lipid bilayer membrane [17]. We hasten to mention that this parameter choice is not necessarily optimized yet. However, this implementation appears to reproduce many experimental properties known for sodium dodecyl sulfate (for which $t = 12$, cf. Table 5.1) self-assembly.

The interaction term (second term in Equation 5.6) is normalized in such a way that the potential vanishes in the bulk of the water-phase.

The third term of Equation 5.6 accounts for the electrostatic interactions, which we implemented on the level of the Poisson–Boltzmann theory for electrified interfaces. The electrostatic potential $\psi(\mathbf{r})$ is found from the Poisson equation which for the current geometry reads:

$$\frac{\partial}{\partial z}\epsilon\frac{\partial \psi}{\partial z} + \frac{1}{r}\frac{\partial}{\partial r}r\epsilon\frac{\partial \psi}{\partial r} = -\frac{q(\mathbf{r})}{\epsilon_0} \qquad (5.7)$$

which was computed for the discrete set of coordinates. In Equation 5.7 the local dielectric permittivity is assumed to be the volume-fraction weighted average: $\epsilon(\mathbf{r}) = \sum_A \varphi_A(\mathbf{r})\epsilon_A$. The local charge density is computed similarly:

$$q(\mathbf{r}) = e\sum_A \varphi_A(\mathbf{r}) v_A$$

The final contribution to the segment potential of Equation 5.6 is needed because the dielectric permittivity is not fixed, but a function of the molecular distributions. The polarization is proportional to the electric field and the gain (hence the minus sign) in energy is once again proportional to the electric field (hence the square). The factor $1/2$ is automatically obtained if the charging is carried out isothermally and reversibly. That procedure accounts properly for the entropy and, hence, yields a Gibbs energy. Note that this polarization term also contributes to the segment potential for the segments that are uncharged except for the vacancies that have $\epsilon_V \equiv 1$ (hence the $\epsilon_A - 1$).

In addition to the hemi-micelle we also need parameters for the isolated water–vapor interface. For this we need a sufficient number of lattice layers in the homogeneous vapor phase and in the homogeneous aqueous bulk. We have implemented reflecting boundary conditions in the z-direction to minimize artifacts of the system boundaries. The main aim is to model a micelle at the interface. This micelle should not interact with other micelles and therefore we need a sufficient number of shells in the radial direction. Again we have implemented reflecting boundaries at the upper ring (at $r = M_r$). Mathematically reflecting boundaries are implemented by forcing all gradients in quantities that are a function of $\mathbf{r}$ to vanish at these boundaries.

5.3.3
From Potentials to Volume Fractions

In this section we will assume that all segment potentials are available and give the procedure as to how from this information the volume fraction profiles are determined. Referring to Table 5.1, where all molecular species are shown, we introduce segment ranking numbers $s_i = 1, 2, \ldots, N_i$ for molecule i, where N is a measure of the molar volume in number of sites and chain architecture operators $\delta_{i,s}^A$, which assume the value unity when segment s of molecule i is of type A, and zero otherwise. At the segment-level we need Boltzmann factors $G_A(\mathbf{r}) = \exp - u_A(\mathbf{r})/kT$. In the chain statistics it is convenient to generalize these Boltzmann factors such that they depend on the molecule number i and the ranking number s. To this end we introduce chain architecture operators: $G_i(\mathbf{r}, s) = \sum_A \delta_{i,s}^A G_A(\mathbf{r})$. To account for the conformational degrees of freedom and to allow for an efficient evaluation of the volume fraction profiles, we generate two complementary so-called chain end distribution functions $G_i(\mathbf{r}, s|1)$ and $G_i(\mathbf{r}, s|N)$. These quantities have the physical meaning of collecting the combined statistical weight of all allowed chain conformations starting with segment $s_i = 1$ or segment $s_i = N_i$, respectively, and ending at segment s at position $\mathbf{r}$. These end-point distributions follow from two propagators, one running to higher segment ranking numbers and the other to lower ones:

$$G_i(\mathbf{r}, s|1) = G_i(\mathbf{r}, s)\langle G_i(\mathbf{r}, s-1|1)\rangle \tag{5.8}$$

$$G_i(\mathbf{r}, s|N) = G_i(\mathbf{r}, s)\langle G_i(\mathbf{r}, s+1|N)\rangle \tag{5.9}$$

The propagators are started by $G_i(\mathbf{r}, 0|1) = G_i(\mathbf{r}, N+1|N) = 1$ for all coordinates. Note that at the reflecting boundaries the no-gradient conditions also apply to the end-point distribution functions. These complementary propagator equations can be used for monomeric species ($N=1$) and for linear chains. The extension to branched chains is straightforward but slightly involved. We refer the reader to Ref. [13] for details.

The volume fractions follow from the composition equation, which combines the two complementary end-point distributions that start on opposite chain ends and end at the same segment at the same coordinate:

$$\varphi_i(\mathbf{r}, s) = C_i \frac{G_i(\mathbf{r}, s|1) G_i(\mathbf{r}, s|N)}{G_i(\mathbf{r}, s)} \tag{5.10}$$

where the division by $G_i(\mathbf{r}, s)$ is to correct for the fact that the potential field for the joining segment s is accounted for in both end-point distribution functions. In Equation 5.10, C_i is a normalization factor. It can be shown that:

$$C_i = \frac{\varphi_i^b}{N_i} = \frac{\theta_i}{N_i Q_i} \tag{5.11}$$

with φ_i^b the volume fraction of component i in the bulk (of the water phase), θ_i is the amount of component i in the system, $\theta_i = \sum_\mathbf{r} = L(\mathbf{r})\varphi_i(\mathbf{r})$, and Q_i is the single chain partition function, $Q_i = \sum_\mathbf{r} L(\mathbf{r}) G_i(\mathbf{r}, N|1)$. For all components except the solvent either the bulk volume fraction (grand canonical) or the amount (canonical) is an input quantity. Using Equation 5.11 we can compute φ_i^b (canonically) or θ_i (grand canonically). For the solvent (water), we employed the (in)compressibility relation in the bulk $\varphi_1^b = 1 - \sum_{i \neq 1} \varphi_i^b$ to be used in the normalization of the water distribution. The volume fraction profiles that depend on the segment type can be obtained from the ranking number-depending quantities, that is $\varphi_A(\mathbf{r}) = \sum_i \sum_s \varphi_i(\mathbf{r}, s) \delta_{i,s}^A$.

Above we have introduced a special component in the system that was termed a vacancy. In our system the number of vacancies is countable and from this point of view it is natural to assign a chemical potential to this "component." As a consequence there is no pV term in the thermodynamic analysis and we have five "molecular" components. Alternatively, one could choose to convert the chemical potential of the vacancies into a pressure and renormalize the chemical potentials of all other components accordingly. In this latter case there are only four molecular components, and in addition a $-pV$ term occurs in the thermodynamic analysis. The first approach is the classical incompressible (lattice) solution and the second approach is known as a lattice gas (even when we model a condensed liquid solution). In Table 5.2 we have introduced Flory–Huggins interactions also for the vacancies and in this framework the choice for an incompressible lattice solution is the more logical one. We will follow this approach.

5.3.4
Grand Potential

Once the mutually consistent volume fraction and segment potential profiles are known, which typically requires an iterative numerical search procedure, one can evaluate thermodynamic quantities. From the above it is clear that the grand potential $\Omega = F - \sum_i n_i \mu_i$ is the central quantity of interest. From its definition it is easily seen that Ω collects all the free energy contributions that are in the system in excess of the bulk free energy. The implication is that all inhomogeneous regions in the system will contribute, that is, $\Omega = \sum_r \omega(\mathbf{r}) L(\mathbf{r})$, with the grand potential density $\omega(\mathbf{r})$ given by:

$$\frac{\omega(\mathbf{r})}{kT} = -\sum_A \varphi_A(\mathbf{r}) \frac{u_A(\mathbf{r})}{kT} - \sum_i \frac{\varphi_i(\mathbf{r}) - \varphi_i^b}{N_i} + \frac{1}{2} \frac{q(\mathbf{r}) \psi(\mathbf{r})}{kT}$$

$$+ \frac{1}{2} \sum_A \sum_B \chi_{AB} \left\{ \varphi_A(\mathbf{r}) \left(\langle \varphi_B(\mathbf{r}) \rangle - \varphi_B^b \right) - \varphi_A^b \left(\varphi_B(\mathbf{r}) - \varphi_B^b \right) \right\} \quad (5.12)$$

In the case that the μ_is are fixed for all components except that of the surfactant (and water), that is when the ionic strength and the free volume bulk concentration are fixed, the Gibbs equation reads:

$$\frac{d\Omega}{d\mu_2} = -n_2^\sigma \quad (5.13)$$

where n_2^σ is the excess number of surfactants (component 2, cf. Table 5.1) at the air–water interface with respect to a suitably chosen Gibbs plane. Here and below, the Gibbs plane z_G is found such that the excess of V vanishes. By plotting the volume fraction of vapor segments with respect to the z-coordinate, a gradual transition from its maximum value, $\varphi_5 (z = 1)$, to the value in the bulk (in the water phase), $\varphi_5^b = \varphi_5(z = M_z)$, is found. The Gibbs plane is located at the z coordinate at which the area enclosed between the $\varphi_5(z)$ curve and the line $\varphi_5 = \varphi_5 (z = 1)$ equals the area enclosed between the $\varphi_5(z)$ curve and the line $\varphi_5 = \varphi_5 (z = M_z)$. Mathematically:

$$z_G = \frac{\sum_z \left(\varphi_5(z) - \varphi_5^b \right)}{\varphi_5(1) - \varphi_5^b} \quad (5.14)$$

Now the excess of surfactant (with respect to the Gibbs plane) is given by:

$$n_2^\sigma = \frac{1}{N_2} \sum_r L(r) \theta_2(r) \quad (5.15)$$

where:

$$\theta_2(r) = \sum_1^{z_G} (\varphi_2(z, r) - \varphi_2(1, r)) + \sum_{z_G + 1}^{M_z} (\varphi_2(z, r) - \varphi_2(M_z, r)) \quad (5.16)$$

5.3 Fundamentals of SCF Theory and the Molecular Model

In Equation 5.16 M_z is the layer number far in the bulk water phase, and $z = 1$ is a layer well in the vapor phase. Below, we will be interested in cases where the interfacial surfactants are not homogeneously distributed, that is, in the case that some interfacial micelles exist. For computational reasons such micelles are positioned at small r-values, that is at values of r that are small compared to the total number of shells in the r-direction (given by M_r). At regions far from the micelle (i.e., for coordinates $r \approx M_r$) the adsorbed amount is laterally homogeneous and the limiting value $\theta_2^b = \theta_2(M_r)$. The excess number of surfactants g^σ in the adsorbed micelle is now given by:

$$g^\sigma = \frac{1}{N_2} \sum_r L(r)(\theta_2(r) - \theta_2^b) \tag{5.17}$$

In the regions far from the central micelle, where the adsorbed amount is not a function of r, it is also possible to compute the limiting interfacial tension $\gamma^b = \sum_z \omega(z, M_r)$. This value is used to evaluate the surface pressure at which the surface micelles start to exist. We also use this tension to help evaluate the translationally-restricted grand potential for the central interfacial micelle by:

$$\Omega_m = \Omega - \sum_r L(r) \gamma^b \tag{5.18}$$

Coexistence of micelles of different size at certain surfactant chemical potential is always expected. The size distribution can be derived from the SCF calculations. The central quantity is the excess Helmholtz energy of the micelle, F^σ [5, 18]. At fixed values of the chemical potentials (μ_i), T and p, the excess Helmholtz energy has a minimum at the most likely micelle size. Very close to this minimum, and in first approximation, the excess Helmholtz energy is found to be a quadratic function of the micellar size:

$$F^\sigma(g) - F^\sigma(g^\ddagger) + K_g(g - g^\ddagger)^2 \tag{5.19}$$

where $F^\sigma(g^\ddagger) = \mathcal{E}_m(g^\ddagger)$ and $g^\ddagger$ are the translationally-restricted grand potential and the aggregation number of the most probable micelle, respectively.

Then, the size distribution is Gaussian:

$$\varphi_m(g) \propto \exp\left(-\frac{\mathcal{E}_m(g^\ddagger)}{kT}\right) \exp\left(-\frac{(g-g^\ddagger)^2}{2\sigma}\right) \tag{5.20}$$

with the width of the distribution given by:

$$\sigma^2 = -kT \frac{g^\ddagger}{(\partial \mathcal{E}_m / \partial g)_{g^\ddagger}} \tag{5.21}$$

Similarly, the hemi-micelle size distribution can be generated using Ω_m and g^σ instead of $\mathcal{E}_m$ and g, respectively.

5.4
Results

The primary results from the SCF calculations are the segment distributions of each type of segment in the system. Figure 5.2 shows results for a typical $C_{18}X_5$ hemi-micelle, adsorbed at the air–water interface at an ionic volume fraction of $\varphi_4^b = 0.01$. With the aim of showing where the interface is located, in Figure 5.2a we present the volume fraction profile of vapor molecules minus that for water molecules $[\varphi_5(r) - \varphi_1(r)]$. At low z values, such a quantity goes to unity (vapor phase, dark), and at high z values it goes to minus unity (water phase, white). As in the hemi-micelle there is both little water and little free volume, we can also extract the shape of the aggregate. It appears that the overall shape of the micelle is lens-type with a maximum diameter in the radial direction, D_{max}, located at $z = 11$, and a maximum height in the z direction, d_{max}, located at $r = 0$. A detailed analysis of the hemi-micelle structure is given below. In Figure 5.2b the surfactant head distribution is plotted. The heads form the hemi-micelle corona facing the water phase. The tails (not shown) then form the core of the lens, and from Figures 5.2a and b it can be seen that the tails face the vapor phase (as expected). The counterions (not shown) are mainly found in the corona. A negligible amount is found in the center of the apolar core as well as in the vapor phase. The counterion density decreases exponentially with the decay length given by the Debye length, from its highest value (corona) down to the bulk value (water phase).

In the following, we present results from the thermodynamic and structural analysis of these adsorbed lenses. By taking into account findings from charged surfactant self-assembly in solution, it is expected that various parameters influence the stability and structure of adsorbed charged hemi-micelles at the air–water interface. For instance, it is known that surfactant characteristics, such as the tail

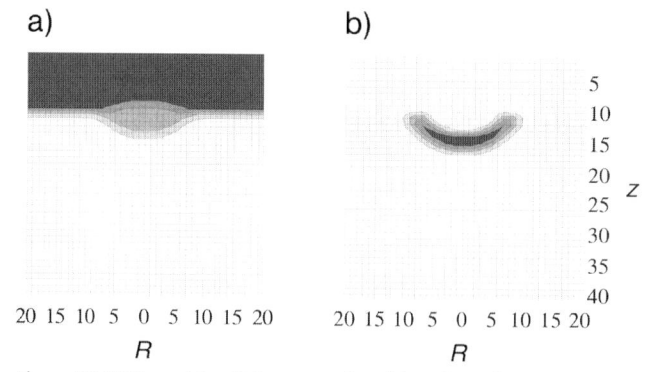

Figure 5.2 2D "equal density" contour plot of the volume fraction profile of vapor minus water (a) and surfactant heads (b) for a $C_{18}X_5$ hemi-micelle, adsorbed at the vapor–water interface, $\varphi_2^b = 0.92 \times 10^{-5}$, $g^\sigma = 23$, $\gamma^* = 0.6487$. P_1, $N_1\varphi_4^b = 0.01$. (a) Dark: $\varphi_5 - \varphi_1 = 1$ (vapor phase), white: $\varphi_5 - \varphi_1 = -1$ (water phase), $\Delta(\varphi_5 - \varphi_1) = 0.6$; (b) dark: $\varphi_X = 0.15$, white: $\varphi_X = 0$, $\Delta\varphi_X = 0.03$.

length, as well as the nature of the electrolyte solution (i.e., ionic strength, counterion size) have an important effect on the CMC and the aggregation number of charged spherical micelles in solution. Therefore, we have selected these parameters for a more detailed investigation.

5.4.1
Stability Analysis

The translationally-restricted grand potential for the central interfacial hemi-micelle as a function of its size $\Omega_m(g^\sigma)$, called stability curve, is presented in Figure 5.3a for surfactants with different tail lengths.

Similarly to the spherical micelle problem, at the CSAC the concentration of hemi-micelles, ϕ_m, is a minimum. From Equation 5.3, Ω_m must be positive, and also $\partial \Omega_m / \partial g^\sigma < 0$ for stable equilibrium. As $\phi_m = \exp(-\Omega_m/kT)$, and as the surface fraction of micelles should be small ($\phi_m \ll 1$) for isolated micelles, we can restrict ourselves to focus on hemi-micelles with relatively large Ω_m. For short-tail surfactants ($t < 15$) such a condition does not occur and we conclude that even for the relatively high ionic strength we need $t \geq 15$ surfactants to obtain dilute surface micelles. Then, the maximum of this curve corresponds to the smallest hemi-micelle that is stable, and the corresponding ϕ_2^b value is taken to be the CSAC. From Equation 5.13, a minimum at the CSAC for the surfactant concentration in the bulk versus the excess number of surfactants in the adsorbed hemi-micelle curve is expected. Examples of such curves for surfactants with different tail lengths are presented in Figure 5.3b. The excess number of surfactants in the hemi-micelles increases with increasing $\phi_m = \exp(-\Omega_m/kT)$. This increase can be as much as a factor of two or three. In Table 5.3, selected data from the stability analysis (CSAC, $g^{\sigma,*} = g^\sigma$ (CSAC)) of the systems presented in this work are shown together with the corresponding values for spherical micelles in solution (CMC, g). From this table it can be seen that aggregation at the surface takes place at substantially lower surfactant concentration than for the

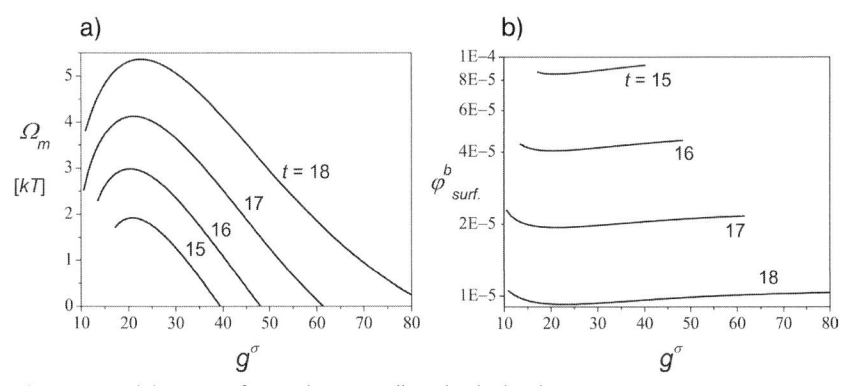

Figure 5.3 Stability curves for C_tX_5 hemi-micelles adsorbed at the vapor–water interface. P_1, N_1, $\phi_4^b = 0.01$. (a) Translationally-restricted grand potential and (b) surfactant concentration in the bulk versus the excess number of surfactants in the hemi-micelle.

5 Modeling the Structure and Stability of Charged Hemi-Micelles at the Air–Water Interface

Table 5.3 Thermodynamic properties of spherical micelles in solution and of hemi-micelles at the air–water interface. The surface pressure has to be multiplied by 105.5 to obtain it in units of mN m^{-1}. The area per molecule is expressed in units of b^2, with $b = 0.5$ nm.

t	Y	φ_4^b (10^{-3})	CMC (10^{-5})	g	CSAC (10^{-5})	$g^{\sigma,*}$	π^* (10^{-3})	a^*
15	1	10	22.97	58	8.49	21	8.63	85.54
16	1	10	10.97	60	4.05	20	5.69	140.04
17	1	10	5.15	76	1.94	21	3.76	224.11
18	1	1	10.18	46	2.58	12	6.86	139.58
		2	6.30	51	1.78	13	5.09	174.92
		3	4.83	68	1.47	16	4.23	207.18
		5	3.51	72	1.19	18	3.35	260.69
		10	2.42	75	0.92	23	2.49	354.22
	2	10	2.80	72	1.09	20	2.67	327.42
	3	10	2.94	71	1.14	19	2.75	318.56
	4	10	3.06	70	1.17	18	2.82	310.13

self-assembly in solution, with the ratio CSAC/CMC in the range 0.25–0.39. It is also found that the number of surfactants in the hemi-micelles is always less than half the number of surfactants in the spherical micelle (in the bulk).

In Figure 5.4 we show the size distribution for spherical micelles in solution (Figure 5.4a), and for hemi-micelles at the air–water interface (Figure 5.4b) at different μ values for $C_{18}X_5$ surfactant. Differences are found between the size distribution for spheres and for adsorbed lenses. The width of the distribution, and

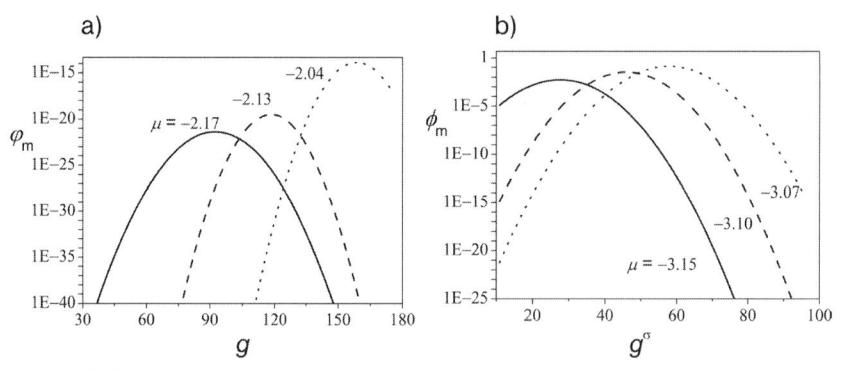

Figure 5.4 Size distribution of spherical micelles in solution (a), and hemi-micelles at the air–vapor interface (b) composed of $C_{18}X_5$ surfactants at different surfactant chemical potential (μ in kT units). In (a) $\varphi_2^b = 2.46 \times 10^{-5}$ (solid line), 2.56×10^{-5} (dashed line) and 2.81×10^{-5} (dotted line). In (b) $\varphi_2^b = 0.93 \times 10^{-5}$ (solid line), 0.98×10^{-5} (dashed line) and 1.00×10^{-5} (dotted line). P_1, N_1, $\varphi_4^b = 0.01$.

the volume fraction of micelles at the maximum, change considerably at changing chemical potential in the case of micelles in solution, but much less in the case of hemi-micelles.

From the SCF calculations it is further possible to obtain the surface tension γ. The difference from that of the pristine interface determines the surface pressure. We have evaluated the surface pressure at the CSAC, π^*, and the corresponding area per adsorbed surfactant molecule a^*. These quantities are also included in Table 5.3.

The effect of the surfactant tail length on the stability behavior of the adsorbed hemi-micelles can be discussed on the basis of the data presented in Figure 5.3. The maximum in the stability curve (Figure 5.3a) decreases, and the CSAC (Figure 5.3b) increases by decreasing the tail length, but $g^{\sigma,*}$ remains almost constant. By comparison with the behavior in solution (Table 5.3), the trends for the CMC and CSAC are similar, but not that for g and $g^{\sigma,*}$. The inhomogeneities at the interface appear at higher surface pressure and lower area per molecule upon decreasing the tail length (Table 5.3). Traube's rule is qualitatively followed by these results, as by adding three CH_2 groups in the surfactant chain the CSAC decreases by a factor of ten.

The effect of electrolyte on the stability of the $C_{18}X_5$ hemi-micelles adsorbed at the interface can be observed in Figure 5.5. By increasing the ionic strength, the maximum in the stability curve, and $g^{\sigma,*}$ (Figure 5.5a) increase, and the CSAC (Table 5.3) decreases. This is consistent with the behavior in solution (see Table 5.3). However, the effect on the CMC is slightly stronger than that for the CSAC, whereas the effect on $g^{\sigma,*}$ is larger than that on g. Electrostatic repulsion of the heads plays a more important role for the structure of the surface object than for those in solution. The inhomogeneities at the interface appear at lower surface pressure and higher area per molecule upon increasing the ionic strength (Table 5.3).

As the counterions penetrate the head group region to directly compensate the charge of the surfactant head, it is natural to expect that the size of the ion has a large

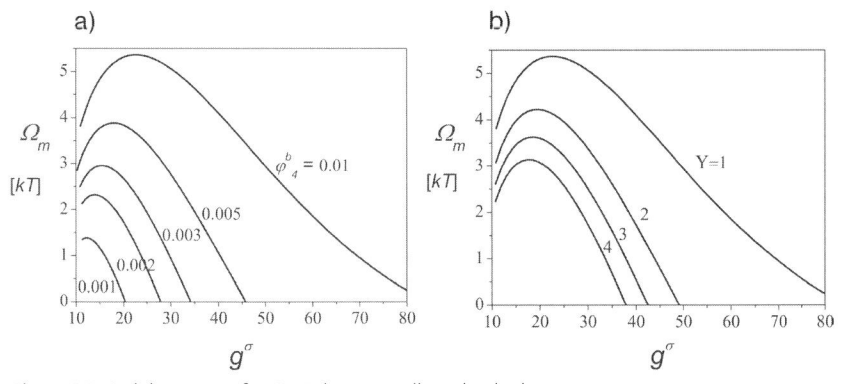

Figure 5.5 Stability curves for $C_{18}X_5$ hemi-micelles adsorbed at the vapor–water interface. (a) Effect of the ionic strength for P_1, N_1; (b) effect of counterion size, given by $Y + 1$ (the number of surrounding water molecules around a central ion Y is indicated) for $\varphi_4^b = 0.01$.

influence on the self-assembly characteristics of the surfactant. This generic effect is important for micelles in solution, but is also expected to be significant for the hemi-micelles at the interface. By increasing the counterion size, which is done by changing the number of W units (given by the parameter Y) around the central ion, the maximum in the stability curve and $g^{\sigma,*}$ (Figure 5.5b) decrease, and the CSAC (Table 5.3) increases. This is in agreement with the trends found for CMC and g. The effect on the CMC is similar to that on the CSAC, whereas that on $g^{\sigma,*}$ is larger than that on g. By the same token, the size of the counterion plays a more important role for the structure of the surface object than for that in solution. The inhomogeneities at the interface appear at higher surface pressure and lower area per molecule upon increasing the counterion size (Table 5.3).

5.4.2
Structural Analysis

From the segment distribution of the surfactant in the system, one can extract information on different structural properties of the hemi-micelle adsorbed on the surface. It was shown before (see Figure 5.2) that the surface aggregate is lens-shaped. An interesting parameter to characterize such a shape is the geometrical ratio defined as D_{max}/d_{max}. Our criterion for establishing the edge of the micelle was $\varphi_2 = 0.250$, which was based on the volume fraction distribution. As the profiles on the edge of the micelle typically change rapidly, the size ratio is well defined.

Another structural parameter based on the surfactant volume fraction profiles is the structural ratio computed from a RMS analysis, δ_R/δ_z with:

$$\delta_R = \sqrt{\frac{\sum_r r^2 [\varphi(11,r) - \varphi(11, M_r)]}{\sum_r [\varphi(11,r) - \varphi(11, M_r)]}} \tag{5.22}$$

and:

$$\delta_z = \sqrt{\frac{\sum_z (z - z_{CM})^2 \varphi(z,1)}{\sum_z \varphi(z,1)}} \tag{5.23}$$

which are consistent with the definitions of D_{max} and d_{max}. The last mentioned quantity needs the z coordinate of the center of mass of the object, z_{CM}, which is given by:

$$z_{CM} = \frac{\sum_{z,r} z [\varphi(z,r) - \varphi(z, M_r)] L(r)}{\sum_{z,r} [\varphi(z,r) - \varphi(z, M_r)] L(r)} \tag{5.24}$$

Finally, it is also interesting to know the z position of the center of mass of the object with respect to the Gibbs plane.

We analyzed all these structural parameters for the first stable hemi-micelle in each of the systems presented above to determine the effects of the tail length, ionic strength and counterion size on such a structure. In all cases the center of mass of the surface object is located in the water phase, above the Gibbs plane, which is at

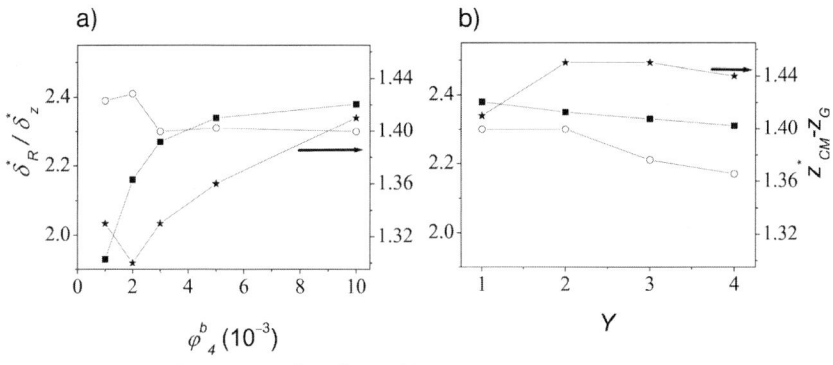

Figure 5.6 Structural properties of the first stable $C_{18}X_5$ hemi-micelles adsorbed at the vapor–water interface. Structural ratio from moment analysis (■), and from geometry (○); deviation of the centre of mass from the Gibbs plane (∗). (a) Effect of the ionic strength for P_1, N_1; (b) effect of counterion size for $\varphi_4^b = 0.01$.

$z_G = 10.1$. The aspect ratio varied between 1.9 and 2.5 which means that the hemi-micelle has a lens shape.

The tail length was found not to influence these structural parameters; this trend was expected from the $g^{o,*}$ behavior. The parameters D_{max}/d_{max}, δ_R/δ_z and $z_{CM} - z_G$ are approximately constant and around 2.3, 2.4 and 1.4, respectively. Then, the geometrical ratio and the ratio from moments are very close to each other.

The effect of the properties of the electrolyte solution are shown in Figure 5.6. The mass distribution of surfactant changes significantly by changing the ionic strength, which gives rise to the decrease of the structural ratio from the RMS analysis, and to the approach of the center of mass to the Gibbs plane upon decreasing the ionic strength (Figure 5.6a). The geometrical ratio does not significantly depend on the ionic strength. From Figure 5.6b it can be seen that the counterion size does not play an important role on the structural properties. Only the geometrical ratio decreases slightly with the ion size.

5.5
Conclusions

Lens-type hemi-micelles are predicted to arise at the air–water interface at relatively low bulk concentrations (with respect to the CMC) of charged surfactants with sufficiently long tail. The heads are exposed to the water phase while the tails form the core that enters the vapor phase. The stability of these hemi-micelles increases with tail length and ionic strength, but decreases with the counterion size. However, the excess number of surfactants in the adsorbed hemi-micelles, and also their structures, depend more strongly on the surfactant head properties than on the surfactant tail size, as is the case for spherical micelles. The effect of the tail length on the aggregation number of the first stable hemi-micelle at the interface is almost negligible when compared with the corresponding trend for the micelle in solution. Electrostatic repulsion of the heads plays, therefore, a more important role for the structure of the surface object than for that in solution. The tail length is found to be the most important parameter influencing the first appearance of inhomogeneities at the interface, which takes place at higher surface pressure and lower area per molecule when this tail is shorter.

Acknowledgment

A.B. Jódar-Reyes thanks the University of Extremadura for financial support through the "Plan de Iniciación a la Investigación, Desarrollo Tecnológico e Innovación."

References

1 Taylor, D.J.F., Thomas, R.K. and Penfold, J. (2007) *Advances in Colloid and Interface Science*, **132**, 69.
2 Kjaer, K., Als-Nielsen, J., Helm, C.A., et al. (1987) *Physical Review Letters*, **8**, 2224.
3 Lyklema, J. (2000) Langmuir monolayers, Ch 3, *Fundamentals of Interface and Colloid Science (FICS)*, Vol III, Academic Press.
4 Lyklema, J. (2000) Gibbs monolayers, Ch 4, *Fundamentals of Interface and Colloid Science (FICS)*, Vol III, Academic Press.
5 Leermakers, F.A.M., Eriksson, J.C. and Lyklema, J. (2005) Association colloids and their equilibrium modelling, Ch 4, *Fundamentals of Interface and Colloid Science*, Vol V, Elsevier, Amsterdam.
6 Lyklema, J. (2000) Interfacial tension, molecular interpretation, Ch 2, *Fundamentals of Interface and Colloid Science (FICS)*, Vol III, Academic Press.
7 Lyklema, J. and Leermakers, F.A.M. (2007) Ion specificity in colloidal systems (ed. Th.R. Tadros), Ch 8 *Colloid Stability and Application in Pharmacy*, Vol 3, Wiley-VCH Verlag GmbH.
8 Laughlin, R.G. (1994) *The Aqueous Phase Behavior of Surfactants*, Academic Press.
9 Hill, T.L. (1991 and 1992) *Thermodynamics of Small Systems, Part 1 and Part 2*, Dover Publications.
10 Hall, D.G. and Pethica, B.A. (1976) *Non-ionic Surfactants*, Ch 16, Marcel Dekker.
11 Scheutjens, J.M.H.M. and Fleer, G.J. (1979) *The Journal of Physical Chemistry*, **83**, 1619.
12 Scheutjens, J.M.H.M. and Fleer, G.J. (1980) *The Journal of Physical Chemistry*, **84**, 178.
13 Meijer, L.A., Leermakers, F.A.M. and Lyklema, J. (1999) *Journal of Chemical Physics*, **110**, 6560–6579.
14 Jódar-Reyes, A.B., Ortega-Vinuesa, J.L., Martín-Rodríguez, A. and Leermakers, F.A.M. (2003) *Langmuir*, **19**, 878.
15 Böhmer, M.R., Koopal, L.K., Janssen, R., et al. (1992) *Langmuir*, **8**, 2228.
16 Jódar-Reyes, A.B., Ortega-Vinuesa, J.L., Martin-Rodriguez, A. and Leermakers, F.A.M. (2002) *Langmuir*, **18**, 8706–8713.
17 Leermakers, F.A.M., Rabinovich, A.L. and Balabaev, N.K. (2003) *Physical Review E*, **67**, 011910.
18 Jódar-Reyes, A.B. and Leermakers, F.A.M. (2006) *The Journal of Physical Chemistry*, **110**, 6300–6311.

6
Foam, Emulsion and Wetting Films Stabilized by Polymeric Surfactants

Dotchi Exerowa and Dimo Platikanov

6.1
Introduction

Thin liquid films have proven their advantages in the study of interaction forces in foam, emulsion, and wetting films stabilized by various types of surfactants: see, for example, Refs. [1–3]. DLVO and non-DLVO surface forces that stabilize these films have been established; in many cases the relation between surface forces and film stability has also been found. Recently, several authors have reported results of model experiments with thin liquid films (foam, emulsion and wetting films) stabilized by polymeric surfactants. In our laboratories all three types of films from aqueous solutions of A–B–A triblock copolymers [4–6] or AB_n hydrophobically modified inulin [8–10] have been studied. The corresponding disperse systems (foams, emulsions, suspensions) stabilized by AB_n polymeric surfactants have also been studied extensively; see, for example, Refs. [11–14]. It was supposed that the stabilizing forces are steric surface forces but they have not been directly proven and quantitatively studied.

In most other papers, polyelectrolytes have been used [15–21]; see also the comprehensive review of Ref. [22]. Amphiphilic diblock copolymers have also been used for foam film studies [23–25]. Emulsion films stabilized by natural polymeric surfactants from bitumen [26] and thin liquid films from protein (biopolymer) aqueous solutions [27] should also be noted.

Here we review the results obtained [4–10] with two types of non-ionic polymeric surfactants: (i) A–B–A triblock copolymers and (ii) novel graft polymeric surfactants based on inulin.

Two PEO–PPO–PEO three-block copolymers of the Synperonic series – F108 and P104 [28, 29] – have been employed. PPO represents the middle hydrophobic block B and both hydrophilic PEO form the two terminal chains A. These commercial, non-ionic, water-soluble polymeric surfactants were used as obtained from BASF. They are pure though not monodisperse. The molecular masses and average EO/PO contents are known from the manufacturer and yield approximate

Highlights in Colloid Science. Edited by Dimo Platikanov and Dotchi Exerowa
Copyright © 2009 WILEY-VCH Verlag GmbH & Co. KGaA, Weinheim
ISBN: 978-3-527-32037-0

chemical formulae: $EO_{128}PO_{54}EO_{128}$ for F108 ($M = 14\,000$ Da) and $EO_{31}PO_{54}EO_{31}$ for P104 ($M = 5900$ Da).

Recently, novel graft AB_n co-polymers have been designed [30] with A consisting of inulin (linear polyfructose backbone) that has been modified by introducing several alkyl groups B [14]. We have used one of these AB_n polymeric surfactants, namely INUTEC SP1, from the company ORAFTI (Belgium). Very stable oil in water (O/W) emulsions stabilized by INUTEC SP1 have been reported [11, 12]. These emulsions were stable for more than one year up to 50 °C at NaCl concentrations of up to $2\,\mathrm{mol\,dm^{-3}}$ or $1\,\mathrm{mol\,dm^{-3}}$ $MgSO_4$. According to Ref. [12] this very high stability is due to the nature of the hydrophilic polyfructose chain. Such a stability at high electrolyte concentrations has not been observed with polymeric surfactants containing poly(ethylene oxide) (PEO) chains. The authors of Ref. [12] explain this fact with the change of the sign of the Flory–Huggins interaction parameter (χ). It is well known that the Flory–Huggins interaction parameter is connected to the solubility of the chains in the medium [31]. In water both inulin and PEO are strongly hydrated by water molecules, hence $\chi < 1/2$ under these conditions. When increasing the electrolyte concentration and the temperature, χ changes. It seems that the inulin-stabilizing chain can retain its hydration to much higher temperatures and electrolyte concentrations than the PEO chains and this is probably the reason for the high emulsion stability observed.

The discussed stabilization effect of these emulsifiers is unique and it is very important to understand which are the interaction forces in the single emulsion film and how they stabilize the film. It is also interesting to understand the influence of the adjacent phase (oil or water, air, solid) on the interaction forces in the film. Correspondingly, emulsion, foam and wetting films have been studied [4–10]. The model of a microscopic thin liquid film (radius $\sim 100\,\mu m$) allows one to obtain films at very low concentrations of polymeric surfactant and to study their formation and stability as well as to establish and to distinguish the surface (interaction) forces in them.

6.2
Microinterferometric Method for Investigation of Thin Liquid Films

The microinterferometric method has widely been used by many authors to investigate both symmetric thin liquid films (foam and emulsion films) and asymmetric ones (wetting films). This method has been described in several papers, for example, Refs. [2, 3, 32, 33, 37] as well as in a book [1]. Here we give only a brief outline.

The measuring cell, in which the microscopic thin liquid films are formed and studied, is the basic part of micro interferometric apparatus. Figure 6.1 presents the main details of three measuring cells. In the Scheludko–Exerowa cell (Figure 6.1a) the film is formed in the middle of a biconcave drop at constant capillary pressure. This is a horizontal round film of radius r of about 50–100 μm. A small portion of the liquid is sucked out of the drop through the capillary using a micro-metrically driven

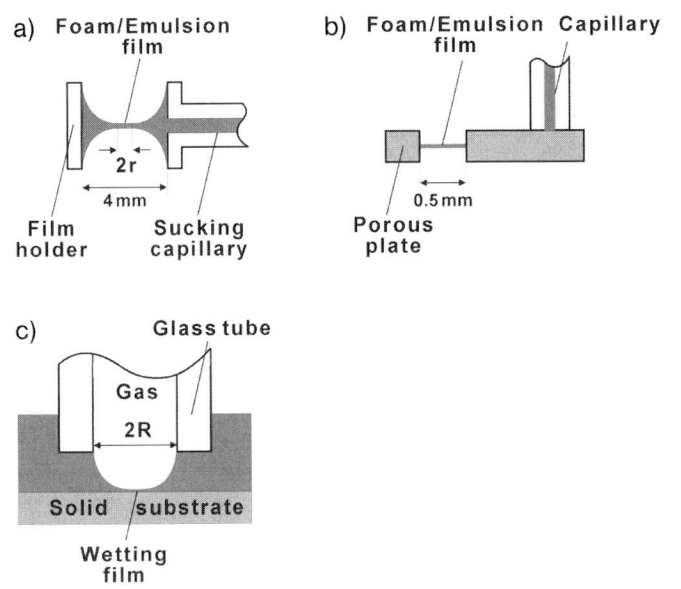

Figure 6.1 Main parts of three experimental cells used to investigate microscopic thin liquid films.

pump. The two liquid/gas (or liquid/liquid) interfaces approach each other and ultimately a foam (or emulsion) film is created. The film thins out, reaching one of several possible states: equilibrium thickness, critical thickness of rupture, black spots and black film formation.

In the Exerowa–Scheludko porous plate cell [34, 35] the foam (or emulsion) film is formed in the cylindrical hole in a porous sinter glass plate (Figure 6.1b). The meniscus goes into the pores and allows one to change the capillary pressure (ΔP_c), which at the equilibrium film thickness h is equal to the disjoining pressure Π. This provides for a direct measurement of the $\Pi(h)$ isotherms, which usually can be performed over a wide range of thicknesses and pressures. This is the so-called Thin Liquid Film – Pressure Balance Technique, which allows direct measurement of the interaction forces in the film. The gas pressure inside the glass chamber of the cell is increased (up to 10^5 Pa) with a membrane pump. Pressure values lower than 10^3 Pa are measured with a water manometer (± 5 Pa). Higher pressures are read from a standard membrane manometer ($\pm 0.6\%$).

The Platikanov cell was designed [36] for investigation of thin wetting films on a solid surface (Figure 6.1c). A smooth plane–parallel quartz glass plate provides a hydrophilic (or hydrophobic if pretreated with trimethylchlorosilane) solid surface. The cell is filed with the aqueous solution. A cylindrical tube is fixed normal to the glass plate, its orifice being submerged into the liquid, close to the solid surface (the distance between this surface and the orifice is somewhat less than the inner radius of the tube). Using a micro-metrically driven pump the pressure of the gas in the tube is slightly increased, the liquid/gas meniscus approaches the solid surface and a microscopic, round, thin, wetting film is formed.

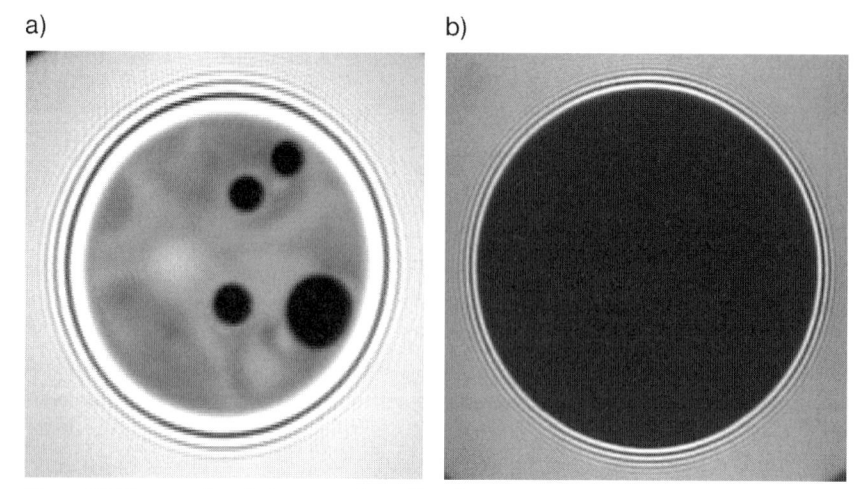

Figure 6.2 Photographs of microscopic foam films from INUTEC SP1 aqueous solution: (a) foam film with black spots; (b) black foam film.

The thickness of all types of films is monitored by measuring the reflection of monochromatic light using the micro-interferometric method. Assuming that the film is optically homogeneous and it has a refractive index equal to that of the bulk solution, a film thickness, h_w, referred to as "equivalent film thickness," is calculated. Figure 6.2 shows photographs of a microscopic foam film with black spots and a black foam film from INUTEC SP1 aqueous solution. This technique provides experimental data about the equilibrium or critical film thickness, contact angles, lifetimes, disjoining pressures, and so on of all three types of films. Details of the measurements are described elsewhere [1–3, 32, 33, 37].

6.3
Intercation Forces in Foam Films

6.3.1
Foam Films Stabilized by A-B-A Block Copolymers. Brush-to-Brush Interaction

A general test for the presence of electrostatic interaction in a film is the influence of the electrolyte concentration. The equivalent film thickness h_w of foam films from Pluronic F108 aqueous solutions decreases with increasing electrolyte concentration (C_{el}) [4]. Above the so-called critical concentration, $C_{el,cr} = 3 \times 10^{-3}\,\mathrm{mol\,dm^{-3}}$ NaCl, h_w remains constant. There is a plateau in the $h_w(C_{el})$ curve. The concentration $C_{el,cr}$ distinguishes the electrostatic interaction from the steric interaction and shows the transition between them.

Direct experimental measurement of disjoining pressure vs thickness isotherms, $\Pi(h_w)$, is very informative in establishing surface forces in films from polymeric

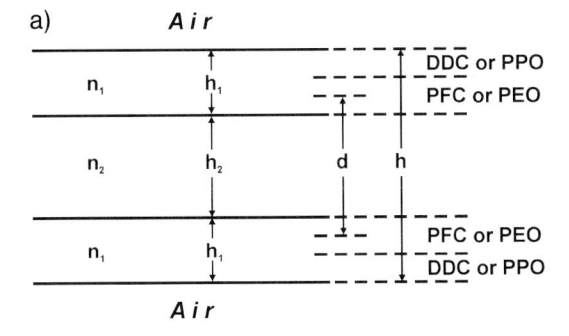

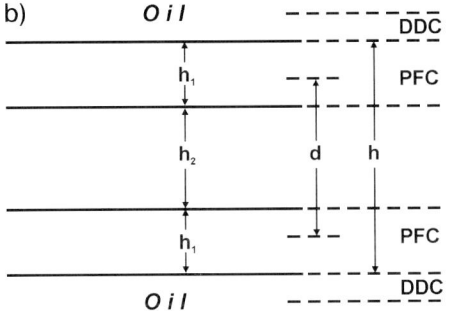

Figure 6.3 Schematic representation of the three-layer film model ("sandwich model"): (a) foam film; (b) emulsion film.

surfactants. Calculation of the surface forces acting in a foam film is based on the three-layer film model [38, 4] or "sandwich model" (Figure 6.3a). It consists of two adsorption layers of thickness h_1 each and a water core h_2 with indexes of refraction n_1 and n_2, respectively:

$$h = 2h_1 + h_2$$
$$h_2 = h_w - 2h_1 \frac{(n_1^2 - 1)}{(n_2^2 - 1)} \tag{6.1}$$

For PEO–PPO–PEO triblock copolymers it was found that the poly(ethylene oxide) chain is in the "brush" configuration if $D < R_J$ (D is the lateral distance between the chains, R_J is the radius of gyration). When $h < 2h_1$ a brush-to-brush interaction appears, which can be described by the scaling theory of de Gennes [39, 40], according to which the steric disjoining pressure is given by:

$$\Pi_{st} = \left(\frac{kT}{D^3}\right)(H^{9/4} - H^{3/4}) \tag{6.2}$$
$$H = h/2h_1$$

The good fit of Equation 6.2 with experimental data for Pluronic F108 can be seen in Figure 6.4, which shows the disjoining pressure isotherm, $\Pi(h_w)$, at 23 °C of a 7×10^{-6} mol dm^{-3} F108 + 10^{-4} mol dm^{-3} NaCl aqueous solution [4], with C_{el} being

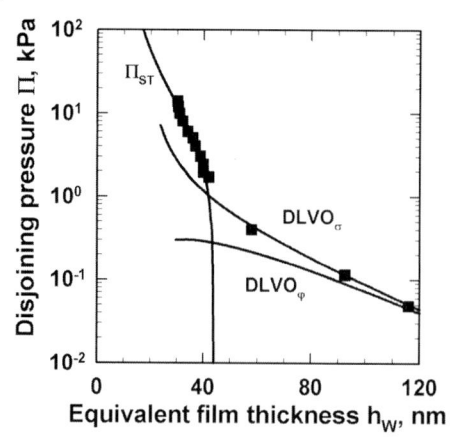

Figure 6.4 Total disjoining pressure versus film thickness for foam films from F108 solution; solid lines are the best fits of DLVO-theory and of Equation 6.2 for Π_{st}.

less than $C_{el,cr}$. The square symbols are the experimentally measured Π-values and the three lines present the $\Pi(h_w)$ dependence calculated according to DLVO theory [41, 42], at constant potential φ or constant charge σ, and Equation 6.2. In the thickness range 50–120 nm the DLVO theory at constant charge describes well the $\Pi(h_w)$ isotherm, while at smaller thickness of 20–40 nm a good fit of the theory [39, 40] of de Gennes is observed. This curve is fitted with $h_1 = 11$ nm in correspondence with the three-layer model. This is proof of the existence of steric interaction between the adsorption layers of A–B–A block copolymers and *brush-to-brush interaction* leading to formation of a film, close to a bilayer film.

By changing the electrolyte concentration we have shown that electrostatic repulsion is operative. If the surface charge arises from preferential adsorption of OH^- ions [43] it should be possible to destroy it by decreasing the pH of the bulk solution at constant (and rather low) ionic strength [5].

The effect of pH on the equivalent film thickness h_w at four different values of the ionic strength ($I = C_{el}$ for 1–1 electrolyte) is depicted in Figure 6.5 for foam films from F108 aqueous solutions [4]. The overall trend is the same in all cases: with decreasing pH, h_w decreases from a plateau value for pH > 5 to a common lower value of 40 nm. The plateau value decreases with increasing the ionic strength. The lower the ionic strength the steeper is the h_w(pH) dependence. Below a critical (steric) value, $pH_{cr,st} = 3.2$, h_w is independent of pH. At approximately pH = 4 all curves converge though they stop at different pH (the lowest possible pH at a given ionic strength). At the highest ionic strength (10^{-2} M), which corresponds to $C_{NaCl} > C_{el,cr}$, the electrostatic repulsion is totally screened and no influence of pH on film thickness is observed.

The results shown in Figure 6.5 are a manifestation of the peculiarity of the water–air interface and give an insight into the mechanism of charge creation there. When the pH decreases, the bulk concentration of H^+ ions increases and, consequently, their adsorption at the solution–air interface increases. Here they

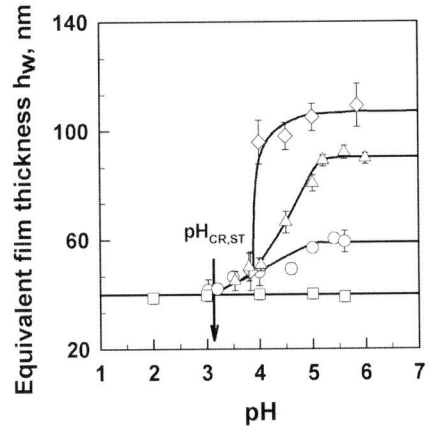

Figure 6.5 Dependence of the film thickness on pH for foam films from F108 solution. Ionic strength (M): (◇) 1.5×10^{-4}; (△) 3×10^{-4}; (○) 10^{-3}; (□) 10^{-2}.

recombine with the excess of potential-forming OH⁻ ions and thus the negative charge is destroyed. The potential of a diffuse double layer decreases and eventually vanishes at the critical pH. Thus the critical value $pH_{cr,st}$ identifies the isoelectric point of the solution–air interface, where electrostatic interaction drops to zero. Importantly, the pH gives the opportunity to vary and eventually suppress the electrostatic interaction at relatively low ionic strength.

6.3.2
Foam Films Stabilized by Hydrophobically Modified Inulin Polymeric Surfactants. Loop-to-Loop Interaction

We consider here another steric interaction – the loop-to-loop interaction. It is observed in foam films from aqueous solutions of hydrophobically modified inulin INUTEC SP1. The alkyl chains are strongly adsorbed at the air–water interface, leaving loops of polyfructose in the aqueous phase. The loops remain hydrated in the presence of high electrolyte concentrations [12, 14].

With increasing electrolyte concentration the film thickness decreases down to the critical value $C_{el,cr} = 2 \times 10^{-2}$ mol dm^{-3} [7]. At $C_{el} > C_{el,cr}$ the h_w remains constant, close to 16 nm. The left-hand part of the $h_w(C_{el})$ dependence indicates that there is an electrostatic component of disjoining pressure while the plateau indicates the existence of non-DLVO forces due to the steric interaction between the adsorbed polymer layers. Similar are the $h_w(C_{el})$ curves of foam films stabilized by A–B–A copolymers, non-ionic surfactants, non-ionic phospholipids, and so on [1–4, 33].

The $\Pi(h_w)$ isotherms were obtained at various electrolyte concentrations, from 10^{-4} to 10^{-3} mol dm^{-3} (below $C_{el,cr}$) as well as 0.5, 1 and 2 mol dm^{-3} (above $C_{el,cr}$). Here, we shall apply the three-layer foam film model (Figure 6.3a) to INUTEC SP1 foam films. The adsorption layers are composed of the hydrophobic dodecyl chains

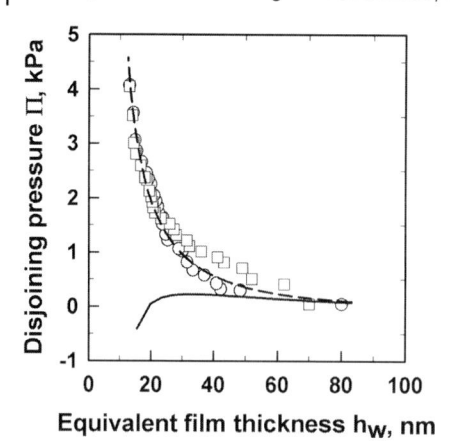

Figure 6.6 Measured disjoining pressure versus film thickness for foam films from INUTEC SP1 solution: (○) run 1 and (□) run 2. Dashed and solid lines — best fits of DLVO-theory, at constant charge and constant potential, respectively.

(DDC) and hydrophilic polyfructose chains (PFC), that is, polyfructose loops of the INUTEC SP1 molecules. According to Equation 6.1 the total film thickness (h) differs from the equivalent film thickness (h_w) that is experimentally measured, the difference being $h_{w,corr} = 2h_1(n_1^2-1)/(n_2^2-1)$ [38]. A value of 1.40 is taken for n_1 (between 1.42 for bulk dodecane and 1.38 for supersaturated aqueous solutions of inulin [44]). The value of h_1 is chosen to be 5 nm, based on data from atomic force microscopy and dynamic light scattering measurements [11, 13]; $h_{w,corr} = 12 \pm 0.7$ nm has been obtained.

Figure 6.6 shows the results [7] for foam films obtained from 2×10^{-5} mol dm^{-3} INUTEC SP1 + 10^{-4} mol dm^{-3} NaCl aqueous solutions at 23 °C. The different symbols indicate two different experimental runs. The disjoining pressure started to increase rapidly when the film thickness decreased below 40 nm. The film remains stable up to a disjoining pressure of about 4 kPa and stable film could be obtained in the range 15–80 nm. The mechanism of film rupture at a disjoining pressure above 4 kPa is difficult to explain at present. One may speculate that it is related to the magnitude of stabilizing forces. Similar rupture behavior has been reported for foam films from ionic and non-ionic surfactants; see, for example, Ref. [1].

Similar is the case of 10^{-3} mol dm^{-3} NaCl. The pressure raise starts before 30 nm and rupture occurs at a film thickness of about 12 nm, which corresponds approximately to a bilayer due to the loop-to-loop interaction [7].

The disjoining pressure isotherms for polymer stabilized foam films can be fitted with theory by considering all interaction forces, namely van der Waals attraction (Π_{vw}), double layer repulsion (Π_{el}) and steric interaction (Π_{st}). At $C_{el} < C_{el,cr}$, Π_{el} predominates over Π_{st}, at least at large film thickness, and in this case the disjoining pressure isotherms can be fitted using the classical DLVO theory, that is, $\Pi = \Pi_{el} + \Pi_{vw}$. In contrast, at $C_{el} > C_{el,cr}$, Π_{st} predominates over Π_{el}, and in this case $\Pi = \Pi_{vw} + \Pi_{st}$.

Theoretical analysis of the $\Pi(h_w)$ isotherms in Figure 6.6 is based on the DLVO theory [41, 42] and a three-layer model [38]. When only DLVO forces are operating, Π_{el} was evaluated by solving the complete Poison–Boltzmann equation, using the numerical procedure given by Chan *et al.* [45]. The solution allows either constant potential or constant charge boundary conditions to be considered as limiting cases of Π_{el}. Π_{vw} was calculated using the empirical equation of Donners *et al.* [46], which is based on the exact Liftshitz theory [47].

The electrostatic charge at the air–water interface containing adsorbed INUTEC SP1 molecules is most likely due to the adsorption of OH^- ions [1, 43]. Under these conditions, the distance d between the planes of the onset of Π_{el} needs additional modeling: the diffuse double layer boundary is supposed to be located in the middle of the adsorbed layers; hence, $d = h - h_1$. This is a rather rough approximation, but it has been proven to be satisfactory in practice [1, 4, 5].

The results of the DLVO theory fit to the experimental results are shown in Figure 6.6. The cases of constant potential and constant charge boundary conditions are plotted by solid and dashed lines, respectively. The diffuse double layer (DDL) potential at infinity was used as a fitting parameter. Keeping in mind the reproducibility of the experimental data and approximations made, the fit when using the constant charge model seems to be satisfactory. The corresponding DDL potentials at infinity are -28 and $-14\,mV$ for 10^{-4} and $10^{-3}\,mol\,dm^{-3}$ NaCl, respectively. As it can be seen, all of the experimental data for these relatively thick films are close to the theoretical calculations based on the constant charge model. Both the constant charge DLVO regime and the fitted values of the DDL potential at infinity are in good agreement with those reported for non-ionic surfactants with relatively large hydrophilic heads. Thus, one may conclude that at $C_{el} < C_{el,cr}$ only DLVO forces are responsible for film stability.

At higher electrolyte concentrations ($C_{el} > C_{el,cr}$) the non-DLVO interaction predominates. These are the results [7] of $\Pi(h)$ measurements for 0.5, 1 and 2 $mol\,dm^{-3}$ NaCl at constant INUTEC SP1 concentration ($2 \times 10^{-5}\,mol\,dm^{-3}$) at 23 °C – presented in Figure 6.7 as the difference of each measured Π-value minus the theoretically calculated Π_{vw}. At such high electrolyte concentrations, the variation of disjoining pressure with film thickness follows roughly the same trend, namely a gradual (but not rapid!) increase in disjoining pressure with reduction in film thickness. In contrast, at $C_{el} < C_{el,cr}$ the increase in disjoining pressure was much more rapid, the disjoining pressure being of the DLVO-type. With $C_{el} > C_{el,cr}$, the $\Pi(h)$ isotherms are roughly the same for each of the three electrolyte concentrations studied.

At $C_{el} > C_{el,cr}$ the double layer repulsion is practically suppressed and this is manifested by the fact that the film thickness no longer changes with increasing C_{el} above $C_{el,cr}$ (Figure 6.7). Under these conditions, $\Pi_{el} = 0$, and the experimental results cannot be explained in the framework of DLVO theory only. If one assumes that the only contribution to Π is Π_{vw} there are very large deviations from the experimental data. This directly implies that there is an additional repulsive contribution to the disjoining pressure, namely Π_{st}.

It has been suggested [12] that the adsorbed INUTEC SP1 molecule produces large "loops" of polyfructose between two adjacent dodecyl chains and strong repulsion

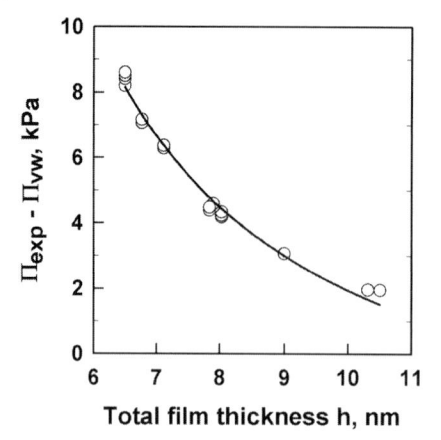

Figure 6.7 Measured disjoining pressure minus the calculated Π_{vw} as a function of film thickness for foam films from INUTEC SP1 solutions; the solid line is the best fit of Equation 6.2.

occurs when the adsorbed layers begin to overlap. This additional repulsion, usually referred to as steric repulsion (Π_{st}) [48], can be estimated from the data plotted in Figure 6.7, if we suppose that $\Pi_{st} = \Pi_{exp} - \Pi_{vw}$. Moreover, it has been recognized that hydrophilic headgroups longer than 1 nm may be treated as "quasibrushes" [49]. If this is also applicable to the "loops," one can use the theory [39, 40] for interacting polymer "brushes" to describe "loop-to-loop" steric interactions between adsorbed layers of INUTEC SP1.

Figure 6.7 shows the best fit of the values ($\Pi_{exp} - \Pi_{vw}$) vs h, that is $\Pi_{st}(h)$, with Equation 6.2 of de Gennes. The resulting value for h_1 from this fit is 6.5 nm, which is only slightly larger than the value used in the three-layer model (5 nm). Moreover, the disagreement becomes less significant if an adsorbed layer of 6 nm is assumed in the three-layer model. In this case, the fit (not shown) yields 6.1 nm for h_1.

In summary, it has been established, using microscopic foam films, that films from aqueous solutions of polymeric surfactants (A–B–A block copolymers and AB_n graft polymers based on inulin) are stabilized by both electrostatic, DLVO and steric, non-DLVO surface forces. Application of the scaling theory of de Gennes [39, 40] to describe the brush-to-brush and loop-to-loop interactions is interesting.

6.4
Interaction Forces in Emulsion Films

6.4.1
Emulsion Films Stabilized by A–B–A Block Copolymers: Brush-to-Brush Interaction and Transition to the Newton Black Film

The study of polymeric surfactants as emulsion stabilizers, the interaction between the emulsion droplets with adsorbed polymer layers, is an interesting topic. Our

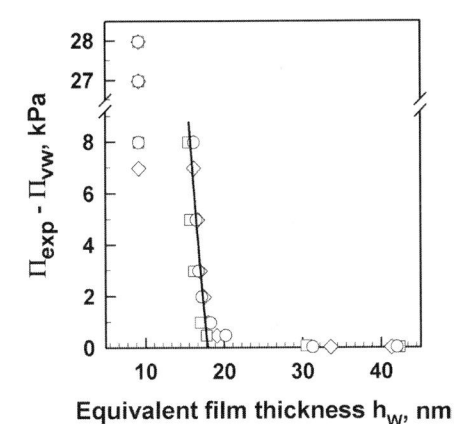

Figure 6.8 Measured disjoining pressure minus the calculated Π_{vw} as a function of film thickness for emulsion films from F108 solutions at C_{NaCl} (M): (◇) 0.05, (□) 1 and (○) 2; the solid line is the best fit of Equation 6.2.

approach to the study of the interaction between two oil droplets in an aqueous solution of polymeric surfactant, namely the study of thin microscopic aqueous films between two oil phases (O/W emulsion films), is quite useful. It enables us to draw a correlation between emulsion films and emulsion, similarly to the approach taken to foam films and foams [8].

The A–B–A block copolymers Pluronic F108 and P104 have been used as polymeric sufactants in 7×10^{-6} mol dm^{-3} (lower than the CAC of 10^{-5} mol dm^{-3}) aqueous solution. The oil phase was isoparaphinic oil Izopar M manufactured by Exxon Mobil Chemicals.

The disjoining pressure/film thickness isotherms, $\Pi(h)$, have also been measured experimentally for emulsion films. Figure 6.8 depicts $\Pi(h_w)$ isotherms [6] at 23 °C for emulsion films from Pluronic F108 aqueous solutions containing high electrolyte concentrations (0.05, 1 and 2 mol dm^{-3} NaCl), that is, $C_{el} > C_{el,cr}$. A small alteration in pressure (of about 1 kPa) results in a change of film thickness of 40–20 nm. A further increase in pressure at about 8 kPa leads to a transition to a *Newton black film*, 9 nm thick, which remains constant up to the maximum pressure applied. Close to this transition the $\Pi(h)$ isotherm is steep and the change in film thickness is negligible. No effect of electrolyte concentration on the experimentally measured isotherms is observed. At $C_{el} > C_{el,cr}$ the electrostatic component of disjoining pressure is suppressed. The other interaction forces could be assumed to be steric repulsion and Van der Walls attraction. It is interesting to compare the experimental data with the theory for steric repulsion of de Gennes [39, 40]. It could also be assumed that the total film thickness is equal to the equivalent thickness measured ($h = h_w$) under these experimental conditions. This applies to emulsion films when the hydrophobic parts of the polymeric surfactant are positioned in the oil phase while the hydrophilic, as is well known, are strongly hydrated. The solid line in Figure 6.8 indicates the fit between Π_{st} (= $\Pi_{exp} - \Pi_{vw}$) vs h. A value of 17.9 nm is obtained for the fitting parameter $2h_1$ (Equation 6.2), and hence $h_1 \approx 9$ nm. A rough estimate of h_1 may be obtained on the basis of the simple scaling theory [39]:

$$h_1 = a^{5/3} N S^{-1/3} \tag{6.3}$$

where a is the monomer size, N is the degree of polymerization and S is the area per molecule; for $S = 0.8$ nm^2, $a = 0.2$ nm, and $N = 150$, $h_1 \approx 14$ nm. Thus, two values for h_1 have been obtained: 14 and 9 nm. Although they do not coincide completely, we could conclude that close to the NBF transition the emulsion films are stabilized by *brush-to-brush* steric interactions.

6.4.2
Emulsion Films Stabilized by Hydrophobically Modified Inulin: Loop-to-Loop Interaction and Transition to the Newton Black Film

For INUTEC SP1, the graft polymeric surfactant based on hydrophobically modified inulin, the critical electrolyte concentration, $C_{el,cr}$, that separates DLVO from non-DLVO interactions is 5×10^{-2} mol dm^{-3} [8]. The effect of the electrolyte on the electrostatic component of disjoining pressure is clearly seen, and the film thickness remains constant ($h_w = 11$ nm) above $C_{el,cr}$ where steric interactions are acting.

Figure 6.9 shows the disjoining pressure isotherms $\Pi(h_w)$ at 22 °C for emulsion films from 2×10^{-5} mol dm^{-3} INUTEC SP1 + 2×10^{-4} mol dm^{-3} NaCl aqueous solution [8]. Two independent experiments are shown, presented with different symbols. The agreement between them is not the best one. This can be attributed to the meta-stability of the films. Initially, there is a gradual decrease of the film thickness with increasing pressure, after which there is a jump (marked with arrows in the figure) toward the Newton black film (NBF) formation. This jump is

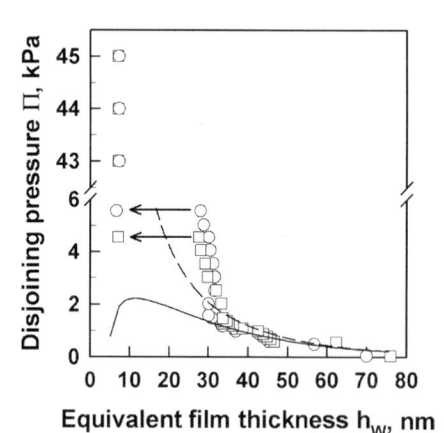

Figure 6.9 Measured disjoining pressure versus film thickness for emulsion films from INUTEC SP1 solution: (○) run 1 and (□) run 2. Dashed and solid lines: best fits of DLVO-theory, at constant charge and constant potential, respectively. Arrows denote the jump transition to NBF.

accompanied by an abrupt reduction in film thickness from about 30 nm to about 7 nm. This very small thickness is due to the formation of the NBF, which remains stable up to 45 kPa – the maximum pressure that can be applied in our experiments. The jump in film thickness occurs at a pressure that is quite different (between 4.5 and 5.5 kPa) in the two experiments presented in Figure 6.9. This variation in pressure is due to the nature of the film, which, as mentioned, is quite meta-stable. Notably, this transition to NBF is observed in all cases.

Because the results in the right-hand branch of the $\Pi(h_w)$ curve (Figure 6.9) are obtained at $C_{el} < C_{el,cr}$ one must analyze the isotherms by taking into account the double-layer repulsion and van der Waals attraction forces as given by the DLVO theory [41, 42]. For this purpose a three-layer model (Figure 6.3b) has been used that is similar to that described for foam films (Figure 6.3a): an aqueous core between two adsorbed layers, each composed of the hydrophobic dodecyl chains (DDC) and hydrophilic polyfructose chains (PFC) of the INUTEC SP1 molecules [7, 8]. However, in the case of emulsion films the DDC are in the adjacent oil phase (Figure 6.3b). Keeping in mind that the NBF is a bilayer structure (e.g. Refs [1, 8]), and that the PFC are probably strongly hydrated [12, 14], the thickness of the PFC layer, h_{PFC}, can be estimated to be 3.6 nm (half that of the NBF, as can be seen in Figure 6.9). Under these conditions, it seems reasonable to approximate the total thickness h by the experimentally derived equivalent thickness h_w.

By considering the DLVO forces alone, the total disjoining pressure is $\Pi = \Pi_{el} + \Pi_{vw}$. As discussed before, the planes of the onset of Π_{el} were chosen to be situated in the middle of the PFC layers, that is $d = h - 3.6$ nm (Figure 6.3b), and Π_{el} was computed following a numeric procedure based on the complete Poisson–Boltzmann equation and considering both constant charge and constant potential cases [45]. The van der Waals interactions were calculated according to the equation $\Pi_{vw} = -A/6\pi h^3$, where A is the effective Hamaker constant. For Isopar M (a mixture of C_{11}–C_{15} iso-alkanes) an average value of 5×10^{-21} J was taken for A on the basis of literature data for these alkanes [50].

Figure 6.9 shows a good fit of the DLVO curves to the experimental data for $h_w > 30$ nm, the experimental points being between the constant charge and constant potential cases. This fit gives a diffuse double layer potential at infinity of about -50 mV (this evaluation should be considered qualitative due to the approximations made). In the transition region ($h_w < 30$ nm), the experimental results cannot be fitted to the DLVO calculation since in this case meta-stable films are formed. Notably, by increasing the pressure, the film jumps from a meta-stable state to a stable NBF that does not rupture up to a pressure of 45 kPa. One can conclude that the high stability of these emulsion films, stabilized by INUTEC SP1 polymeric surfactant, is due to the formation of NBF. The latter is a bilayer film that is stabilized by short-range steric interactions of the strongly hydrated loops and tails. Evidence of this high stability can also be demonstrated by using a higher electrolyte concentration, as discussed below.

Figure 6.10 shows the $\Pi(h_w)$ isotherm for emulsion films obtained from 2×10^{-5} mol dm^{-3} INUTEC SP1 + 0.05 or (0.5, 1, and 2) mol dm^{-3} NaCl aqueous solutions [8]. In these four cases of high electrolyte concentrations ($C_{el} > C_{el,cr}$) the disjoining pressure isotherms were very similar and it was even difficult to discern

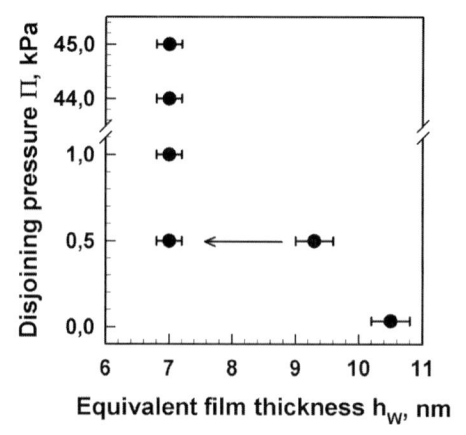

Figure 6.10 Disjoining pressure versus film thickness for O/W emulsion films from INUTEC SP1 aqueous solutions: averaged results from measurements at high NaCl concentrations.

one from another within experimental reproducibility. Consequently, Figure 6.10 represents averaged data for these four NaCl-concentrations. Presumably, there is no contribution from electrostatic repulsion in the films. However, the film thickness still shows a jump (indicated by an arrow) from 9–11 nm to about 7 nm within a relatively small increase in the pressure. Such a transition in film thickness is quite small (<4 nm), implying the presence of a very weak barrier (probably due to a weak steric repulsion) that can be easily overcome. Remarkably, however, the NBF remains very stable, even at 2 mol dm^{-3} NaCl, and no rupture was observed up to the highest pressure (45 kPa) applied.

Figure 6.11 depicts the $\Pi(h_w)$ isotherms for emulsion films from 2×10^{-5} mol dm^{-3} INUTEC SP1 + 2×10^{-4} mol dm^{-3} salt aqueous solutions, the salt being NaCl

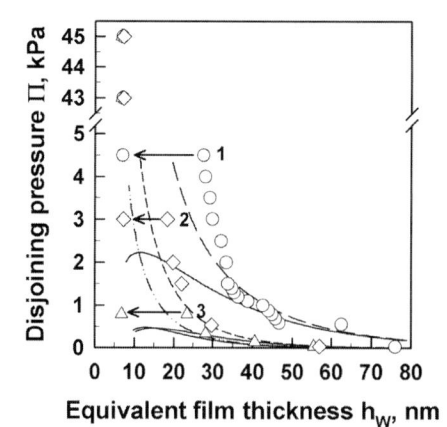

Figure 6.11 Measured disjoining pressure versus film thickness for O/W emulsion films from INUTEC SP1 + 2×10^{-4} mol dm^{-3} NaCl (○), Na$_2$SO$_4$ (◇) and MgSO$_4$ (△) aqueous solutions. Arrows denote the jump transition to NBF.

or Na_2SO_4 or $MgSO_4$ [9]. In all three cases, the pressure increase leads to a decrease in thickness, reaching a value different for each of the three salts, and after that a jump to NBF occurs. Notably, the NBF thickness is the same (7.0 ± 2 nm) for the three electrolytes studied – the points are plotted one on top of the other. Under the maximum pressure of 45 kPa the NBF are stable and do not rupture. This high stability of NBF is worth studying further by other methods. The pressure under which the transition occurs (i.e. the transition barrier) is different for each salt: for NaCl it is about 4.5 kPa, for Na_2SO_4 it is 3 kPa, and for $MgSO_4$ it is 1 kPa. This is an expected result since the transition barrier is related to the electrostatic disjoining pressure Π_{el}. Thus, for a 1 : 1 salt, the barrier is the highest and it decreases for the 1 : 2 salt and for the 2 : 2 salt. Probably, this is due to the decrease in potential of the diffuse electric layer. These potentials can be evaluated from the experimental data preceding the barriers of transition to NBF. These experimental Π-values are determined by the double layer repulsion and van der Waals attraction only [8, 9]; hence one must analyze the $\Pi(h_w)$ isotherms with Π_{el} and Π_{vw} as given by DLVO-theory [41, 42], making use of an appropriate three-layer emulsion film model (Figure 6.3b), that is, $\Pi(h) = \Pi_{el}(d) + \Pi_{vw}(h)$. $\Pi_{el}(d)$ was computed following the numerical procedures for 1 : 1, 1 : 2 and 2 : 2 electrolytes as proposed in Refs. [9, 45, 51]. Both constant charge and constant potential models were considered. The van der Waals interactions were again calculated according to the Hamaker equation.

The results of these DLVO computations are shown in Figure 6.11 by corresponding pairs of solid and dashed lines, which represent the constant potential and constant charge cases respectively. As can be seen there is good fit to the experimental data when the thickness is larger than about 30 nm for NaCl and 25 nm for Na_2SO_4 and $MgSO_4$. For the diffuse double layer potential at infinity this fit gives values of -50, -18 and -15 mV for the cases of NaCl, Na_2SO_4 and $MgSO_4$, respectively. Within the approximations made, it seems that the experimental results are between the constant potential and constant charge cases. In the transition region (h_w less than about 20–30 nm) the experimental data cannot be fitted to the DLVO-theory since either metastable films are formed or the deficiencies of DLVO-theory at small thickness are manifested. The tendency of decreasing diffuse double layer potential in the sequence NaCl, Na_2SO_4 and $MgSO_4$ is confirmed at least qualitatively.

Disjoining pressure isotherms for the cases of NaCl, Na_2SO_4, and $MgSO_4$ have also been measured at high salt concentrations (0.05, 1, 2 mol dm^{-3}) well above $C_{el,cr}$ [9]. In all cases a transition to NBF is observed at thickness of about 10 nm regardless of the electrolyte type. At $C_{el} > C_{el,cr}$ the electrostatic disjoining pressure is eliminated and only steric forces stabilize the film. This experiment clearly indicates that the steric forces do not affect the barrier in the $\Pi(h)$ isotherm, that is, the steric disjoining pressure is not influenced by the electrolyte type. The pressure under which the transition occurs is about 0.5 kPa, that is, the barrier is very low and all films transform into a NBF that is 7.0 ± 0.2 nm thick. The same NBF thickness has been measured at lower electrolyte concentrations.

The formation of such Newton black films, that is, bilayer films stabilized by polymeric surfactants deserves special attention. NBF formation in emulsion films from non-ionic polymeric surfactants was first established in Ref. [8]. For the case of

NBF stabilized by simple surfactants a hole–nucleation theory [1, 33, 52] has been developed, based on short-range interactions in a two-dimensional ordered system. For NBF from polymeric surfactants this still has to be done – an interesting task to understand the reasons for NBF stability on the basis of short-range forces.

6.4.3
Comparison of Film Stability and the Stability of a Real Emulsion

The stability of the emulsion NBF can explain the high stability of emulsions obtained using INUTEC SP1 at both high temperature and high electrolyte concentration [12, 14]. Emulsions of 50/50 Isopar M/water were prepared at an INUTEC SP1 concentration of 2% based on the oil phase (1% on the total). These emulsions remain stable at temperatures as high as 50 °C and in the presence of [NaCl] as high as 2 mol dm^{-3}.

Let us now compare the measured disjoining pressure [see the $\Pi(h)$ isotherm!] at the transition of the emulsion film to NBF and the capillary pressure of the droplets in an emulsion. For droplets of about 10 μm the capillary pressure is about 3.6 kPa, whereas for 1 μm droplets it is 36 kPa. Figure 6.12 presents the transition disjoining pressures for the cases of Pluronic F108 and INUTEC SP1 (Figure 6.8 and Figure 6.10 respectively), and the capillary pressures for the same cases [6]. The arrows indicate the transitions to NBF for both polymers. The transition for F108 is at 8–9 kPa while that for INUTEC SP1 it is about 1 kPa. Thus, the transition for INUTEC SP1 is realized at a disjoining pressure lower than the capillary pressure of the droplets in a real emulsion. This means that all films in the emulsion are NBF. However, this is not the case with emulsion films from F108 – the transition to NBF occurs at a disjoining pressure below the capillary pressure of the small (1 μm) emulsion droplets and above the capillary pressure for droplets of 10 μm diameter. This is an indication that not all emulsion films have been transformed into NBF.

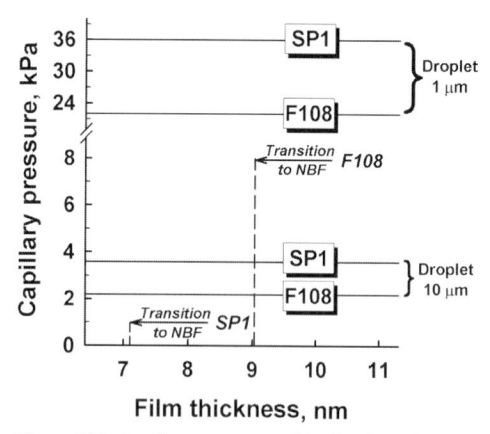

Figure 6.12 Capillary pressures of the droplets of O/W emulsions stabilized by F108 or SP1; transitions to NBF of the corresponding emulsion films are denoted by arrows.

6.5
Wetting Films Stabilized by Hydrophobically Modified Inulin Polymeric Surfactant

The polymeric surfactant INUTEC SP1 has been applied for stabilization of latex dispersions [11], and other dispersed solid particles. The high stability obtained is due to the multi-point attachment to the solid surface by several alkyl chains, leaving strongly hydrated loops and tails of linear polyfructose that provide enhanced steric stabilization. Evidence for the high repulsion obtained by this polymeric surfactant was obtained from measurements using atomic force microscopy [13]. An alternative method for investigating the stabilization of solid surfaces by this polymeric surfactant can also be obtained from the measurement of wetting films on solid surfaces. Here we discuss only the wetting films obtained on a hydrophilic silica surface from aqueous solutions of INUTEC SP1.

The solid substrate used was a quartz glass smooth flat plate that was very carefully washed with acid mixtures and doubly distilled water. A drop of pure water put on this surface completely spread and no contact angle could be measured ($\theta = 0°$), hence such a substrate had a hydrophilic surface. A new experimental cell, based on the Platikanov cell (Figure 6.1c), for the formation and study of the wetting films [36, 53, 54] has been constructed [10]. The thickness h of the microscopic thin wetting films was calculated from the light intensity data, measured by the microinterferometric method [1–3, 32, 33] and using the equations derived in Ref. [37]. All measurements were performed at 22 °C. To insure that an equilibrium thickness is reached all measurements started 90 min after the solution was placed into the cell and for each solution studied up to ten single h-measurements were performed.

Figure 6.13 shows the variation of equilibrium film thickness h_{eq} with INUTEC SP1 concentration C_{SP1}, both in water and at various NaCl concentrations [10]. At $C_{NaCl} < 10^{-1}$ mol dm^{-3} h_{eq} decreases with increasing INUTEC SP1 concentration, reaching a minimum at 10^{-6} mol dm^{-3}, after which h_{eq} increases with further increase in C_{SP1}. However, at and above 10^{-1} mol dm^{-3} NaCl this minimum at 10^{-6} mol dm^{-3} becomes less pronounced and the increase of h_{eq} above 10^{-6} mol dm^{-3} of INUTEC SP1 is slower. All the dependencies in Figure 6.13 shift to smaller h_{eq} values as the concentration of NaCl increases.

The general trend of the variation of film thickness with INUTEC SP1 concentration is difficult to explain. Although not too large, the change of film thickness is significant and repeatable. The following qualitative picture could be drawn up on basis these data. It may be rationalized in terms of the adsorption and orientation of the polymeric surfactant at the solid–liquid and liquid–air interface. At low C_{SP1} the INUTEC SP1

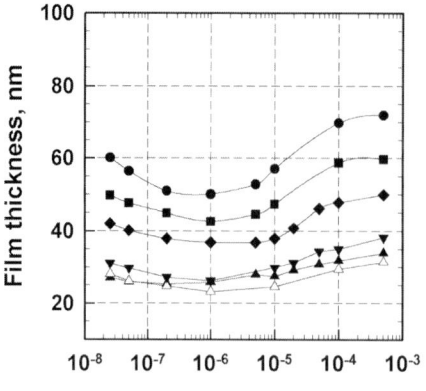

Figure 6.13 Dependence of film thickness of wetting films on INUTEC SP1 concentration in the presence of different NaCl concentrations between 0 (●) and 2 (△) mol dm^{-3}.

molecules adsorb with the hydrophilic polyfructose loops and tails pointing towards the silica substrate, leaving the alkyl chains in solution. In contrast, at the air–water interface, the alkyl chains are in the air, leaving the hydrophilic polyfructose loops and tails in the solution [31]. Contact angle measurements [55] showed the presence of a maximum in the contact angle on a hydrophilic silica surface at 10^{-6} mol dm^{-3} INUTEC SP1. This maximum coincides with the minimum in film thickness. However, it is difficult to visualize why the film thickness should reach a minimum at this concentration. At high C_{SP1} the INUTEC SP1 molecules form a bilayer with the hydrophilic polyfructose loops and tails now pointing to the aqueous phase, as illustrated schematically earlier (Figure 7 in Ref. [55]). This bilayer formation can explain the increase in film thickness with increasing INUTEC SP1 concentration, which is also accompanied by a decrease of the contact angle [55].

Figure 6.14 presents the variation of equilibrium film thickness with NaCl concentration at three different INUTEC SP1 concentrations [10]. The general trend is similar in each case – a decrease in film thickness with increasing electrolyte concentration, reaching a break point in the $h_{eq}(C_{el})$ curve at a critical electrolyte concentration, $C_{el,cr}$. Above that concentration the $h_{eq}(C_{el})$ dependency shows only a slow decrease with increasing C_{el}. The $C_{el,cr}$ value, at which almost constant h_{eq} is reached, is lower for Na$_2$SO$_4$ solutions (10^{-4} mol dm^{-3}) than with NaCl solutions (10^{-3} mol dm^{-3}). This behavior (decrease of a film thickness with electrolyte concentration) reflects the compression of the electrical double layer at higher C_{el} as described in the DLVO-theory [41, 42].

The Debye length $1/\kappa$ has been calculated from the expression:

$$\kappa^2 = 10.514 \times 10^5 \Sigma C_{el,i} z_i^2 / \varepsilon RT \tag{6.4}$$

for the different concentrations of NaCl and Na$_2$SO$_4$. The $1/\kappa$ values are significant at $C_{el} < C_{el,cr}$ and they determine the increase in the film thickness with increasing

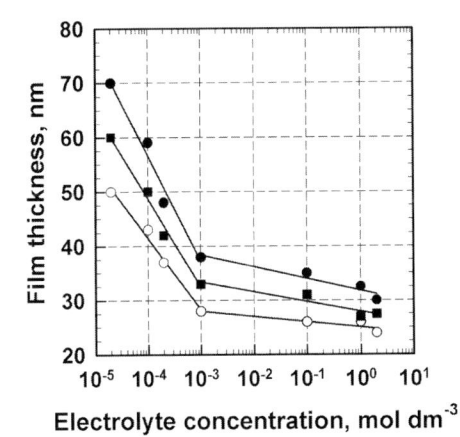

Figure 6.14 Dependence of equilibrium film thickness of wetting films on the NaCl concentration at three different INUTEC SP1 concentrations.

Debye length (increasing electrostatic disjoining pressure). At $C_{el} \geq C_{el,cr}$ the $1/\kappa$ values are rather small, that is, the electrostatic disjoining pressure is negligible. Interestingly, the ratio of the critical electrolyte concentrations ($C_{el,cr}$) for NaCl and Na$_2$SO$_4$ is $C_{NaCl} : C_{Na_2SO_4} \approx 20$. This is close to the value 2^6 predicted by the Schulze–Hardy rule [56], which, however, was derived for 1–1 and 2–2 electrolytes, while Na$_2$SO$_4$ is a 1–2 electrolyte.

The force balance through the wetting film includes, besides the capillary pressure, three components of the disjoining pressure Π – electrostatic Π_{el}, Van der Waals' Π_{VW} and steric Π_{st}. The electrostatic disjoining pressure $\Pi_{el} \approx 0$ at $C_{el} \geq C_{el,cr}$ and an almost constant $h_{eq} \approx 30$ nm is established. The Van der Waals' disjoining pressure $\Pi_{VW} = -A/6\pi h^3 \approx 20$ Nm^{-2} for $h = 30$ nm and Hamaker constant $A = -10^{-20}$ J, given for the system fused quartz/aqueous film/air in Ref. [57]. Obviously, Π_{VW} can also be neglected at all C_{el} (for $h > 30$ nm Π_{VW} is even smaller). Hence in the range $C_{el} \geq C_{el,cr}$ the capillary pressure will be balanced by the steric disjoining pressure Π_{st} only.

Similar behavior has been observed earlier with emulsion films [8]. Above a critical electrolyte concentration the significant compression of the double layer means that the steric repulsion of the loops and tails of the INUTEC SP1 determines the film thickness. The slow decrease in film thickness with increasing electrolyte concentration above $C_{el,cr}$ may be caused by some dehydration of the polyfructose loops and tails.

6.6 Conclusion

A series of studies of microscopic thin liquid films, both symmetric (foam and O/W emulsion films) and asymmetric (wetting films), from aqueous solutions of non-ionic polymeric surfactants, using the microinterferometric technique, has given important information about their properties and stability. The results permitted us to establish the type of interaction forces acting in them and to determine their stability under various conditions – capillary pressure, electrolyte concentration, surfactant concentration and temperature. Two types of polymeric surfactants – PEO–PPO–PEO three-block copolymers (ABA type) and hydrophobically modified inulin graft polymer (AB_n type) – have been explored.

All three types of thin liquid films from both ABA and AB_n polymeric surfactants are stabilized by DLVO-forces at low electrolyte concentrations and by non-DLVO-forces at higher electrolyte concentrations. The latter are steric surface forces of the type brush-to-brush and loop-to-loop interactions (according to de Gennes). These steric forces act in O/W emulsion films as well, but there transitions to Newton black films (NBF) have also been established. A difference between foam and O/W emulsion films has been observed. The barrier in the $\Pi(h)$ isotherm for an emulsion film is much lower and the transition to NBF can occur. The NBFs from polymeric surfactants are very stable, as are the emulsions obtained from the same solutions. Actually, two types of bilayer emulsion films are obtained, those stabilized by brush-to-brush or loop-to-loop steric interactions and the others – by short-range interactions, also steric, in a two-dimensional ordered system. The minor difference in the experimentally measured thickness (about 2 nm) is not sufficient to characterize the state of these films.

In contrast, the wetting films are relatively thicker and their thickness depends on the concentration of the AB_n polymeric surfactant. This behavior is due to the different adsorption and orientation of the polymeric surfactant molecules at the solid–liquid and liquid–air interface of the asymmetric wetting film; the results suggest the formation of adsorption bilayers at the solid interface, and the steric repulsion of the loops and tails of the polymeric surfactant determined the film thickness.

Obtaining Newton black emulsion films stabilized by polymeric surfactants requires special attention. Short-range interactions in NBF between polymeric surfactant molecules determine their high stability. For simple surfactants a hole–nucleation theory for stability of NBF, considering the film as a two-dimensional ordered system and assuming formation of holes that rupture the film, has been developed [1, 33, 52]. A similar theoretical consideration is yet to be developed for emulsion NBF from aqueous solutions of polymeric surfactants. It can be expected that the short-range interactions between hydrated hydrophilic chains determine NBF stability. We hope that in the future their nature will be elucidated.

References

1 Exerowa, D. and Kruglyakov, P.M. (1998) *Foam and Foam Films*, Elsevier, Amsterdam.

2 Platikanov, D. and Exerowa, D. (2005) Thin liquid films, in *Fundamentals of Interface and Colloid Science* (ed. J. Lyklema), Vol. 5, Elsevier, Amsterdam.

3 Platikanov, D. and Exerowa, D. (2006) Symmetric thin liquid films with fluid interfaces, in *Emulsions and Emulsion*

Stability (ed. J. Sjoblom), CRC Taylor & Francis, Boca Raton.

4 Sedev, R. and Exerowa, D. (1999) *Advances in Colloid and Interface Science*, **83**, 111.

5 Exerowa, D., Ivanova, R. and Sedev, R. (1998) *Progress in Colloid and Polymer Science*, **109**, 29.

6 Exerowa, D., Gochev, G., Kolarov, T., *et al.* (2008) *Colloids and Surfaces* (submitted).

7 Exerowa, D., Kolarov, T., Pigov, I., *et al.* (2006) *Langmuir*, **22**, 5013.

8 Exerowa, D., Gochev, G., Kolarov, T., *et al.* (2007) *Langmuir*, **23**, 1711.

9 Gochev, G., Kolarov, T., Levecke, B., *et al.* (2007) *Langmuir*, **23**, 6091.

10 Nedyalkov, M., Alexandrova, L., Platikanov, D., *et al.* (2007) *Colloid and Polymer Science*, **285**, 1713.

11 Esquena, J., Dominguez, F.J., Solans, C., *et al.* (2003) *Langmuir*, **19**, 10463.

12 Tadros, Th.F., Vandamme, A., Levecke, B., *et al.* (2004) *Advances in Colloid and Interface Science*, **108–109**, 207.

13 Nestor, J., Esquena, J., Solans, C., *et al.* (2007) *Journal of Colloid and Interface Science*, **311**, 430.

14 Tadros, Th.F. (2007) in *Colloid Stability. The Role of Surface Forces*, Vol. I, Wiley-VCH Verlag GmbH, Weinheim, p. 235.

15 Kolaric, B., Jaeger, W. and Klitzing, R.v. (2000) *Journal of Physical Chemistry. B*, **104**, 5096.

16 Klitzing, R.v., Espert, A., Asnacios, A., *et al.* (1999) *Colloids and Surfaces A: Physicochemical and Engineering Aspects*, **149**, 131.

17 Kolaric, B., Jaeger, W., Hedicke, G. and Klitzing, R.v. (2003) *Journal of Physical Chemistry. B*, **107**, 8152.

18 Klitzing, R.v. (2005) *Advances in Colloid and Interface Science*, **114/115**, 253.

19 Klitzing, R.v., Kolaric, B., Jaeger, W. and Brandt, A. (2002) *Physical Chemistry Chemical Physics*, **4**, 1907.

20 Ciunel, K., Armelin, M., Findenegg, G. and Klitzing, R.v. (2005) *Langmuir*, **21**, 4790.

21 Klitzing, R.v. (2000) *Tenside Surfactants Detergents*, **37**, 338.

22 Qu, D., Haenni-Ciunel, K., Rapoport, D. and Klitzing, R.v. (2007) in *Colloid Stability and Application in Pharmacy* (ed. Th.F. Tadros), Wiley-VCH Verlag GmbH, Weinheim, p. 307.

23 Rippner, B., Boshkova, K., Claesson, P. and Arnebrant, T. (2002) *Langmuir*, **18**, 5213.

24 Schillen, K., Claesson, P.M., Malmsten, M. *et al.* (1997) *Journal of Physical Chemistry*, **101**, 4238.

25 Kolaric, B., Foerster, S. and Klitzing, R.v. (2001) *Progress in Colloid and Polymer Science*, **117**, 195.

26 Khristov, Kh., Taylor, S.D., Czarnecki, J. and Masliyah, J. (2000) *Colloids and Surfaces A: Physicochemical and Engineering Aspects*, **174**, 183.

27 Yampolskaya, G. and Platikanov, D. (2006) *Advances in Colloid and Interface Science*, **128/130**, 159.

28 Nace F V.M.(ed.) (1996) *Nonionic Surfactants: Polyoxyalkylene Block Copolymers*, Marcel Dekker, New York.

29 Alexandridis, P. and Lindman, B.(eds) (1997) *Amphiphilic Block Copolymers: Self-Assembly and Applications*, Elsevier, Amsterdam.

30 Stevens, C.V., Merrigi, A., Peristeropoulo, M., *et al.* (2001) *Biomacromolecules*, **2**, 1256.

31 Flory, P.J. (1953) *Principles of Polymer Chemistry*, Cornell University Press, New York.

32 Scheludko, A. (1967) *Advances in Colloid and Interface Science*, **1**, 391.

33 Exerowa, D., Kashchiev, D. and Platikanov, D. (1992) *Advances in Colloid and Interface Science*, **40**, 201.

34 Exerowa, D. and Scheludko, A. (1971) *Comptes Rendus de L'Academie Bulgare des Sciences*, **24**, 47.

35 Mysels, K.J. and Jones, M.N. (1966) *Faraday Discussions of the Chemical Society*, **42**, 42.

36 Platikanov, D. (1964) *Journal of Physical Chemistry*, **68**, 3619.

37 Scheludko, A. and Platikanov, D. (1961) *Kolloid-Zeitschrift*, **175**, 150.

38 Duyvis, E.M. (1962) Thesis, University Uthrecht.
39 de Gennes, P.G. (1979) *Scaling Concepts in Polymer Physics*, Cornell University Press, Ithaca, New York.
40 de Gennes, P.G. (1987) *Advances in Colloid and Interface Science*, **27**, 189.
41 Derjaguin, B.V. and Landau, L.D. (1941) *Acta Physicochimica USSR*, **14**, 633.
42 Verwey, E.J.W. and Overbeek, J.Th.G. (1948) *The Theory of the Stability of Lyophobic Colloids*, Elsevier, Amsterdam.
43 Exerowa, D. (1969) *Kolloid-Zeitschrift*, **232**, 703.
44 Bot, A., Erle, U., Veeker, R. and Agterof, W.G.M. (2004) *Food Hydrocolloids*, **18**, 547.
45 Chan, D.Y., Pashley, R.M. and White, R.L. (1980) *Journal of Colloid and Interface Science*, **77**, 283.
46 Donners, W.Y., Rijnbout, B. and Vrij, A. (1977) *Journal of Colloid and Interface Science*, **60**, 540.
47 Dzyaloshinski, I.E., Lifshitz, E.M. and Pitaevski, L.P. (1960) *Advances in Physics*, **10**, 165.
48 Napper, D.H. (1983) *Polymeric Stabilization of Colloidal Dispersions*, Academic Press, New York.
49 Israelachvili, N. and Wennerstroem, H. (1992) *Journal of Physical Chemistry*, **96**, 520.
50 Lyklema, J. (1991) *Fundamentals of Interface and Colloid Science*, Vol. I, Academic Press, London, p. A9.2.
51 Kuo, Y.-C. and Hsu, J.-P. (1993) *Journal of Colloid and Interface Science*, **156**, 250.
52 Kashchiev, D. and Exerowa, D. (1980) *Journal of Colloid and Interface Science*, **77**, 501.
53 Zorin, Z., Platikanov, D. and Kolarov, T. (1987) *Colloids and Surfaces A: Physicochemical and Engineering Aspects*, **22**, 147.
54 Diakova, B., Filiatre, C., Platikanov, D. *et al.* (2002) *Advances in Colloid and Interface Science*, **96**, 193.
55 Nedyalkov, M., Alexandrova, L., Platikanov, D. *et al.* (2008) *Colloid and Polymer Science*, **286**, 713.
56 Lyklema, J. (2005) *Fundamentals of Interface and Colloid Science*, Vol. IV, Elsevier, Amsterdam, p. 3.102.
57 Israelachvilli, J. (1991) *Intermolecular and Surface Forces*, Academic Press, New York.

7
Conditions for the Existence of a Stable Colloidal Liquid
Gerard J. Fleer and Remco Tuinier

7.1
Introduction

Studying the phase behavior of colloids is relevant for two main reasons. Firstly, colloids are found in many industrial and biological systems. To name a few examples: living cells, food, ink and air all contain colloids. Understanding the properties of colloidal dispersions may lead to technological developments in engineering practical suspensions. Secondly, investigating well-defined colloidal systems helps to gain more insight into atomic and molecular systems. This is because the relevant time scale of particle motion for colloids is measurable, and individual colloid particles can be observed microscopically. The colloid–atom analogy, already advanced by Einstein [1], has been shown to be of great importance. Colloids are large enough to consider the solvent as a continuum. Colloids are, however, still small enough to exhibit Brownian motion [2], first demonstrated experimentally by Perrin [3]. Also, the phase behavior of colloidal dispersions and atomic and molecular fluids shows common features [4]. In the high-temperature limit atomic systems undergo a fluid–solid transition [5] around 50% in volume fraction, just like hard-sphere colloids [6].

In atomic and molecular systems the *range* and *strength* (inverse temperature) of the attraction is set by quantum mechanics. In many cases a Lennard-Jones potential describes the pair interaction quite well [7]. The phase behavior of atomic and molecular systems is often represented in a pressure versus temperature diagram. Quite often the distance between *triple point* (tp) and *critical point* (cp) is significant so that there is a wide region where a liquid exists: the liquid window is then wide.

Upon changing the range and strength of the attraction the distance between tp and cp changes. At a certain range and strength this distance reduces to zero; then we reach the *critical endpoint* (cep), which constitutes the boundary condition for the existence of a stable liquid. The simplest case for which this can be illustrated is a

Highlights in Colloid Science. Edited by Dimo Platikanov and Dotchi Exerowa
Copyright © 2009 WILEY-VCH Verlag GmbH & Co. KGaA, Weinheim
ISBN: 978-3-527-32037-0

one-component system of hard spheres with an added exponential (Yukawa-type) attraction where range and strength can be adjusted [8, 9]. In this chapter we abbreviate this system as "Yuk."

Also in colloidal dispersions the range and strength of the attraction can, in principle, be varied. A suitable way of doing this is by adding nonadsorbing polymer chains to the colloids, where the depletion effect [10–12] gives rise to attraction. The width of the depletion zone (the depletion thickness) determines the range; the polymer concentration fixes the strength. Now we have a three-component system of colloids, polymer and solvent. Experimental phase diagrams for such colloid/polymer mixtures are now available [13, 14], and theory has also been developed. The first theory taking into account polymer partitioning over the coexisting phases [15] assumes that the depletion thickness is equal to the coil size of the polymer and is "fixed" (independent of the polymer concentration); we denote this model by the abbreviation "fix." It is adequate for the situation where the coils are much smaller than the colloids (the so-called *colloid limit*, where the polymer concentrations along the binodals are below overlap). However, for polymer concentrations above overlap the depletion zone is compressed and the depletion thickness varies (decreases) with increasing polymer concentration [16]. Recently, we introduced a "var" model that takes into account this variation of the depletion thickness with polymer concentration [17]. It describes not only the colloid limit but also the *protein limit* of small (protein-like) colloids and long polymer chains (and the crossover between these two limits).

In this chapter we compare the predictions of these three models, "Yuk," "fix" and "var." We first present phase diagrams in terms of the strength and the range as a function of the colloid concentration in these three models. We find that the critical endpoint is essentially the same in the three cases: the range in the cep is about one-third of the colloid radius, the strength is about $2kT$. Outside the cep the three models behave quite differently.

We then investigate how these results translate into $p(T)$ diagrams. For the Yukawa system classical textbook diagrams are found, but for the polymer/colloid systems the behavior is slightly different, yet well understood.

7.2
Theory

7.2.1
Free Energy

For the three systems discussed here, the free energy F is split into a hard-sphere part F_0 and an attractive part F_a. It is most convenient to consider the normalized free energy $f = Fv/V$, where $v = 4\pi a^3/3$ is the volume of a (colloidal) particle and V is the volume of the system. We thus write:

$$f = f_0 + f_a \tag{7.1}$$

The hard-sphere part depends only on the (colloid) volume fraction η and is the same in the three cases. We have to differentiate between a fluid (F) phase [either gas (G) or liquid (L)] and a crystalline solid (S). Expressions are available in the literature [18, 19]:

$$\frac{f_0}{kT} = \begin{cases} \eta[\ln\eta - 1 + 4x + x^2] & F \\ \eta[2.1306 - 3\ln(\eta^{-1} - \eta_{cp}^{-1})] & S \end{cases} \quad (7.2)$$

where $x = \eta/(1-\eta)$ and $\eta_{cp} = (\pi/6)\sqrt{2} = 0.741$ is the volume fraction at close-packing.

For the attractive part f_a we need the specifics of the system, as discussed in the next sections. For all three systems we consider first the pair potential $W(H)$, where H is the interparticle distance: H is zero for particles in contact. We shall use only the relative particle separation $h = H/a$, where a is the particle radius. In all cases $W(h) = \infty$ for $h < 0$. The pair potential is characterized by the strength $\varepsilon = -W(0)$ and by the relative range $q = \kappa^{-1}/a$, where κ^{-1} is the range of the attraction. From an appropriate expression for $W(h)$ as a function of ε and q the attractive free energy can be derived, and from $f = f_0 + f_a$ as a function of η, ε and q all thermodynamic properties (including the phase behavior) follow from standard thermodynamics. For example, the chemical potential μ of the spherical particles and the pressure p of the system are given by:

$$\mu = \partial f / \partial \eta \quad pv = -f + \eta\mu \quad (7.3)$$

where again $v = 4\pi a^3/3$ is the particle volume. The quantities μ and pv are thus a function of three parameters: the colloid volume fraction η, the strength ε, and the range q of the attraction. Explicit analytical expressions for these dependencies are available for the three systems discussed in this chapter. Both μ and pv have a hard-sphere part and an attractive part: $\mu = \mu_0 + \mu_a$ and $pv = (pv)_0 + (pv)_a$. The hard-sphere part of pv is given by:

$$\frac{(pv)_0}{kT} = \begin{cases} \eta + 4x^2 + 2x^3 & F \\ 3/(\eta^{-1} - \eta_{cp}^{-1}) & S \end{cases} \quad (7.4)$$

When the attractive contribution to f, μ and pv is zero, we have a pure hard-sphere system where the only possible phase coexistence is FS at $\eta_F^0 = 0.492$ and $\eta_S^0 = 0.542$, according to Equations 7.2 and 7.3, with results nearly identical to that of computer simulations [5]. At this coexistence $\mu_0^0 = 15.463 \, kT$ and $(pv)_0^0 = 6.081 \, kT$. When there is an attractive component the phase behavior is richer, with the possibility of two-phase GL, GS and LS coexistence and a GLS triple point, just like in simple atomic or molecular systems.

Once expressions for μ and pv (and the first and second derivatives μ' and μ'' with respect to η) have been found, the complete phase diagram for each system may be calculated. Binodal points and triple points follow from equal μ and pv in two or three phases, respectively. Critical GL points are obtained from $\mu' = \mu'' = 0$. Of central importance for the existence of a stable liquid is the critical endpoint (cep); its coordinates follow from $\mu' = \mu'' = 0$, supplemented with equal μ and pv in the (critical) fluid (F) and the coexisting solid (S).

7.2.2
Yukawa Attraction

The simplest possible situation is a one-component system of hard spheres with an added attractive interaction. For attractive Yukawa spheres the pair potential $W(h)$ for $h > 0$ is a simple exponential:

$$W(h) = -\frac{\varepsilon}{1+h/2} e^{-kh}, \quad W(0) = -\varepsilon, \quad k = 1/q \tag{7.5}$$

In this equation we introduced the inverse (relative) range $k = 1/q$.

Earlier [9] we simplified the expressions for f_a as given by the first-order mean spherical approximation by Tang et al. [20, 21], which is based upon pair-wise additivity of the interaction. The result is:

$$f_a = -\varepsilon \eta^2 \frac{P_1}{P_2}, \quad P_1 = a_0 + a_1 \eta, \quad P_2 = b_0 + b_1 \eta + b_2 \eta^2 \tag{7.6}$$

where P_n is a polynomial in η of degree n. The coefficients a_i and b_i are only a function of the range:

$$a_0 = 2k + 4k^2, \quad a_1 = 4k + 2k^2 \tag{7.7}$$

$$b_0 = 2k^3/3, \quad b_1 = (1+2k)e^{-2k} - 1 + 2k^2, \quad b_2 = (2+k)e^{-2k} - 2 + 3k - 3k^2 + b_0 \tag{7.8}$$

Inserting Equation 7.6 into Equation 7.1 gives $f(\eta, \varepsilon, q)$, and Equation 7.3 then provides $\mu(\eta, \varepsilon, q)$ and $pv(\eta, \varepsilon, q)$. Explicit analytical forms for these latter quantities were presented in a recent publication [8]. The pressure in the system is given by:

$$pv = (pv)_0 - \varepsilon \eta^2 \frac{P_3}{P_4} \tag{7.9}$$

where the P_n's are again polynomials in η of degree n, with coefficients that are only a function of k; for the precise expressions we refer to Ref. [9]. The important point is that P_3 and P_4 are positive so the attractive part $(pv)_a$ is negative, as expected.

7.2.3
FVT ("Fix")

The abbreviation FVT stands for free-volume theory, which describes the thermodynamics of a mixture of colloids and nonadsorbing polymer in a solvent. In the original form of FVT [15] the solvent was disregarded and only two components (colloids and polymers) were considered in an effective two-component system. Recently [17] we reformulated the model and included the solvent as a separate component – so we have a three-component system. Surprisingly, the expression for f_a turns out to be simpler for three components than for two. Actually, f_a is the semigrand potential ω_a obtained from $\Omega = F - N_p\mu_p$, where N_p is the number of polymer chains and μ_p their chemical potential. For simplicity we denote ω_a as f_a.

In this system the range of the interaction is given by the depletion thickness δ, which is the thickness of the depletion zone (which is void of polymer) around a spherical colloidal particle. Hence, the relative range is $q = \delta/a$. In the present section we assume that δ is not affected by the polymer concentration: q is fixed, and we use the abbreviation "fix." This "fix" assumption is a fair approximation for dilute polymer solutions (below overlap), where δ is of order of the radius of gyration R of the polymer coils, and it is appropriate for the so-called colloid limit, where $\delta \approx R$ is much smaller than the particle radius a: q is well below unity.

For two colloidal particles in a solution of nonadsorbing polymer the pair potential for "fix" is given by the product of the osmotic pressure Π of the polymer solution and the geometrical overlap volume v_{ov} of two spherical depletion layers [11, 22, 23]:

$$W(h) = -\Pi v_{ov}(h), \quad v_{ov}(h)/v = (q-h/2)^2(3/2+q+h/4) \quad (h<2q) \quad (7.10)$$

For the contact potential $W(0)$ we insert $h = 0$ and abbreviate $v_{ov}(0)$ to $v_{ov} = vq^2(q+3/2)$:

$$\varepsilon = -W(0) = \Pi v_{ov} = \Pi v q^2(q+3/2) \quad (7.11)$$

The parameter Πv is the work to insert a bare colloidal particle (without depletion layer) into the polymer solution. For the "fix" model it is the most convenient thermodynamic parameter to describe the polymer properties. It is directly coupled to ε and q, as shown in Equation 7.11.

In terms of the parameter Πv the expression for f_a in FVT takes a very simple form [8]:

$$f_a = -\alpha \Pi v \quad (7.12)$$

where α is the fraction of the free volume in the system. According to standard scaled-particle theory [15, 24] it is given by:

$$\alpha = (1-\eta)\exp[-Ax - Bx^2 - Cx^3] \quad (7.13)$$

where again $x = \eta/(1-\eta)$. The coefficients A, B and C depend only on the relative range q:

$$A = (1+q)^3 - 1, \quad B = 3q^2(q+3/2), \quad C = 3q^3 \quad (7.14)$$

Note that the coefficient B equals $3v_{ov}/v$ according to Equation 7.10 (with $h=0$).

Inserting Equation 7.12 into Equation 7.1 and applying Equation 7.3 gives explicit analytical expressions for μ and pv as a function of η, Πv and q or, equivalently, as a function of η, ε, and q, from which the full phase diagram $\Pi v(\eta)$ or $\varepsilon(\eta)$ for any q may be calculated. Details of these expressions are given elsewhere [8, 25]. The pressure is given by:

$$pv = (pv)_0 + \Pi v g \quad (7.15)$$

where $g = \alpha - \eta \partial \alpha / \partial \eta$ is a known analytical function of q and η; it runs from unity at $\eta = 0$ to zero at high η. As a consequence, in FVT the pressure is higher than for hard spheres (unlike in the Yukawa system where it is lower). The background is, obviously, that the polymeric component contributes to the pressure. For a dilute gas ($\eta \to 0, g \to 1$) Equation 7.15 reduces to $pv = \Pi v$ (here the colloid contribution is negligible), whereas in a concentrated L or S system ($g \to 0$) the polymer contribution vanishes and $pv = (pv)_0$.

Thus far we did not specify the relation between δ and R or, equivalently, between $q = \delta/a$ and the coil/colloid size ratio $q_R = R/a$. For this relation we have to account for curvature effects: the depletion thickness δ around a sphere is somewhat smaller than the depletion thickness $1.071 R$ [26] next to a flat plate. It turns out [17, 25] that the relation between q and q_R is a simple power law:

$$q = 0.866 q_R^{0.88} \tag{7.16}$$

We will need this relation in the next section.

7.2.4
GFVT ("Var")

In this section we summarize a generalized free-volume theory (GFVT) that takes into account the variation of the attraction range with the polymer concentration [17], hence the notation "var." We saw above that in dilute polymer solutions, corresponding to the colloid limit, the relative range is given by $q = 0.866 q_R^{0.88}$. However, with increasing polymer concentration the depletion thickness at a flat plate decreases from $1.071 R$ in the dilute limit to the correlation length (blob size) ξ in the semi-dilute limit. This correlation length is independent of R and scales with the polymer concentration φ as $\xi \sim \varphi^{-\gamma}$ [16]. Here $\gamma = 0.77$ is the De Gennes exponent for excluded-volume chains in a good solvent; it is related to the Flory exponent $\nu = 0.588$ in $R \sim N^\nu$ (where N is the chain length) through $1/\gamma + 1/\nu = 3$. For theta solvents the De Gennes exponent γ equals 1; we do not discuss this situation although a full description is possible [25]. The prefactor in $\xi \sim \varphi^{-\gamma}$ is also known [17, 27]: for a good solvent $\xi/R = 0.539(\varphi/\varphi^*)^{-\gamma}$, where φ^* is the overlap concentration.

For the variable (i.e. concentration-dependent) relative range we use the symbol q_v; it varies from $q_v = q = 0.866 q_R^{0.88}$ in the dilute limit to $q_v = q_{sd} = 0.866(\xi/a)^{0.88}$ in the semi-dilute limit. The dependence of q_v on q_R and φ is most easily formulated in terms of a parameter Y, which is defined as:

$$Y = (\varphi/\varphi^*) q_R^{-1/\gamma} \tag{7.17}$$

For small (protein-like) colloidal particles in semi-dilute solutions (this is the so-called *protein limit* where $q_R \approx q$ is well above unity) the parameter Y becomes independent of q or q_R [17, 25]. The background is that φ/φ^* scales as $(\xi/R)^{-1/\gamma}$, where ξ does not depend on R: $\varphi/\varphi^* \sim R^{1/\gamma} \sim q_R^{1/\gamma}$, which cancels against the factor $q_R^{-1/\gamma}$ in Equation 7.17.

In terms of Y and q_R (which is directly related to q, see Equation 7.16) the dependence of q_v on the polymer concentration may be written as [17, 25, 27]:

$$q_v = 0.866\{q_R^{-2}+3.95\,Y^{2\gamma}\}^{-0.44} \tag{7.18}$$

In the dilute limit (= colloid limit) the Y term may be omitted and $q_v = q$ reduces to Equation 7.16. In the semi-dilute limit (= protein limit) the term q_R^{-2} is negligible and $q_v = 0.47\,Y^{-0.68}$, which is independent of q or q_R because Y is independent of these quantities.

In the GFVT model Equation 7.11 for FVT is generalized to [22, 23]:

$$\varepsilon = \int_0^{\Pi} v_{ov}\,d\Pi = \int_0^{Y} q_v^2(q_v + 3/2)\partial\Pi v/\partial Y\,dY \tag{7.19}$$

where $\partial\Pi v/\partial Y$ is given by [17, 25, 27]:

$$\frac{\partial\Pi v/\partial Y}{kT} = q_R^{-1/\nu} + 3.77\,Y^{3\gamma-1} \tag{7.20}$$

Finally, Equation 7.12 is generalized to [17, 25]:

$$f_a = -\int_0^{\Pi v} \alpha\,d\Pi v = -\int_0^{Y} \alpha\,\partial\Pi v/\partial Y\,dY \tag{7.21}$$

In Equation 7.21 we use Equation 7.13 for α; however, in Equation 7.14 we have to replace the "fixed" q by the "variable" q_v. Unlike in FVT, where α depends only on q and η, in GFVT α is also a function of the polymer concentration: $\alpha = \alpha(q, \eta, Y)$. Inserting Equation 7.21 into Equation 7.1 gives $f(q, \eta, Y)$ from which, through Equation 7.3, $\mu(q, \eta, Y)$ and $pv(q, \eta, Y)$ follow; again explicit analytical expressions are known [25], but these are not given here. Using Equations 7.18 and 7.19 we can now find phase diagrams $\varepsilon(\eta, q)$; we show some results in Figures 7.1 and 7.2 below.

Analogously to Equation 7.15 the pressure is given by:

$$pv = (pv)_0 + \int_0^{\Pi v} g\,d\Pi v = (pv)_0 + \int_0^{Y} g\,\partial\Pi v/\partial Y\,dY \tag{7.22}$$

As in the FVT model pv is higher than $(pv)_0$. The limits are also the same: $pv = \Pi v$ in a dilute gas, and pv approaches $(pv)_0$ in a concentrated liquid or solid where the polymer content is low.

7.3
Phase Diagrams $\varepsilon(\eta)$

Figure 7.1 shows phase diagrams in terms of the strength ε/kT as a function of the colloid concentration η for the three systems "Yuk," "fix" and "var." This figure applies

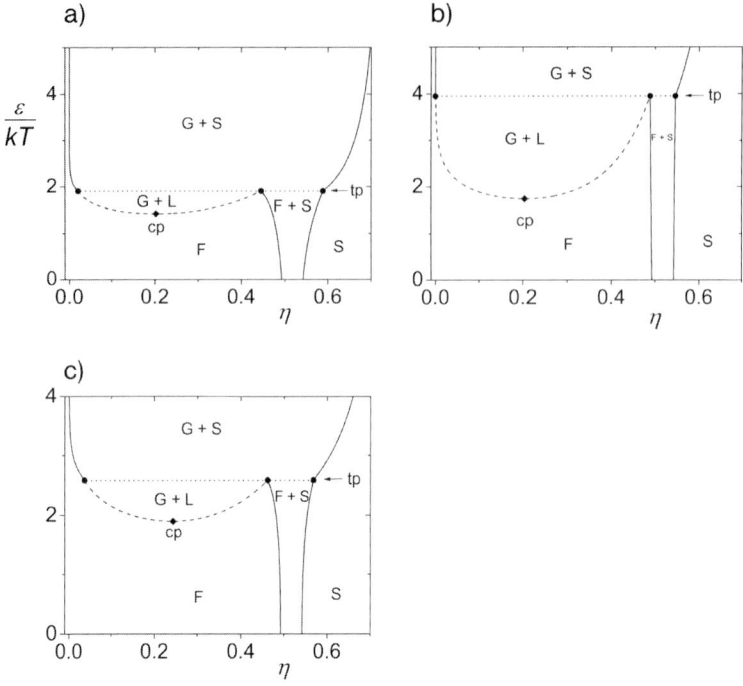

Figure 7.1 Phase diagrams $\varepsilon(\eta)$ for "Yuk" (a), "fix" (b) and "var" (c), and for $q = 0.552$. The FS binodals ($\varepsilon < \varepsilon^{\text{tp}}$) and GS binodals ($\varepsilon > \varepsilon^{\text{tp}}$) are the solid curves, the GL binodal ($\varepsilon^{\text{cp}} < \varepsilon < \varepsilon^{\text{tp}}$) is dashed. The horizontal dotted line connecting the circles is the triple point at $\varepsilon = \varepsilon^{\text{tp}}$, the filled diamond is the GL critical point at $\varepsilon = \varepsilon^{\text{cp}}$. The demixing regions are indicated as G + S, G + L, and F + S. The one-phase regions F and S are also shown.

to one particular (relative) range $q = 0.552$; the range is just above half the particle radius. The solid part at the bottom of the three diagrams is the FS binodal. For $\varepsilon = 0$ we have hard-sphere coexistence at $\eta_F = \eta_F^0 = 0.492$ and $\eta_S = \eta_S^0 = 0.542$. With increasing ε the FS demixing gap widens somewhat, with η_F slightly decreasing and η_S slightly increasing. For "fix" this widening is hardly noticeable at this value of q.

At a certain value ε^{tp} (1.91kT for "Yuk," 3.95kT for "fix," 2.59kT for "var"), the F branch jumps discontinuously from η_L^{tp} (0.445, 0.488 and 0.461, respectively) to η_G^{tp} (0.020, 0.00044 and 0.036) at constant ε; at this value ($\varepsilon = \varepsilon^{\text{tp}}$) the S branch shows a discontinuity in slope at η_S^{tp} (0.589, 0.546 and 0.569). This is the GLS triple point at $q = 0.552$, which is characterized by four parameters: ε^{tp} and three coexisting compositions. For $\varepsilon > \varepsilon^{\text{tp}}$ there is only GS demixing, as shown by the GS binodals in the upper part of the diagrams: a (very) dilute gas phase is in equilibrium with a (very) concentrated solid phase.

At the triple point ($\varepsilon = \varepsilon^{\text{tp}}$) there is GL demixing in the region $\eta_G^{\text{tp}} < \eta < \eta_L^{\text{tp}}$. For $\varepsilon > \varepsilon^{\text{tp}}$ no liquid is possible, but for $\varepsilon < \varepsilon^{\text{tp}}$ there exists stable liquid, over an η window that narrows as ε decreases. At the critical point cp (♦ in Figure 7.1) this window is

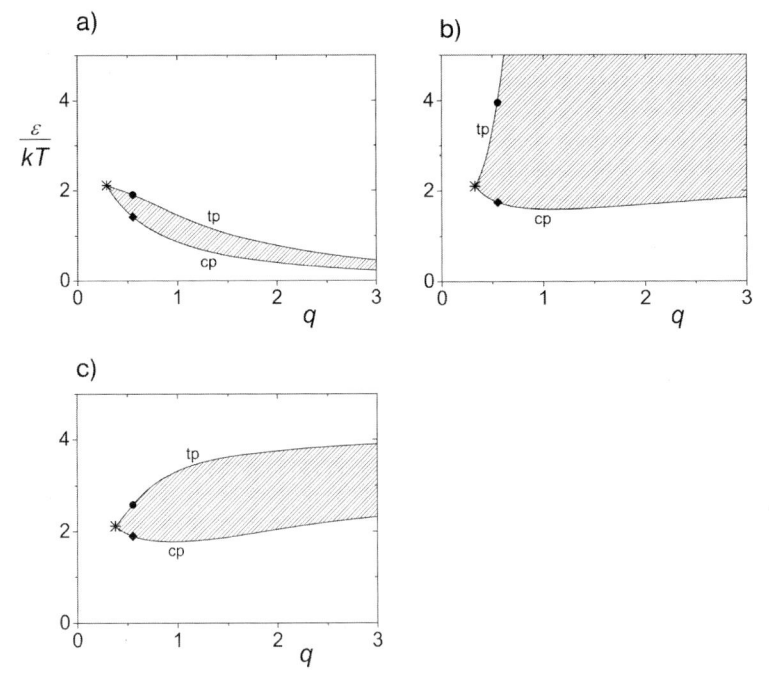

Figure 7.2 Width of the liquid window (hatched) in terms of the strength ε as a function of the relative range q, for "Yuk" (a), "fix" (b), and "var" (c). In each diagram the upper curve is $\varepsilon^{tp}(q)$, and the lower curve is $\varepsilon^{cp}(q)$. The two curves merge at the cep (asterisk); the coordinates q^{cep} and ε^{cep} are given in Equation 7.23. The filled circles (tp) and diamonds (cp) correspond to $q = 0.552$ and are the same as in Figure 7.1.

reduced to zero. Liquid is thus only stable over the range $\varepsilon^{cp} < \varepsilon < \varepsilon^{tp}$, in the region indicated as G + L in Figure 7.1. The cp coordinates ε^{cp}/kT, η^{cp} are 1.12, 0.202 for "Yuk," 1.74, 0.204 for "fix" and 1.90, 0.243 for "var."

For $q = 0.552$ the liquid window is very narrow for "Yuk": $1.42kT < \varepsilon < 1.91kT$. For "fix" it is much wider ($1.74kT < \varepsilon < 3.95kT$), and for "var" we have an intermediate situation ($1.90kT < \varepsilon < 2.59kT$). The η range for a stable liquid is not much different in the three cases.

The next two figures show how the coordinates of the triple point tp and of the critical point cp change with the relative range q. Figure 7.2 gives the data $\varepsilon(q)$: the upper curve in each diagram is the triple curve $\varepsilon^{tp}(q)$, the lower one gives the critical curve $\varepsilon^{cp}(q)$. The region in between (hatched) is the liquid window. The symbols in this figure (● for tp, ♦ for cp) correspond to $q = 0.552$, and are the same as in Figure 7.1.

We first consider what happens when q is made smaller than 0.552, the value in Figure 7.1. From Figure 7.2 it is clear that in all cases the liquid window narrows; the critical and triple curves approach each other. At a certain value (q^{cep}) the critical curve coincides with the triple point (which then actually is a double point as the G and L phases merge into one critical F phase that coexists with a solid). This is the critical endpoint (cep), indicated by the asterisks in the three diagrams. The four coordinates of the cep are:

$$q^{cep} = \begin{cases} 0.289 \\ 0.328 \\ 0.337 \end{cases} \quad \frac{\varepsilon^{cep}}{kT} = \begin{cases} 2.119 \\ 2.104 \\ 2.122 \end{cases} \quad \eta_F^{cep} = \begin{cases} 0.272 \\ 0.318 \\ 0.317 \end{cases} \quad \eta_S^{cep} = \begin{cases} 0.621 \text{ "Yuk"} \\ 0.583 \text{ "fix"} \\ 0.594 \text{ "var"} \end{cases}$$

(7.23)

It is striking to see how close the three cep's are in the different models. In all cases the relative range is about 1/3 and the strength is $2.1kT$. Nearly the same numbers are found from yet another system: a Lennard-Jones-type fluid [28]. This suggests that a universal principle applies: stable liquid is only possible when the range is longer than one-third of the particle radius. For a shorter range a crystalline solid is the preferred thermodynamic state for the condensed phase. Liquid structures require a relatively long attraction range to allow for the positional disorder.

As to the strength: for a Yukawa system ε should be less than $2kT$ for a stable liquid: both ε^{tp} and ε^{cp} decrease with increasing q. The width of the liquid window is small (at most about $0.6kT$). For "fix" the liquid window diverges, but this model breaks down for $q > 0.5$ because then the polymer concentrations along the GL binodals are above overlap and the assumption of a constant depletion thickness fails. For the more realistic "var" model the liquid region is again narrow: for high q it spans a range of roughly $1.7kT$ in strength, which is nevertheless considerably wider than in a Yukawa system. The difference is that multiple interactions are taken into account, whereas "Yuk" is restricted to pair interactions [8].

Figure 7.3 illustrates the variation with q of the colloid concentrations η^{cp} in the critical point (dashed curve) and of the three colloid concentrations η_G^{tp}, η_L^{tp}, and η_S^{tp} in the triple point. The asterisks indicate the cep: its coordinates for the three models are given in Equation 7.23. The filled circles (tp) and diamonds (cp) correspond to $q = 0.552$ and are the same as in Figure 7.1.

Also in terms of η_F^{cep} and η_S^{cep} (asterisks), there is not much difference between the three diagrams (Equation 7.23). Outside the cep η_S^{tp} decreases weakly, but it soon levels off to reach a value that is close to $\eta_S^0 = 0.54$ for hard-sphere coexistence. The variation of η_G^{tp} and η_L^{tp} with q is very strong for q just above q^{cep} but for high q the liquid part η_L^{tp} approaches $\eta_L^0 = 0.49$, whereas η_G^{tp} becomes essentially zero. For η^{cp} we see in all cases a decrease with q towards a final level: 0.13 for "Yuk," zero for "fix" and 0.11 for "var." The Yukawa limit 0.13 is the same as predicted by mean-field theory [29] for liquid–vapor coexistence.

7.4
Phase Diagrams pv/kT Versus ε/kT

The pressure in the system for any value of ε and q is calculated from Equation 7.9 for "Yuk," from Equations 7.11 and 7.15 for "fix," and from Equations 7.20 and 7.22 for

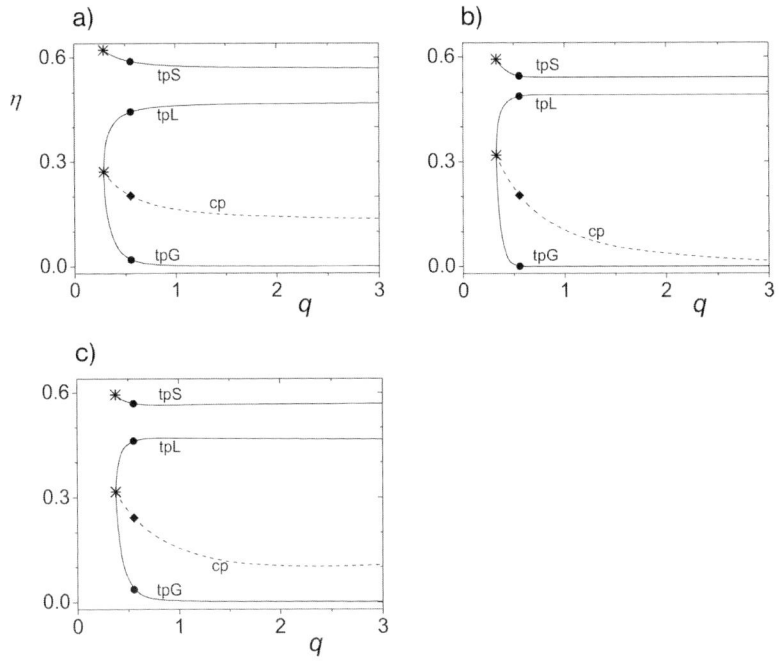

Figure 7.3 Variation of η^{cp} in the critical point cp (dashed curves) and of the three colloid concentrations η_G^{tp}, η_L^{tp}, and η_S^{tp} in the triple point tp (solid curves, labels tpG, tpL and tpS, respectively), for "Yuk" (a), "fix" (b) and "var" (c). The asterisks are the fluid and solid parts of the cep, with coordinates as given in Equation 7.23. The filled circles (tp) and diamonds (cp) in each diagram correspond to $q=0.552$ and are the same as in Figure 7.1.

"var." In Figure 7.4 we plot pv/kT along the GS, GL and LS binodals as a function of ε/kT for the three systems. The dashed curves in each diagram are for $q=0.429$, the solid curves for $q=0.789$. For each q, the three curves join in the triple point (●); triple points are indicated for a wide q range (q^{cep} and up), as indicated in the caption. Each GL binodal ends in the critical point (◆), given for the same q range. The filled circles (tp) and diamonds (cp) merge in the cep (asterisk).

There are considerable differences between the three diagrams. In "Yuk" the pressure at tp and cp is very low, and the tp pressure is lower than the cp pressure. Starting from tp, the LS and GL binodals go up, and GS goes down. The LS binodal is very steep and approaches, eventually, the hard-sphere coexistence pressure $(pv)_0$ 6.08kT for $\varepsilon \to 0$; this happens far outside the scale of the graph. For "fix" and "var" the pressure is high, and the tp pressure is higher than that in cp. Here a suitable reference level is $(pv)_0$ for the entire diagram.

The "fix" diagram is most easily understood, especially for high q (0.789 in the diagram, solid curves) where the condensed phases contain very little polymer (exclusion limit). Then we have for the triple point $(pv)^{tp} = (pv)_0 = 6.08kT$ because

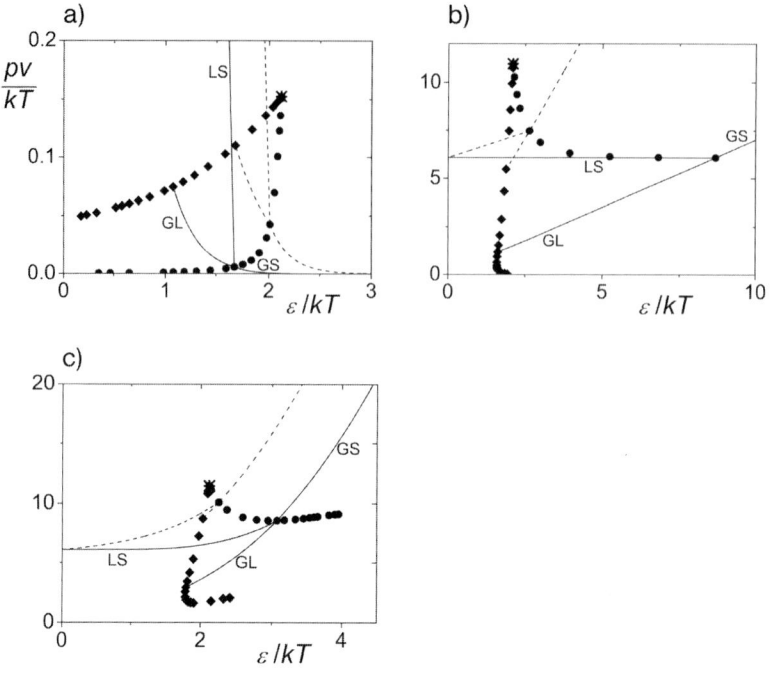

Figure 7.4 GL, GS and LS binodals in terms of pv/kT versus ε/kT, for "Yuk" (a), "fix" (b), and "var" (c). Dashed curves are for $q=0.429$, solid ones for $q=0.789$. Filled circles are triple points and diamonds are critical points for the following q values for "Yuk" (a): 0.289 ($=q^{cep}$), 0.29, 0.3, 0.31, 0.33, 0.37, 0.429, 0.471, 0.552, 0.633, 0.712, 0.789, 0.866, 1.017, 1.164, 1.31, 1.453, 1.594, 2.277, 2.933 and 3.57. For "fix" (b) the first six values are different: 0.3275 ($=q^{cep}$), 0.328, 0.33, 0.34, 0.36 and 0.38, and for "var" (c) the first two values are 0.377 ($=q^{cep}$) and 0.387; in both cases the values for 0.429 and above are the same as for "Yuk.".

g in Equation 7.15 vanishes. For the LS binodal the same applies, so this binodal is a horizontal line at $pv=6.08kT$. For the GS and GL binodals we have $pv=\Pi v$ since in the dilute gas phase $g=1$ and the colloid pressure is negligible. Moreover, Πv at a given q is directly proportional to ε (Equation 7.11). Hence, for GS and GL pv as a function of ε is a straight line that extrapolates to the origin; its slope is $q^{-2}(q+3/2)^{-1}$. The GS binodal goes up from tp, the GL binodal goes down but the latter ends at cp. For smaller q (0.429 in the diagram, dashed curves) the exclusion limit is not fully reached and $(pv)^{tp}$ is somewhat higher than $6.08kT$. As a consequence, the LS binodal is no longer horizontal, but GS and GL are still straight.

For "var" the polymer exclusion is less than in "fix" because of the compression of the depletion layers. Therefore, LS is not fully horizontal, but still $pv=6.08kT$ for small ε. GS is not straight, but for high ε the same limit is approached as for "fix." In addition, GL is also no longer a straight line. Nevertheless, the GL and GS binodals around tp follow a very smooth course.

7.5
Phase Diagrams pv/ε Versus kT/ε

In Figure 7.4 we plotted pv/kT against ε/kT. This is the most convenient representation for constant temperature: the parameters pv/kT and ε/kT are then varied by changing p and ε, respectively. However, we can replot the data and divide pv/kT by ε/kT to obtain pv/ε and plot this against kT/ε: if we now keep constant ε we get a $p(T)$ diagram analogous to the well-known textbook diagrams for simple atomic and molecular substances. Figure 7.5 shows data from Figure 7.4 replotted in this way.

The Yukawa diagram (Figure 7.5a), for a one-component system, has all the features of a classical $p(T)$ diagram: a very steep LS curve, a GL binodal where the pressure increases with temperature in an Arrhenius-type fashion up to the critical point, and a GS curve at low temperature and pressure.

The "fix" diagram for a three-component system colloid plus polymer plus solvent looks rather different, but its features are well understood from Figure 7.4. The horizontal (high q) LS curve in Figure 7.4 ($pv = 6.08kT$ for any ε/kT) is now transferred into a straight line pv/ε against kT/ε in Figure 7.5b with slope 6.08 (which is considerably smaller than for "Yuk"). The linearly increasing GL and GS branches in Figure 7.4 become a (nearly) horizontal line in Figure 7.5 at the level

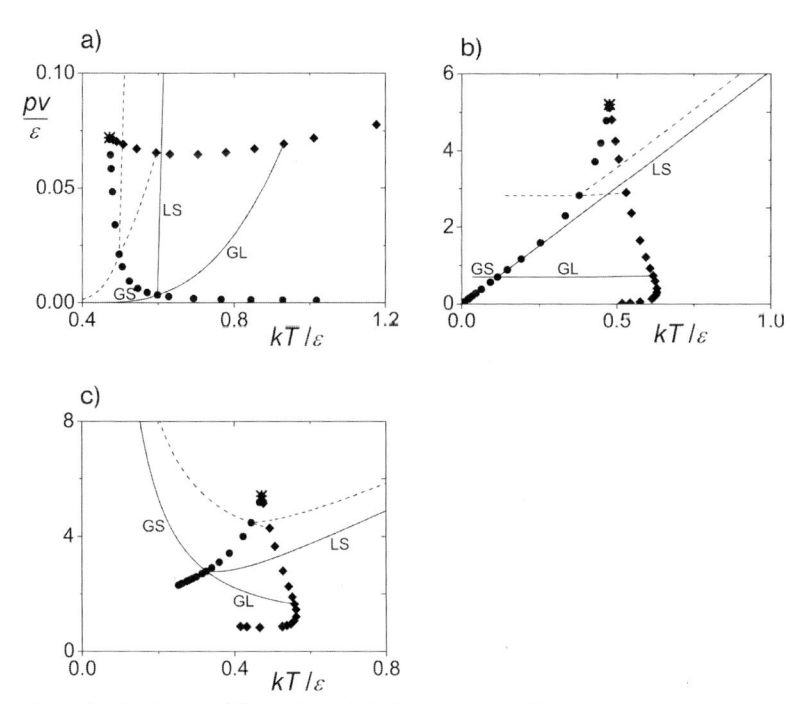

Figure 7.5 (a–c) Data of Figure 7.4 replotted as pv/ε versus kT/ε, to show the analogy with standard $p(T)$ diagrams for simple fluids. Solid and dashed curves and filled circles (tp) and diamonds (cp) are the same as in Figure 7.4.

$pv = (pv)^{\mathrm{tp}} \approx (pv)^{\mathrm{cp}}$. In this representation $(pv)^{\mathrm{cp}}$ is (slightly) above $(pv)^{\mathrm{tp}}$ but the difference is small.

The more realistic "var" diagram resembles the "fix" diagram to some extent, but the LS branch is even less steep, and the horizontal "fix" curves for GS and GL change into decreasing $p(T)$ curves: in this case the cp pressure is clearly below the tp pressure.

7.6
Concluding Remarks

In this chapter we have shown that the critical endpoint (cep), which corresponds to the lowest (relative) attraction range q where a stable colloidal liquid exists, does not depend on the details of the pair potential. It is situated at $q^{\mathrm{cep}} = 1/3$ and an interaction strength $\varepsilon^{\mathrm{cep}} = 2\,kT$ for four different systems: hard spheres with a Yukawa attraction ("Yuk"), a Lennard-Jones type (LJ) fluid (both one-component systems), and two three-component colloid/polymer/solvent systems with either a fixed depletion thickness ("fix") or a depletion thickness that varies with the polymer concentration ("var"). In the latter case the depletion thickness decreases from (roughly) the coil radius R in dilute solutions to the blob size ξ in semi-dilute solutions.

Although the cep is almost universal, for $q > q^{\mathrm{cep}}$ the $\varepsilon(\eta)$ phase diagram depends on the specifics of the system. For "Yuk" the liquid window is very narrow, spanning only a strength range of (at most) $0.6kT$. For "fix" the liquid window is very wide but this prediction is not realistic as this range in ε occurs at polymer concentrations where the assumption of a fixed depletion thickness breaks down. For the more accurate "var" model the liquid window is again relatively narrow (though wider than for "Yuk"): for high q it spans a range in strength of roughly $1.7kT$, corresponding to a factor of two in (external) polymer concentration.

We have also investigated the shape of $p(T)$ pressure versus temperature diagrams. For the one-component Yukawa system this diagram is analogous to textbook examples for simple atomic or molecular systems, with a very steep LS branch and an increasing Arrhenius-type GL branch ending in the fluid critical point. For the colloid/polymer systems it is rather different because both the colloids and the polymer chains contribute to the pressure: now the LS branch is much less steep and the GL branch in "var" decreases because the triple-point pressure is smaller than the pressure in the critical point.

References

1 Einstein, A. (1956) *Investigations on the Theory of the Brownian Motion*, Dover Publications.
2 von Smoluchowski, M. (1906) *Annals of Physics*, **21**, 756.
3 Perrin, J. (1916) *Atoms*, Constable & Company, London.
4 Lekkerkerker, H.N.W. (1997) *Physica A*, **244**, 227.

5. Hoover, W.G. and Ree, F.M. (1986) *Journal of Chemical Physics*, **49**, 3609.
6. Pusey, P.N. and van Megen, W. (1986) *Nature*, **320**, 340.
7. Hansen, J.-P. and McDonald, I.R. (2006) *Theory of Simple Liquids*, 3rd edn, Elsevier.
8. Fleer, G.J. and Tuinier, R. (2007) *Physica A*, **379**, 52.
9. Tuinier, R. and Fleer, G.J. (2006) *Journal of Physical Chemistry B*, **110**, 20540.
10. Asakura, S. and Oosawa, F. (1954) *Journal of Chemical Physics*, **22**, 1255.
11. Vrij, A. (1976) *Pure and Applied Chemistry*, **48**, 471.
12. Milling, A. and Biggs, S. (1995) *Journal of Colloid and Interface Science*, **170**, 604.
13. Ilett, S.M., Orrock, A., Poon, W.C.K. and Pusey, P.N. (1995) *Physical Review E*, **51**, 1344.
14. Faers, M.A. and Luckham, P.F. (1997) *Langmuir*, **15**, 2922.
15. Lekkerkerker, H.N.W., Poon, W.C.K., Pusey, P.N. Stroobants, A. and Warren, P.B. (1992) *Europhysics Letters*, **20**, 559.
16. de Gennes, P.G. (1979) *Scaling Concepts in Polymer Physics, Cornell*, University Press, Ithaca.
17. Fleer, G.J. and Tuinier, R. (2007) *Physical Review E*, **76**, 041802.
18. Carnahan, N.F. and Starling, K.E. (1969) *Journal of Chemical Physics*, **51**, 635.
19. Hall, C.K. (1972) *Journal of Chemical Physics*, **52**, 2252.
20. Tang, Y. and Lu, B.C.-Y. (1993) *Journal of Chemical Physics*, **99**, 9828.
21. Tang, Y., Lin, Y.-Z. and Li, Y.-G. (2005) *Journal of Chemical Physics*, **122**, 184505.
22. Tuinier, R., Aarts, D.G.A.L., Wensink, H.H. and Lekkerkerker, H.N.W. (2003) *Physical Chemistry Chemical Physics*, **5**, 3707.
23. Tuinier, R. and Fleer, G.J. (2004) *Macromolecules*, **37**, 8764.
24. Reiss, H., Frisch, H.L. and Lebowitz, J.L. (1959) *Journal of Chemical Physics*, **31**, 369.
25. Fleer, G.J. and Tuinier, R. (2008) Advances in Colloid and Interface Science, in print. Available online through http://dx.doi.org/10.1016/j.cis.2008.07.001.
26. Hanke, A., Eisenriegler, E. and Dietrich, S. (1999) *Phys Rev E*, **59**, 6853.
27. Fleer, G.J., Skvortsov, A.M. and Tuinier, R. (2007) *Macromolecular Theory & Simulations*, **16**, 531.
28. Vliegenhart, G.J., Lodge, J.F. and Lekkerkerker, H.N.W. (1999) *Physica A*, **263**, 378.
29. Hansen, J.-P. and McDonald, I.R. (2006) *Theory of Simple Liquids*, 3rd edn, Elsevier, p. 135.

8
Preparation, Properties and Chemical Modification of Nanosized Cellulose Fibrils
Per Stenius and Martin Andresen

8.1
Introduction

Microfibrillar cellulose (MFC) is a basic structural component of wood fibers, one of the most abundant biological raw materials on the planet. Recently, the potentialities of MFC as a renewable, abundant and biodegradable material have attracted increasing interest. Some examples of uses envisaged are MFC as a reinforcement material in composites, as a component in wood- and paper-based products with enhanced strength properties and built-in advanced functionalities, and as an additive for control of the stability and rheology of emulsions, suspensions and foams.

This chapter summarizes present knowledge about the preparation of MFC directly from wood, and its surface and colloidal properties. In most of the more advanced applications suggested for this material, chemical modification of the surfaces of the fibrils is of major importance, and methods to achieve such modification are described in some detail. Finally, a brief survey of already demonstrated applications of MFC is given.

A form of nanosized cellulose particles that has recently been quite extensively studied is cellulose whiskers or nanocrystalline cellulose [1, 2]. This material, which is manufactured by acid hydrolysis of cellulose fibers (see below), will be discussed only to the extent that knowledge of its properties are of immediate relevance to MFC.

8.2
Microfibrillar Cellulose

Native cellulose is a linear polysaccharide, consisting of β-D-anhydroglucopyranose units bound together by β-(1 → 4)-glycosidic linkages (Figure 8.1). The degree of polymerization (DP) varies from a few hundred to more than 15 000 monomers, depending on the source of origin [3].

Highlights in Colloid Science. Edited by Dimo Platikanov and Dotchi Exerowa
Copyright © 2009 WILEY-VCH Verlag GmbH & Co. KGaA, Weinheim
ISBN: 978-3-527-32037-0

Figure 8.1 β-D-glucopyranose units in cellulose.

During biosynthesis, intermolecular hydrogen bonds promote parallel association of the cellulose chains, resulting in thin, thread-like aggregates of extreme length. These primary fibril structures make up the wall layers and, finally, the whole cell wall of the cellulose fiber (Figure 8.2) [4].

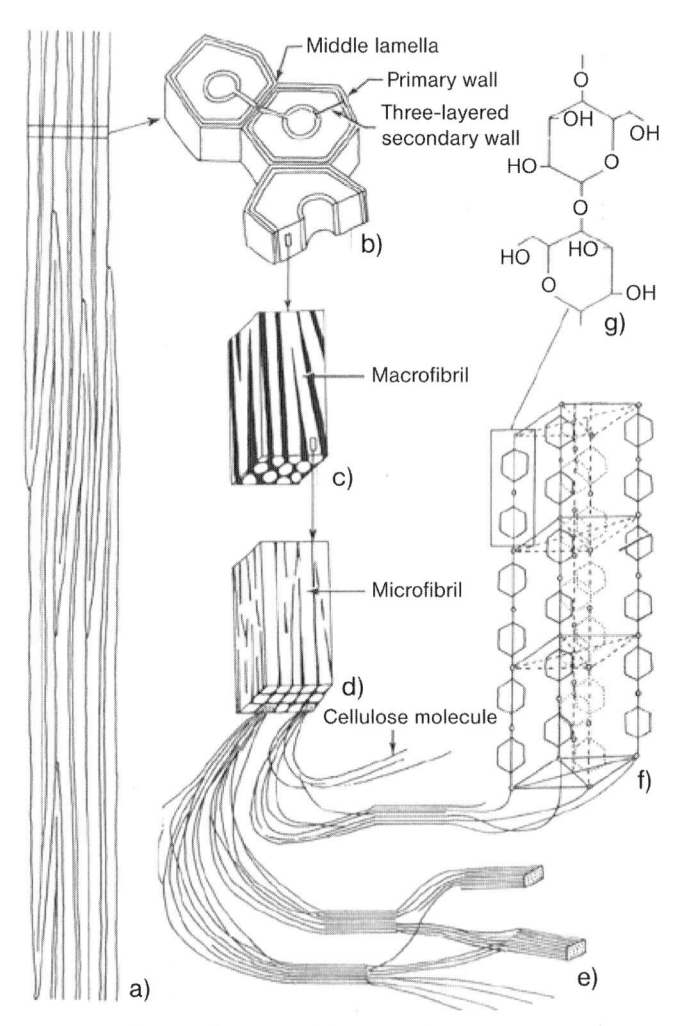

Figure 8.2 Schematic illustration of the structural components in the fiber cell wall (Reproduced with permission from Ref. [4]).

In wood, the cross-sectional diameters of the smallest basic units, often referred to as elementary fibrils, are in the range 2–4 nm. These fibrils are associated in higher systems with diameters of 10–30 nm (microfibrils) [3]. The microfibrils are also known as nanofibrils, protofibrils, and so on. Here the traditional and well-established term microfibril is used. Microfibrils possess a high degree of crystallinity, but along their transverse direction there are localized (paracrystalline) distortions, and they also contain amorphous domains [1, 5, 6].

In the wood fiber cell wall the microfibrils are a reinforcement material that is embedded in a matrix of hemicellulose and lignin. In this sense natural fibers can be regarded as composite materials. Microfibrils can have lengths up to several micrometers, giving very high aspect ratios, which result in gel-like suspension properties [7]. This, combined with high strength and ductility – the tensile strength properties of crystalline microfibrils exceed those of steel –, high specific surface area and a surface that is amenable to chemical functionalization, makes the cellulose microfibrils highly interesting as a nanomaterial.

8.3
Preparation of Microfibrillar and Nanocrystalline Cellulose

There are two main strategies for disintegration of cellulose fibers down to their microfibrillar components: (i) hydrolysis with strong acid and (ii) high shear mechanical treatment.

8.3.1
Acid Hydrolysis

When native cellulose is treated with strong acid, the amorphous domains of the microfibrils are preferentially attacked. The result is a colloidal suspension of rod-like crystallites [2, 8–12]. These nanocrystals (also referred to as microcrystals or whiskers) are shorter and stiffer than the native microfibrils. Hydrolysis with sulfuric acid is commonly used, which leads to aqueous suspensions that are electrostatically stabilized by negative surface charges imparted during hydrolysis [10–14]. Using hydrochloric acid results in nanocrystals with lower surface charge, which therefore tend to aggregate [15–17]. Nanocrystal suspensions have been prepared from wood pulp [10], cotton [11–13], sugar-beet pulp [18], bacterial cellulose (BC) [19] and sea animals (e.g. tunicates) [20]. Figure 8.3 shows cellulose microcrystals obtained by acid hydrolysis [20].

Owing to the relatively low aspect ratio and rod-like character of nanocrystals, their suspensions do not have the same gelling properties as those of more thread-like microfibrils. On thickening, nanocrystal suspensions may separate spontaneously into chiral-nematic liquid crystalline phases [10–13]. The crystallites may grow in size because of the larger freedom of motion of the cellulose molecules after hydrolytic cleavage and, therefore, tend to be larger in dimension than the original microfibrils [2].

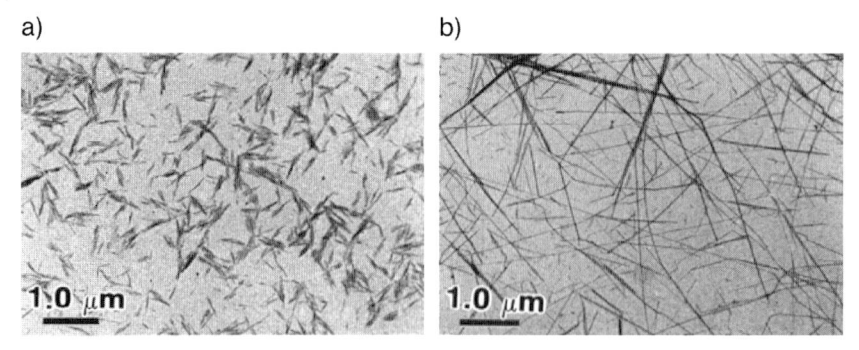

Figure 8.3 Transmission electron micrographs (TEM) of cellulose nanocrystals. The shape of the nanocrystals depends on the origin of the cellulose, with those from tunicate (b) being longer and more needle-like than the ones obtained from cotton (a). (Reproduced with permission from Ref. [20]).

8.3.2
High Shear Mechanical Treatment

In 1983 Herrick, Turbak et al. [21, 22] prepared suspensions containing a large fraction of fibrils by passing dilute slurries of wood pulp several times through a high-pressure homogenizer. They used the term microfibrillated cellulose for this material. At 2% concentration in water, it had the appearance of a pseudoplastic and thixotropic gel, suggesting a highly entangled network of interconnected fibrils. Several later reports describe the preparation of MFC using a similar approach, which can be regarded as the "classic" method for producing high aspect ratio, nanoscale microfibrils from wood [23–27].

However, due to the very high strength of the wood fiber secondary wall, the material resulting from mechanical homogenization of wood fibers also contains a substantial part of larger fibril bundles and even residual fibers and fiber fragments (Figure 8.4).

Homogenization has also been applied to various types of primary wall cellulose (e.g. sugar beet, potato tuber, banana rachis) where the microfibrils are more loosely organized than in the secondary wall of wood cells [28–31]. Hence, the disintegration requires less energy and the fibers are more easily delaminated. However, the material still mainly consists of bundles of microfibrils.

Taniguchi and Okamura used a super grinder to give sufficient shearing stress along the longitudinal fiber axis for delamination of wood fibers [32]. This yielded microfibrillated fibers with diameters in the range 20–90 nm and a fairly heterogeneous material. However, it seems that even with a high energy input, high shear mechanical treatment alone is insufficient for producing well-defined fully dispersed MFC.

On the other hand, by combination of high shear with chemical pretreatment, MFC with well-controlled diameter may be produced at reduced energy consumption.

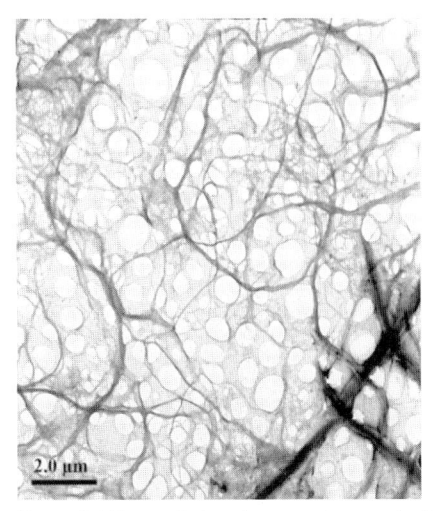

Figure 8.4 Transmission electron micrograph of microfibrillar wood pulp. The material consists of microfibrils and bundles of these, but also occasional fiber fragments and even small amounts of whole, unfibrillated fibers occur. (Reproduced with permission from Ref. [64].)

Thus, Pääkkö et al. [7] combined mild enzymatic hydrolysis with high-pressure homogenization to prepare MFC consisting mainly of elementary fibrils with diameter 5–6 nm and microfibrils of 10–20 nm diameter (Figure 8.5).

This mild hydrolysis retained the high aspect ratio of the fibrils, giving a highly entangled gel-network in aqueous suspension. Another method was presented by Saito and coworkers [33], who used TEMPO-mediated oxidation (Section 8.5.4.1) of never-dried pulp before homogenization. This treatment rendered the microfibril

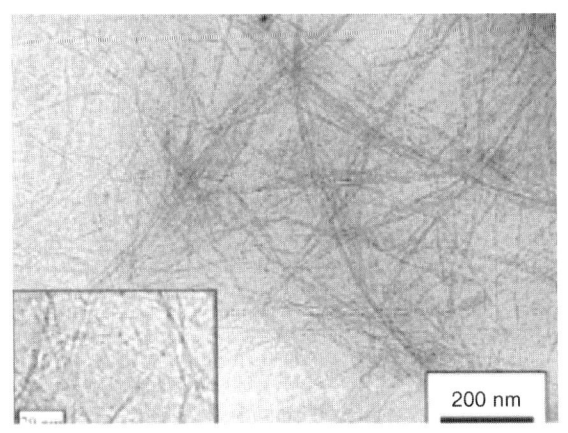

Figure 8.5 Cryo-TEM of MFC gel obtained after refining, enzymatic hydrolysis and homogenization of wood pulp. (Reproduced with permission from Ref. [7].)

surfaces negatively charged, which most likely led to increased electrostatic repulsion between the microfibrils in the fibers. As a result, the pulp could be disintegrated into long individual elementary fibrils with widths of 3–5 nm by simple homogenization in a Waring blender.

8.3.3
Other Routes to Cellulose Microfibrils

Some strains of *Acetobacter* produce BC as ribbon-shaped fibrils with a cross sectional size of 10×50 nm [34]. The gelatinous pellicle membrane formed at the surface of the culture medium under static culture conditions can be disintegrated into microfibrils and microfibril bundles by treatment in a homogenizer. Alternatively, BC may be prepared under agitated culture conditions, to yield dispersions of well-defined, individualized microfibrils, with properties similar to those of MFC [35].

8.4
Methods Used to Characterize Cellulose Microfibrils

8.4.1
Fibril Morphology and Structure

The most commonly used tool by far for sub-micrometer morphology analysis of cellulosic materials is transmission electron microscopy (TEM). TEM typically yields information about length, aspect ratio, shape and aggregation. Metal shadowing or negative staining is commonly used for contrast enhancement. TEM has been applied to visualize cellulose microfibrils from sugar beets [28, 29], wood [7, 25, 27, 33, 36], potato tubers [30, 37] and bacteria [38, 39]. TEM has also been used extensively to study the morphology of cellulose nanocrystals [10, 15, 19, 20, 40, 41].

Scanning electron microscopy (SEM) has also been widely used to examine cellulose microfibrils and nanocrystals [21, 22, 24, 26, 32, 34, 41]. The limited resolution of SEM compared to TEM does, however, make detailed analysis of nanosized fibrils and whiskers challenging. In addition, metal coating applied in SEM most likely broadens the nanosized structures.

Atomic force microscopy (AFM) has the necessary resolution to observe nanosized cellulose fibril structures without the need for metal coatings. AFM has been applied to both MFC [7, 25] and cellulose whiskers [41–44] and to generate surface profiles of films cast from cellulose nanocrystals [44–46]. AFM may, however, overestimate the width of the particles [41, 44] due to the tip-broadening effect (the shape of the tip contributes to the recorded image). One way to overcome this problem is to measure the height of the fibrils, which is not subject to tip-broadening artifacts [43].

Solid-state cross polarization magic angle spinning carbon-13 magnetic resonance (CP/MAS ^{13}C-NMR) spectroscopy has yielded information on the crystallinity, allomorph composition and surface properties of microfibrils [5, 39, 47–49]. Both the surface and the interior structures can be quantified and the dimensions of the microfibrils can be calculated using the so-called crystallinity index [48, 50].

Other methods used to gain information about the structure, size and crystallinity of cellulose microfibrils are X-ray and neutron diffraction analysis [29, 34, 51–54] and various scattering techniques, including small-angle neutron scattering (SANS) and small-angle X-ray scattering (SAXS) [14, 55, 56].

8.4.2
Fibril Surface Chemistry

Fourier-transform infrared spectroscopy (FT-IR) is the most commonly used technique for evaluation of the chemistry of surface-modified cellulose microfibrils and nanocrystals (whiskers) [25, 51, 52, 57–60]. However, FT-IR analysis is not confined to the surface of the material, and is therefore best suited for qualitative evaluations of the chemistry of microfibrillar cellulose.

^{13}C NMR has also been used to obtain information about the chemical composition of microfibril surfaces [39, 50, 58, 61]. As with FT-IR, the method is sensitive to the bulk properties of the sample, and the low signal-to-noise ratio make identification of the chemical structure of the surface difficult.

Probably the most suitable technique for evaluation of cellulose microfibril surface chemistry is X-ray photoelectron spectroscopy (XPS) [62]. The method provides a full elemental analysis, except for hydrogen, of the topmost 1–20 nm of the sample surface. High-resolution XPS C1s measurements give quantitative information on carbon bonding chemistry [63]. XPS has been used extensively to quantify the changes of the chemical composition in the microfibril surface during derivatization [25–27, 60, 64].

Bulk sensitive elemental analysis [19, 38, 39, 52, 53, 60], zeta-potential measurements [27] and various titration techniques [34, 39, 40, 50, 58, 59] have also been employed for chemical characterization of fibril surfaces.

8.5
Modification of Microfibril Surfaces

The hydroxyl groups in the surface of cellulose fibrils offer rich possibilities for chemical modification. Fibrils are partially amorphous and MFC prepared from chemical pulps will be microporous due to the removal of lignin and hemicelluloses. To avoid unacceptably high consumption of reactants, surface modification reactions should therefore be designed so that penetration of chemicals into the fibrils is avoided.

Surface modification of fibrils reported in the literature are mainly hydrophobication reactions aimed at compatibilization with nonpolar polymer matrices. Hence, much of the work concerns modification of whiskers that are easier to mix into composites than MFC. However, methods for introduction of specific functionalities, as part of other applications of MFC, have also been reported.

8.5.1
Esterification Reactions

8.5.1.1 Acetylation

Cellulose acetate is prepared by derivatization of fibers in a mixture of acetic acid anhydride and acetic acid (Figure 8.6). In the "homogeneous" process the cellulose acetate is solubilized into the reaction medium as it is produced. In the "heterogeneous" process, non-swelling diluents, such as toluene or amyl acetate, are added to the reaction medium. Then the cellulose acetate remains insoluble, without any change in fiber morphology [65].

Sassi and Chanzy subjected nanocrystals, from *Valonia* and tunicin, to both homogenous and heterogeneous acetylation [51]. The acetyl content was evaluated by IR spectroscopy. In both methods, acetylation essentially took place at the surface of the crystals. However, in homogenous acetylation, the crystal diameter was drastically reduced, indicating that sufficiently acetylated cellulose chains located in the surface dissolved in the reaction medium. In heterogeneous acetylation the acetylated cellulose remained insoluble and surrounded an unreacted core of cellulose crystals.

Geyer *et al.* [57] have described the heterogeneous acetylation of highly crystalline BC microfibrils. Following the derivatization procedure described above, they achieved a degree of substitution (DS) of 2.5, as determined volumetrically by alkali saponification.

In a similar study by Kim *et al.* [34] heterogeneous surface acetylation of BC resulted in DS values between 0.04 and 2.77, as determined by titration and FT-IR. The DS could be controlled by changing the amount of acetic anhydride added. X-ray diffraction showed that the acetylation proceeded from the surface of the microfibrils, leaving the core portion unreacted. SEM analysis revealed that even low levels of acetylation effectively prevented aggregation of the microfibrils on direct drying from water, as a result of an increase in surface hydrophobicity. Yamamoto *et al.* [39] have reported CP/MAS ^{13}C NMR spectroscopic studies of the effects of solid-phase acetylation on BC structure. The DS ranged from 1.19 to 2.73. Like Kim *et al.*, they found evidence that acetylation proceeded from the surface to the core of each microfibril.

8.5.1.2 Reaction with Anhydrides

Yuan *et al.* [60] have described an elegant method to hydrophobize the surface of tunicin whiskers. Aqueous emulsions of an alkenyl succinic anhydride were mixed with cellulose whisker suspensions, freeze-dried and then heated. Bulk and surface DS was evaluated by FT-IR and XPS, respectively. DS values were low (~0.02), but this

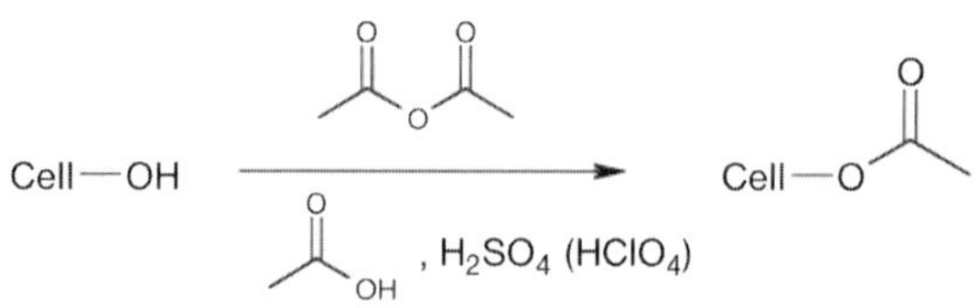

Figure 8.6 Acetylation of cellulose with acetic anhydride–acetic acid.

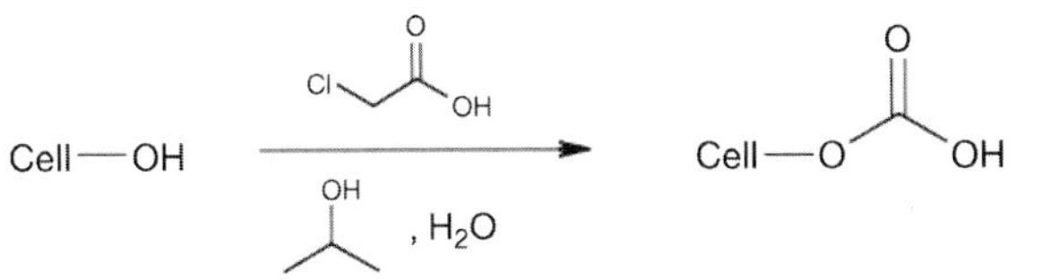

8.5 Modification of Microfibril Surfaces | 143

Figure 8.7 Carboxymethylation of cellulose.

was apparently sufficient to allow good dispersion of the modified fibrils into a wide range of nonpolar solvents, indicating possible application in nanocomposites.

Stenstad et al. [27] grafted both succinic and maleic anhydride to the surface of microfibrillated wood pulp under water-free conditions. This method introduced negative charge onto the fibril surfaces. Grafting of maleic anhydride to the MFC surface introduced vinyl groups that may serve as starting point for further radical-induced polymerization reactions. Grafting was confirmed qualitatively by FT-IR. The DS was 0.8 and 0.3 for succinic and maleic acid, respectively, as evaluated by XPS.

8.5.1.3 Carboxymethylation

Carboxymethyl cellulose (CMC), the most important and widely produced ionic cellulose ether ($\approx$300 000 ton y^{-1}) [66], is usually produced as a slurry of cellulose in a mixture of isopropanol and water, treated with aqueous NaOH and converted with monochloroacetic acid into its sodium salt (Figure 8.7) [67].

Geyer et al. [57] have reported on the heterogeneous carboxymethylation of BC. The introduction of ester groups was verified with FT-IR. The DS was 1.3, as determined gravimetrically by the uranyl method. The substitution pattern was assessed by HPLC, revealing a relatively uniform distribution of carboxymethyl groups within the polymer chain.

8.5.1.4 Isocyanate Grafting

Stenstad et al. [27] grafted hexamethylene di-isocyanate onto the surface of microfibrils obtained from homogenization of wood pulp (Figure 8.8). Under anhydrous conditions, isocyanates react with the surface hydroxyls of the microfibrils to form

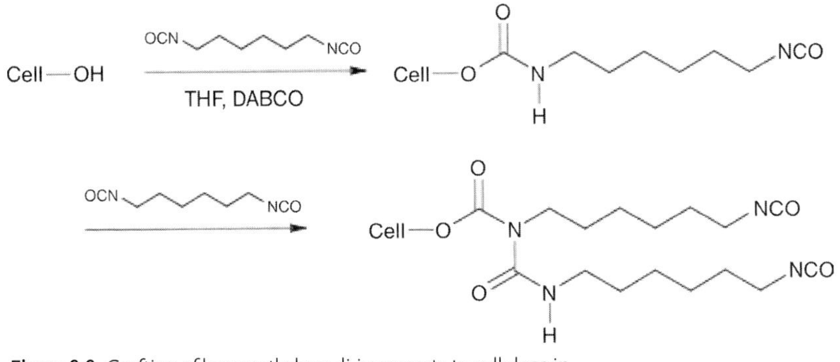

Figure 8.8 Grafting of hexamethylene di-isocyanate to cellulose in dry tetrahydrofuran (THF), using 1,4-diazabicyclo[2.2.2]octane (DABCO) as catalyst.

urethane bonds. The urethane groups reacted further with free di-isocyanate added in excess, yielding a highly crosslinked, hydrophobic polymer layer on the surface of the fibrils (confirmed by FT-IR).

The isocyanate-coated MFC in dry THF could be immediately coated with bisamino-propyl amine or 3-ethylenamino-1-propylamine, thus introducing positive charges to the microfibril surfaces.

8.5.2
Etherification

8.5.2.1 Silylation

A very effective method to hydrophobize the surface of MFC is etherification of the hydroxyl functions with alkyl silanes. Both chlorosilanes and alkoxysilanes react readily with the hydroxyl functions of cellulose.

Grunert and Winter [19] have described topochemical silylation of BC nanocrystals with a typical cross-section 8×50 nm and a few hundred to several thousand nanometers in length The crystals were treated with hexamethyldisilazane in formamide, resulting in an average DS of 0.49.

Goussé and coworkers [52, 53] have reported on the surface silylation of both cellulose nanocrystals and microfibrils in solvents of low polarity. Cellulose whiskers from tunicin and sugar beet microfibrils were silylated in toluene by the addition of a series of alkyl dimethylchlorosilanes (Figure 8.9).

A wide range of degrees of surface substitution (DSS) was achieved, determined by elemental analysis. The reactions started rapidly, but slowed after a few hours to reach a plateau beyond which no further silylation took place. If too much silylating reagent or too a long reaction time were used, the cellulose chains located at the surface of the whiskers became soluble in the reaction medium, in analogy with the observations by Sassi and Chanzy [51] for the acetylation of cellulose nanocrystals. X-ray studies showed that the microfibrillar structure stayed intact for DSS up to 0.4 for the microfibrils and DSS up to1 for the whiskers; at too high a surface substitution, total decrystallization occurred. At controlled DSS values, the derivatized whiskers and microfibrils formed stable suspensions in solvents of low polarity, such as THF. Furthermore, suspensions of silylated microfibrils in organic solvents exhibited pseudoplastic properties.

Recently, Andresen et al. [25] adapted the silylation procedure described by Goussé et al. [52, 53], using chloro(dimethyl)isopropylsilane to modify the surface of MFC from wood pulp. Silylated MFC with DSS between 0.6 and 1 (determined by XPS) formed stable dispersions in THF and toluene. Like Goussé et al., Andresen et al.

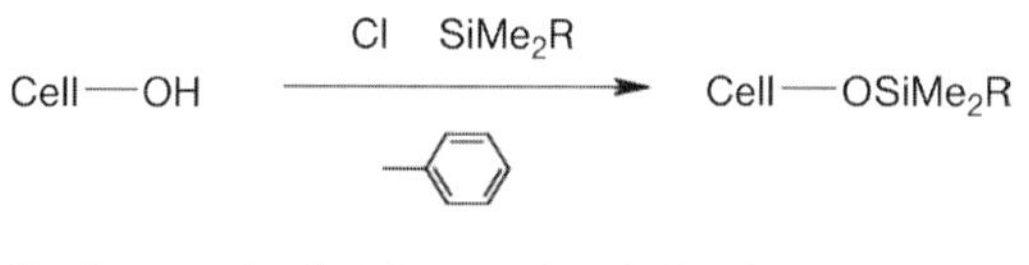

$$\text{Cell—OH} \xrightarrow{\text{Cl—SiMe}_2\text{R, } C_6H_6} \text{Cell—OSiMe}_2\text{R}$$

R = isopropyl, *n*-butyl, *n*-octyl, *n*-dodecyl

Figure 8.9 Silylation of cellulose with alkyl dimethylchlorosilanes.

found that when reaction conditions were too harsh the MFC was partially solubilized and the initial microfibrillar morphology was lost.

8.5.3
Nitration

Yamamoto et al. [39] have reported on the solid-phase nitration of BC microfibrils, using a mixture of anhydrous nitric acid and dichloromethane. DS values, based on elemental analysis of nitrogen, varied between 0.05 and 2.73. CP/MAS ^{13}C NMR revealed that the reactivity of the OH groups in the surface decreased in the order O(6)H > O(2)H > O(3)H.

8.5.4
Oxidation Reactions

8.5.4.1 TEMPO-Mediated Oxidation
Oxidation with NaClO catalyzed by the 2,2,6,6-tetramethyl-1-piperinidyloxy (TEMPO) radical has proven to be a highly selective, fast, and well-controlled method for carboxylation of the alcohol groups in various cellulose materials [61]. Owing to the steric structure of the TEMPO molecule, only the primary C6 hydroxyls are converted into carboxyl groups (Figure 8.10) [68–70].

This method has been used to introduce carboxyl groups onto the surface of MFC from wood and sugar beet pulp, BC microfibrils and nanocrystals based on cotton, sugar beet pulp and tunicin [50, 58, 59, 61, 71]. Using ^{13}C NMR spectroscopy, Isogai and Kato [61] showed that only small amounts of the hydroxyls in the MFC surface were oxidized. Montanari et al. [50] found that TEMPO-mediated carboxylation resulted in DS = 0.23 for sugar beet nanocrystals and 0.15 for cotton nanocrystals (^{13}C NMR and conductometric titration). The crystals formed electrostatically stabilized suspensions in water. Habibi et al. [59] have reported similar results with tunicin cellulose.

Araki et al. prepared sterically stabilized suspensions by grafting poly(ethylene glycol) (PEG) onto the surface of cotton nanocrystals that had previously been carboxylated by TEMPO-mediated oxidation (Figure 8.11) [40].

The bonding of PEG was confirmed by gravimetry, IR and conductometric titration of carboxyl groups. Freeze-dried nanocrystals could be redispersed in water or organic solvents and showed strongly increased stability towards flocculation with polyelectrolyte.

Figure 8.10 TEMPO-mediated oxidation of cellulose.

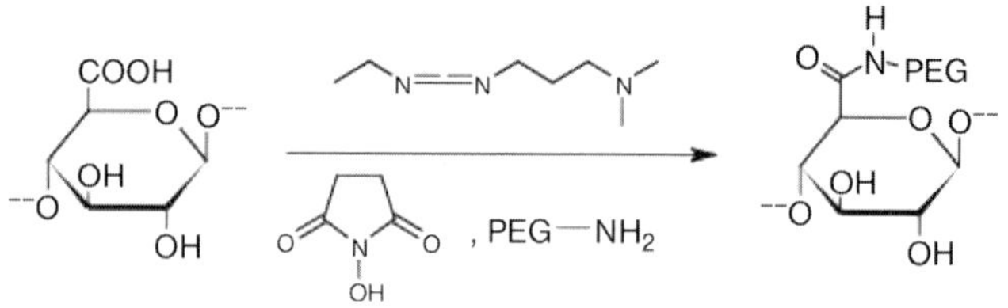

Figure 8.11 PEG-grafting of TEMPO-oxidized cellulose.

8.5.4.2 Cerium Induced Grafting

Stenstad et al. [27] have used cerium(IV) to generate free radicals on the cellulose backbone of wood pulp MFC. In the presence of glycidyl methacrylate (GMA) monomers this initiated grafting of polymerized GMA onto the MFC surface (Figure 8.12), as confirmed qualitatively by FT-IR and quantitatively by XPS. High-resolution C1s XPS showed that the length of the polymer chains could be controlled by varying the amount of GMA. Coating of the MFC with a polymer layer was also documented visually by TEM. Preliminary studies indicated that the cerium treatment did not significantly reduce the mean chain length of the cellulose in the MFC.

8.5.5
Coating with Surfactant

Heux et al. [20] have prepared suspensions of cotton and tunicate nanocrystals in nonpolar solvents, using surfactant as a stabilizing agent. The surfactant, a phosphoric ester of a poly(ethylene oxide) nonylphenyl ether, was added to aqueous

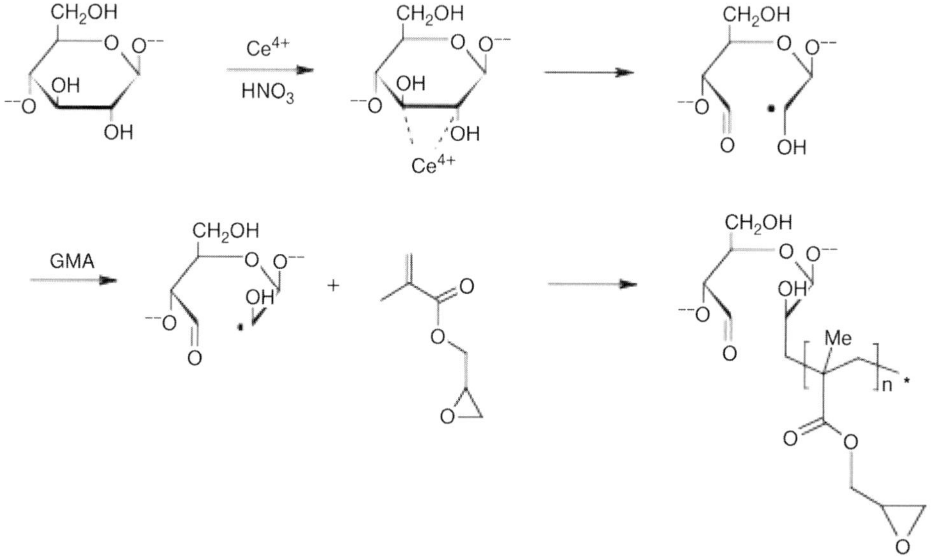

Figure 8.12 Proposed mechanism for the cerium-induced grafting of GMA to the cellulose backbone.

suspensions of the nanocrystals. Freeze-dried suspensions could be re-dispersed in toluene and cyclohexane, forming dilute suspensions that were stable over several weeks. TEM analysis of the redispersed nanocrystals showed that their overall characteristics were similar to the original, unmodified aqueous suspensions. A SANS study indicated a significant increase of the apparent cross section of the modified crystals, suggesting that the surfactant was physically adsorbed, rather than chemically grafted, onto the surface [55].

8.6
Applications of Nanofibrillar Cellulose

8.6.1
Rheology

The use of MFC to modify the rheology of the media into which they are introduced was already proposed in the first reports on their manufacture [22]. Suggested utility areas included foods, paints, cosmetics, drilling fluids and pharmaceutical products. As already mentioned, the formation of highly entangled fibril networks results in suspensions with viscoelastic (gel-like) properties (Figure 8.13), which suggests possible application of MFC as thickener or dispersant [72]. Owing to the hydrophilicity resulting from the hydroxyl functional groups in the surface, MFC does not disperse well in nonpolar solvents. This led to several studies (see above) concerning the hydrophobization of the surface of nanosized cellulose fibrils and nanocrystals, while retaining their initial morphology and crystalline aspects. The rheological

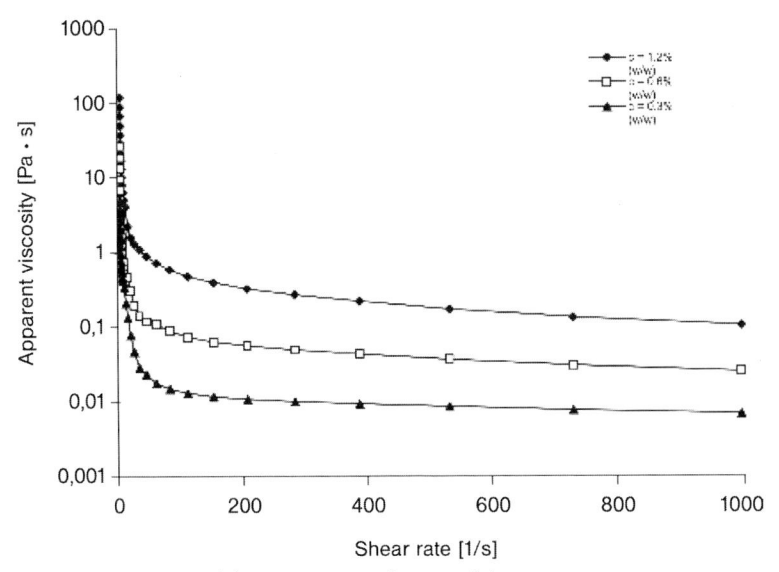

Figure 8.13 Variation of shear viscosity as a function of shear rate for 1.2, 0.6 and 0.3% (w/w) suspensions of native MFC in water.

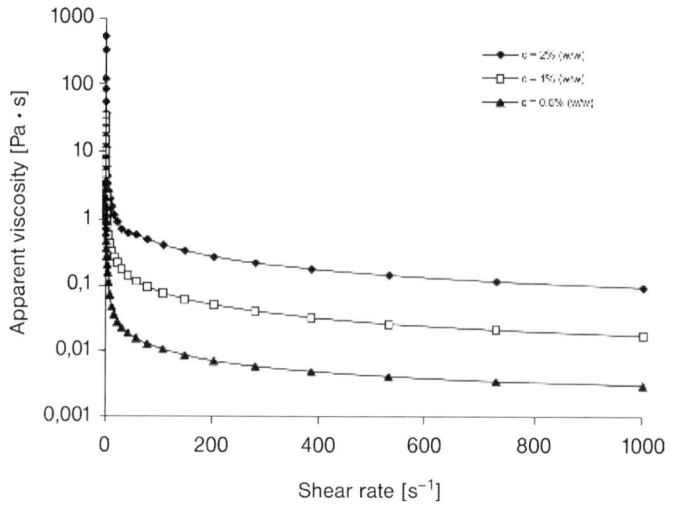

Figure 8.14 Variation of shear viscosity as a function of shear rate for 2, 1 and 0.5% (w/w) suspensions of hydrophobized MFC in toluene.

properties of suspensions of hydrophobized MFC in toluene are exemplified in Figure 8.14.

Dispersions of native MFC in water and hydrophobized MFC in toluene are pseudoplastic, that is, viscosity decreases at increased shear rates. At similar concentrations, the viscosities of MFC dispersed in water are much higher than those of hydrophobized MFC in toluene, indicating that interfibrillar interactions are markedly weakened after the surface modification.

8.6.2
Nanocomposites

The advantageous mechanical properties and the high specific surface area suggest that cellulose microfibrils would have good reinforcing ability in materials, namely nanocomposites. This possibility has given rise to most of the research conducted on fibrillated cellulose materials. Several publications describe the incorporation of both native and surface-modified cellulose microfibrils [24, 37, 73–75] and nanocrystals [18, 19, 76–79] into various polymer matrices. Owing to their much less elongated shape, nanofibril-reinforced composites are simpler to prepare and have been more extensively investigated than those based on microfibrils.

8.6.3
Thin Films

Upon drying, cellulose microfibrils form very thin, highly homogenous films, with tensile strengths far superior to those of regular print grade paper, indicating possible application in paper and paper coatings [32, 80]. Furthermore, it has been shown that water contact angles on films prepared from hydrophobized MFC are close to those of super hydrophobic surfaces, suggesting their possible use in water-repellent and self-cleaning material applications [25].

The ability of rod-like cellulose microcrystals to form self-assembling chiral-nematic phases in suspension has also led to attention for its potential applications. By casting films from suspensions of cellulose microcrystals, cellulose films with the optical properties of chiral nematic liquid crystals can be prepared. The films can be tailored to give different colors of reflected light by altering the salt content of the suspension for a given source of cellulose and set of hydrolysis conditions. Possible areas of application include optical variable films and ink pigments for security papers [81].

8.6.4
Dispersion Stabilizers

Various fibrillated cellulose materials have been reported to stabilize water-in-oil emulsions [82–85]. Recently, Andresen *et al.* [64] have shown that highly stable oil-in-water emulsions with widely varying O/W ratio and droplet size can be prepared with hydrophobized MFC as the sole stabilizer. Figure 8.15 shows emulsions stabilized with the same MFC that was hydrophobized to different degrees.

8.6.5
Biochemical and Biomedical Applications

Cellulose microfibrils were initially proposed to have potential in various biomedical applications, like drug carriers and topical wound dressings [22, 86]. Membranes based on bacterial cellulose microfibrils have been successfully used as wound-

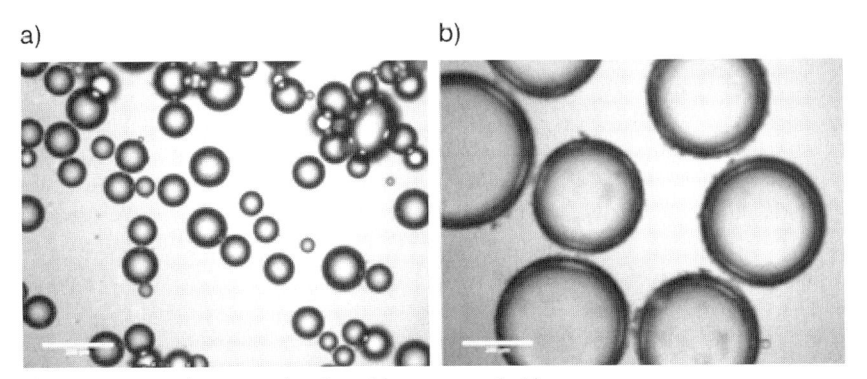

Figure 8.15 Optical micrographs of emulsions prepared with silylated MFC; DS = 0.6 (a) and DS = 1.1 (b). The toluene : water ratio was 1 : 1 and the concentration of silylated MFC was 0.15% (w/v) for both emulsions. (Reproduced with permission from Ref. [64].)

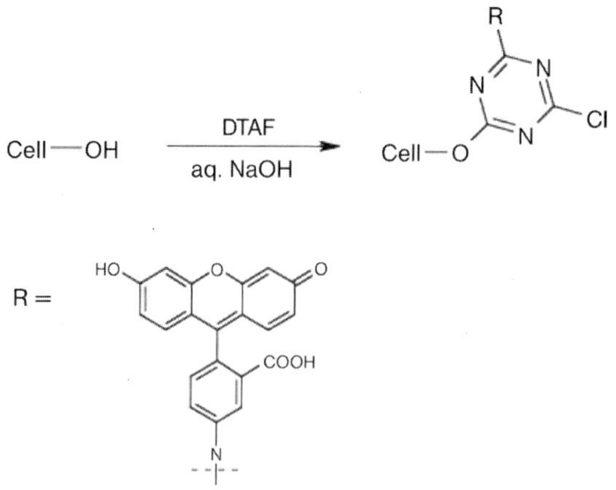

Figure 8.16 Grafting of 5-(4,6-dichlorotriazinyl)aminofluorescein (DTAF) onto cellulose.

healing devices for damaged skin and as small-diameter blood vessel replacements, and are believed to offer many other uses in regenerative medicine, such as guided tissue engineering or periodontal treatments [22, 86].

8.6.5.1 Enzymatic Assay

Helbert and coworkers [38] have prepared suspensions of fluorescent cellulose microfibrils by dissolving 5-(4,6-dichlorotriazinyl)aminofluorescein (DTAF) into a suspension of bacterial cellulose microfibrils in aqueous NaOH (Figure 8.16).

The amount of DTAF grafted onto the microfibril surface depended on the concentration of NaOH, a more alkaline suspension leading to more grafting. The extent of grafting could also be increased by successive grafting operations, with extensive washing in between. This resulted in a DS value of 0.025. Enzymatic assays were carried out with the microfibrils in suspension and on microwell titer plates. Exo- and endocellulase activities were differentiated by comparing the amount of released fluorescence and that of released reducing sugar. These observations may open a way to automation of cellulase assays, which would be of importance to applications in the textile and paper industry that benefits from cellulase treatments.

8.6.5.2 Grafting of an Antimicrobial

Using a simple adsorption-curing process, Andresen *et al.* have grafted the antimicrobial ammonium compound octadecyldimethyl-(3-trimethoxysilylpropyl)ammonium chloride (ODDMAC) onto the surface of MFC from wood pulp (Figure 8.17) [26].

After prehydrolysis to its corresponding silanol, the ODDMAC was adsorbed onto the MFC surface and covalently bound by curing. The reaction was verified by FT-IR and XPS. The bactericidal activity of films prepared from the modified MFC was tested against *Escherichia coli*, *S. aureus* and *P. aeruginosa*. The films showed significant reduction in viable *E. coli* and *S. aureus* when compared to an unmodified reference, even at very low concentrations of ODDMAC in the MFC surface. Owing

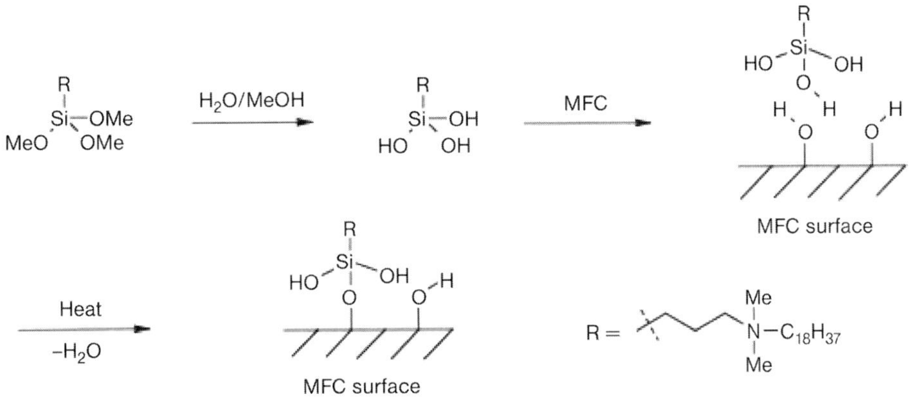

Figure 8.17 Mechanism for the prehydrolysis and adsorption/curing of ODDMAC onto a MFC surface.

to the very high surface area-to-volume ratio of these films, they could prove to be effective for wound healing applications, air filtration uses or as an antimicrobial separation filter for submicron particles.

8.6.6
Paper Products

Obvious applications of MFC are as a reinforcement or coating in paper products with special functionalities. So far, however, little has been reported on these applications in the open literature.

8.7
Concluding Remarks

Wood microfibrils are unique organic colloids. They are manufactured from an abundantly available, biodegradable and renewable raw material consisting of particles of extreme strength and aspect ratio. The applicability of MFC as a reinforcing component in composites, as a stabilizer of dispersions, as a coating material and as an additive to control the rheological properties of dispersions have so far been demonstrated. The colloidal properties of MFC have not been studied in any detail. Some problems that need to be much better understood are how to manufacture stable, well-dispersed MFC at acceptable costs; how to ensure that chemical modifications for applications that require retainment of the unique strength properties of the fibrils do not cause unacceptable scissoring of the cellulose chains, and how to chemically modify the MFC by mild treatments, preferably in aqueous environment. As we have shown, good progress is being made to solve these problems, but there is still room for much research in this area. Of particular importance seems to be problems directly related to surface and colloid chemistry: to understand how to reduce the energy required to disintegrate wood fibers, and to find methods that limit chemical modification to the MFC surface. The research that has been reviewed above shows that good progress is being made. Notably, very recently published results [7, 75] indicate that enzymatic treatments may offer one important way to solve these problems.

References

1 Azizi Samir, M.A.S., Alloin, F. and Dufresne, A. (2005) *Biomacromolecules*, **6**, 612.
2 Lima, M.M.d.S. and Borsali, R. (2004) *Macromolecular Rapid Communications*, **25**, 771.
3 Fengel, D. and Wegener, G. (1989) *Wood Chemistry, Ultrastructure, Reactions* Reprinted., Walter de Gruyter, Berlin.
4 Esau, K. (2001) *Anatomy of Seed Plants*, 2nd edn, Wiley Interscience, New York.
5 Larsson, P.T., Wickholm, K. and Iversen, T. (1997) *Carbohydrate Research*, **302**, 19.
6 Wickholm, K., Larsson, P.T. and Iversen, T. (1998) *Carbohydrate Research*, **312**, 123.
7 Pääkkö, M., Ankerfors, M., Kosonen, H. et al. (2007) *Biomacromolecules*, **8**, 1934.
8 Marchessault, R.H., Morehead, F.F. and Walter, N.M. (1959) *Nature*, **184**, 632.
9 Marchessault, R.H., Morehead, F.F. and Koch, M.J. (1961) *Journal of Colloid and Interface Science*, **16**, 327.
10 Revol, J.F., Bradford, H., Giasson, J. et al. (1992) *International Journal of Biological Macromolecules*, **14**, 170.
11 Revol, J.F., Godbut, L., Dong, X.M. et al. (1994) *Liquid Crystals*, **16**, 127.
12 Dong, X.M., Kimura, T., Revol, J.F. and Gray, D. (1996) *Langmuir*, **12**, 2076.
13 Dong, X.M., Revol, J.F. and Gray, D. (1998) *Cellulose*, **5**, 19.
14 Orts, W.J., Godbut, L., Marchessault, R.H. and Revol, J.F. (1998) *Macromolecules*, **31**, 5717.
15 Araki, J., Wada, N., Kuga, S. and Okano, T. (1998) *Colloids and Surfaces A: Physicochemical and Engineering Aspects*, **142**, 75.
16 Araki, J., Wada, N., Kuga, S. and Ogano, T. (1999) *Jounal of Wood Science*, **45**, 258.
17 Araki, J., Wada, N., Kuga, S. and Ogano, T. (2000) *Langmuir*, **16**, 2413.
18 Azizi Samir, M.A.S., Alloin, F., Sanches, J.-et al. (2004) *Macromolecules*, **37**, 1386.
19 Grunert, M. and Winter, W.T. (2002) *Journal of Polymers and the Environment*, **10**, 27.
20 Heux, L., Chauve, G. and Bonini, C. (2000) *Langmuir*, **16**, 8210.
21 Herrick, F.W., Casebier, R.L., Hamilton, J.K. and Sandberg, K.R. (1983) *Journal of Applied Polymer Science: Applied Polymer, Symposia*, **37**, 797.
22 Turbak, A.F., Snyder, F.W. and Sandberg, K.R. (1983) *Journal of Applied Polymer Science: Applied Polymer, Symposia*, **37**, 815.
23 Wågberg, L., Winter, L., Ödberg, L. and Lindström, T. (1987) *Colloids Surface*, **27**, 163.
24 Nakagaito, A.N. and Yano, H. (2005) *Applied Physics A*, **80**, 155.
25 Andresen, M., Johansson, L.-S., Tanem, B.S. and Stenius, P. (2006) *Cellulose*, **13**, 655.
26 Andresen, M., Stenstad, P., Møretrø, T. et al. (2007) *Biomacromolecules*, **8**, 2149.
27 Stenstad, P., Andresen, M., Tanem, B.S. and Stenius, P. (2008) *Cellulose*, **15**, 35.
28 Dinand, E., Chanzy, H. and Vignon, M.R. (1996) *Cellulose*, **3**, 183.
29 Dinand, E., Chanzy, H. and Vignon, M.R. (1999) *Food Hydrocolloids*, **13**, 275.
30 Dufresne, A., Dupeyre, D. and Vignon, M.R. (2000) *Journal of Applied Polymer Science*, **76**, 2080.
31 Zuluaga, R., Putaux, J.-L., Restrepo, A. Mondragon, I. and Gañón, P. (2007) *Cellulose*, **14**, 585.
32 Taniguchi, T. and Okamura, K. (1998) *Polymer International*, **47**, 291.

33 Saito, T., Nishiyama, M., Putaux, J.-L. *et al.* (2006) *Biomacromolecules*, **7**, 1687.
34 Kim, D.-Y., Nishiyama, Y. and Kuga, S. (2002) *Cellulose*, **9**, 361364.
35 Ougiya, H., Hioki, N., Watanabe, K. *et al.* (1998) *Bioscience, Biotechnology, and Biochemistry*, **62**, 1880.
36 Saito, T. and Isogai, A. (2006) *Colloids and Surfaces A: Physicochemical and Engineering Aspects*, **289**, 219.
37 Dufresne, A. and Vignon, M.R. (1998) *Macromolecules*, **31**, 2693.
38 Helbert, W., Chanzy, H., Husum, T.L. *et al.* (2003) *Biomacromolecules*, **2**, 481.
39 Yamamoto, H., Horii, F. and Hirai, A. (2006) *Cellulose*, **13**, 327.
40 Araki, J., Wada, N. and Kuga, S. (2001) *Langmuir*, **17**, 21.
41 Kvien, I., Tanem, B.S. and Oksman, K. (2005) *Biomacromolecules*, **6**, 3160.
42 Baker, A.A., Helbert, W., Sugiyama, J. and Miles, M.J. (1997) *Journal of Structural Biology*, **119**, 129.
43 Beck-Candanedo, S., Roman, M. and Gray, D.G. (2004) *Biomacromolecules*, **6**, 1048.
44 Roman, M. and Gray, D.G. (2005) *Langmuir*, **21**, 5555.
45 Edgar, C.D. and Gray, D.G. (2003) *Cellulose*, **4**, 299.
46 Eriksson, M., Notley, S.M. and Wågberg, L. (2007) *Biomacromolecules*, **8**, 912.
47 Larsson, P.T., Hult, E.-L., Wickholm, K. *et al.* (1999) *Solid State Nuclear Magnetic Resonance*, **15**, 31.
48 Heux, L., Dinand, E. and Vignon, M.R. (1999) *Carbohydrate Polymers*, **40**, 115.
49 Masuda, K., Adachi, M., Hirai, A. *et al.* (2003) *Solid State Nuclear Magnetic Resonance*, **23**, 198.
50 Montanari, S., Roumani, M., Heux, L. and Vignon, M.R. (2005) *Macromolecules*, **38**, 1665.
51 Sassi, J.-F. and Chanzy, H. (1995) *Cellulose*, **2**, 111.
52 Goussé, C., Chanzy, H., Ezcoffier, G. *et al.* (2002) *Polymer*, **43**, 2645.
53 Goussé, C., Chanzy, H., Cerrada, M.L. and Fleury, E. (2004) *Polymer*, **45**, 1569.
54 Zuluaga, R., Putaux, J.-L., Restrepo, A. *et al.* (2007) *Cellulose*, **14**, 585.
55 Bonini, C., Heux, L., Cavaillé, J.Y. *et al.* (2002) *Langmuir*, **18**, 3311.
56 Terech, P., Chazeau, L. and Cavaillé, j.Y. (1999) *Macromolecules*, **32**, 1872.
57 Geyer, U., Heinze, T., Stein A. and Klemm, D. (1994) *International Journal of Biological Macromolecules*, **16**, 343.
58 da Silva Perez, D., Montanari, S. and Vignon, M.-R. (2003) *Biomacromolecules*, **4**, 1417.
59 Habibi, Y., Chanzy, H. and Vignon, M.R. (2006) *Celllulose*, **13**, 679.
60 Yuan, H., Nishiyama, M., Wada, M. and Kuga, S. (2006) *Biomacromolecules*, **7**, 696.
61 Isogai, A. and Kato, Y. (1998) *Cellulose*, **5**, 153.
62 Stenius, P. and Vuorinen, T. (1999) in *Analytical Methods in Wood Chemistry, Pulping, and Papermaking* (eds E. Sjöström and R. Alén), Springer-Verlag, Berlin, p. 149.
63 Johansson, L.-S., Campbell, J.M., Fardim, P. *et al.* (2005) *Surface Science*, **584**, 126.
64 Andresen, M. and Stenius, P. (2007) *Journal of Dispersion Science and Technology*, **28**, 837.
65 Tanghe, L.J., Genung, L.B. and Mench, J.W. (1963) in *Methods in Carbohydrate Chemistry* (eds R.L. Whistler, J.W. Green, J.N. Bemiller and L. Wolfrom), Academic Press, New York, p. 201.
66 Heinze, T. and Liebert, T. (2001) *Progress in Polymer Science*, **26**, 1689.
67 Hamatsu, T. and Ito, M. (1995) *Kotingu Jiho*, **200**, 7.
68 de Nooy, A.E., Besemer, A.C. and van Bekkum, H. (1995) *Carbohydrate Research*, **269**, 89.
69 de Nooy, A.E., Besemer, A.C. and van Bekkum, H. (1996) *Synthesis*, 1153.
70 Chang, P.S. and Robyt, J.F. (1996) *Journal of Carbohydrate Chemistry*, **15**, 819.
71 Nge, T.T. and Sugijama, J. (2007) *Journal of Biomedical Materials Research*, **81**, 124.
72 USA US 6,703,497, **2004**, Cellulose microfibrils with modified surface, preparation method and use thereof,

Rhodia Chimie, Ladouce, L., Fleury, E., Gousse, C., Cantiani, R., Chanzy, H., Excoffier, G., 787272, 23 March, 2004.

73 Bruce, D.M., Hobson, R.N., Farrent, J.W. and Hepworth, D.G. (2005) *Composites Part A*, **26**, 1486.

74 Malainine, M.E., Mahrouz, M. and Dufresne, A. (2005) *Composites Science and Technology*, **65**, 1520.

75 López-Rubio, A., Lagaron, J.M., Ankerfors, M. *et al.* (2007) *Carbohydrate Polymers*, **68**, 718.

76 Dubief, D., Samain, E. and Dufresne, A. (1999) *Macromolecules*, **32**, 5765.

77 Ljungberg, N., Bonini, C., Bortolussi, F. *et al.* (2005) *Biomacromolecules*, **6**, 2732.

78 Orts, W.J., Shey, J., Imam, S.H. *et al.* (2005) *Journal of Polymers and the Environment*, **13**, 301.

79 Favier, V., Chanzy, H. and Cavaillé, J.Y. (1995) *Macromolecules*, **28**, 6365.

80 Iguchi, M., Yamanaka, S. and Budhiono, A. (2000) *Journal of Materials Science*, **35**, 261.

81 Fleming, K., Gray, D.G. and Matthews, S. (2001) *Chemistry - A European Journal*, **7**, 1831.

82 Oza, K.P. and Frank, S.G.J. (1986) *Journal of Dispersion Science and Technology*, **7**, 543.

83 Ougiya, H., Watanabe, K., Morinaga, Y. and Yoshinaga, F. (1997) *Bioscience, Biotechnology, and Biochemistry*, **61**, 1541.

84 USA US 6,534,072 B1, **2003**, Composition in the form of an oil-ion-water emulsion containing cellulose fibrils, and its uses, especially cosmetic uses, LÓreal, Tournilhac, F., Lorant, R. 09/573,175, March 18, 2003.

85 Khopade, A.J. and Jain, N.K. (1990) *Drug Development and Industrial Pharmacy*, **24**, 289.

86 Czaja, W.K., Young, D.J., Kawecki, M. and Brown, R.M. (2007) *Biomacromolecules*, **8**, 1.

9
Melting/Freezing Phase Transitions in Confined Systems
Ludmila Boinovich and Alexandre Emelyanenko

9.1
Introduction

The physical properties of materials in a confined state have attracted considerable attention both due to their fundamental significance and to their primary importance for nanotechnology. The term confined state embraces a wide variety of systems: the boundary layers at the interfaces between two bulk phases, the adsorption layers, wetting films, epitaxial structures, emulsions, free-lying nanoparticles and particles embedded in solid matrices, the substances condensed in pores, and so on. As a rule, the transition of a substance from the bulk state to the confined state is accompanied by an essential alteration of many physical properties. In particular, for practically all of the confined systems mentioned above, a shift of the first order phase transition temperature with respect to that in the bulk state was detected experimentally (see [1–4] for reviews). In nanoporous systems, not only a considerable (tens of degrees) depression of the freezing temperature can be observed, but the transition to solid state might even disappear. At the same time there are systems that demonstrate not a depression but an elevation of the solid/liquid phase transition temperature in pores. A similar situation occurs with small particles, adsorbed films and boundary layers at plane interfaces.

Usually, theoretical analysis of the shifts of phase transition temperatures is performed separately for different types of confinement on the bases of different models. Thus, several theoretical approaches have been developed to describe either surface freezing or surface melting, taking into account the peculiarities of the system under consideration. Tkachenko and Rabin [5] have developed the theory of surface freezing of normal alkanes, based on the notion that strong fluctuations along the molecular axis of uniaxially ordered stretched chains of alkanes provide sufficient entropy to stabilize the solid monolayer on a top of the liquid phase. This entropic mechanism is applicable to a class of linear chain molecules of intermediate length, but not to liquids with spherical or branched molecules. Chernov and Mikheev [6] have explained the surface melting, taking into account the exponentially decaying smectic ordering induced by interfaces in a layer of a hard sphere liquid. Such ordering results in an additional damped oscillatory contribution to the interfacial free energy. This additional term shifts the thermodynamic equilibrium, making the existence of a thin liquid layer on top of the bulk solid energetically favorable at temperatures below the bulk melting point. However, the results of above approach cannot be directly applied to electrolytes or liquids with non-spherical molecules.

An approach based on the analysis of the van der Waals contribution to the free energy of the system in the presence of melted layer at the solid surface has been applied to explain the premelting of ice [7]. Below 0 °C the action of van der Waals forces might provide energetic favor for an ice–water interlayer-substrate configuration in comparison to ice–substrate. The calculations performed in Ref. [7] on the basis of the Dzyaloshinsky, Lifshitz and Pitaevsky theory have shown that, depending on the substrate, the equilibrium thickness of a water interlayer might be either finite or diverge under approach to the ice melting temperature.

Many approaches have been developed for theoretical studies of melting/freezing phenomena in small particles, both free-lying and embedded in a matrix of another condensed substance. Seemingly, the first one was proposed by Pawlow [8], who had considered the equilibrium condition between a solid particle, a liquid particle having the same mass, and their saturated vapor. In essence, in Pawlow's treatment the solid/liquid phase transition is associated with the triple point, but at a definite ratio of the radii of solid and liquid particles. This approach, referred to in the modern literature as the homogeneous melting and growth model, was generalized later by Defay *et al.* [9] for the analysis of triple point in systems with arbitrary particle sizes. Another type of model, known as the liquid shell (or liquid-skin) model, assumes the existence of a liquid layer, surrounding the solid particle, and describes the equilibrium of such a system in the presence of a vapor phase [10]. The dependence of the thickness of such a premelted layer on the particle radius in the case of thin layers governed by short range interfacial forces was accounted for by Kofman *et al.* [11]. The inclusion of long-range interactions was given by Dash *et al.* [12].

The liquid nucleation and growth model [13–15] considers the melting of a particle as a kinetic process driven by a definite activation energy. The melting starts by the nucleation of a liquid layer at the solid–vapor interface and progresses into the core of particle. The vibrational model [16] is based on the Lindemann criterion for the analysis of size-dependent melting. The key parameter of the model is the ratio between the amplitude of the atomic thermal vibrations at the surface of nanocrystals and in the bulk crystal for free or embedded particles. The model predicts both the depression of melting points for free nanocrystals and the possibility of superheating the embedded crystals.

One of the first theoretical analyses of freezing/melting in thin pores was undertaken by Batchelor and Foster [17] on the basis of the Clausius–Clapeyron equation for solid and liquid states of substance inside the pores. From the geometric representation they derived an equation that, as shown later by Defay *et al.* [9], describes the shift of the triple point for the system. Since then this equation, as well as so-called Gibbs–Thomson equation – which is easily derived from Batchelor–Foster formula by assuming equality of solid and liquid densities and by replacing the difference of solid/vapor and liquid/vapor surface tensions with solid/liquid surface tension – long served as a basis for comparison between experimentally observed temperatures of melting/freezing transitions in pores and theoretical predictions. The generalized Landau–Ginzburg approach, where spatial inhomogeneity in the order parameter was incorporated [4], allows one to calculate the free energy surface of inhomogeneous fluid inside the pore. Using molecular simulations and free energy calculations, Radhakrishnan et al. [4] have shown that the freezing/melting behavior of fluids of small molecules in pores of simple geometry can be understood in terms of two main parameters: the dimensionless pore width and a parameter that measures the ratio of the fluid–wall to the fluid–fluid attractive interaction. They note that the value of the second parameter determines the qualitative nature of the freezing behavior (e.g. the direction of change in the freezing temperature and the presence or absence of new phases).

Up to this point, the theoretical approaches treated the origin of freezing and melting in various confined geometries as completely different. In this chapter we show that a unified thermodynamic approach based on Clausius–Clapeyron relations for the coexistence of solid and liquid phases jointly, with equations relating the vapor pressures above bulk and confined state of substance, can explain the shift of the temperature of triple point in various confined systems and define the key physico-chemical parameters determining this shift.

Here we consider the melting/freezing phase transitions in systems characterized by one of the following types of confinement:

1. Surface phase transitions at the plane interface between vapor phase and bulk solid or liquid phase;
2. Phase transitions at the surface and in the interior of small spherical particles;
3. Phase transitions in the substance condensed in a porous matrix.

We restrict ourselves, for clarity and simplicity, to the analysis of the systems where each phase can be considered as a single component.

Since in all the above-mentioned systems the phase transition occurs with the presence of the vapor phase, the problem of finding the melting/freezing point reduces to the determination of the triple point for each of the cases considered. The general approach to treating this problem is based on the Clausius–Clapeyron relations, describing the solid–vapor and liquid–vapor lines of equilibrium on the phase diagrams in the pressure versus temperature coordinates:

$$\frac{d \ln p^0_{(S)}}{dT} = \frac{L_{SV}}{RT^2} \quad (9.1)$$

and:

$$\frac{d \ln p^0_{(L)}}{dT} = \frac{L_{LV}}{RT^2} \quad (9.2)$$

where $p^0_{(S)}$ and $p^0_{(L)}$ are the pressures of vapor being in equilibrium with the bulk solid and the bulk liquid phase, respectively; T is the temperature; R is the gas constant; L_{SV}

and L_{LV} are the molar heats of sublimation and evaporation, respectively. Exploitation of these relations traditionally relies on three main assumptions. Firstly, in their derivation the vapor is considered as an ideal gas. Secondly, it is assumed that Equations 9.1 and 9.2 are also valid for metastable states, that is, for the overcooled liquid and the overheated solid. Finally, the temperature dependence of the heats of sublimation and evaporation is neglected.

From Equations 9.1 and 9.2 it follows that:

$$\frac{d \ln p^0_{(L)}/p^0_{(S)}}{dT} = -\frac{q}{RT^2} \tag{9.3}$$

where $q = L_{SV} - L_{LV}$ is the latent heat of melting of the bulk phase. If we integrate this equation from the state corresponding to the triple point of the bulk solid/liquid/vapor coexistence to some other state, associated to the triple point in the confined system, and assume that q is sensibly constant along this path, we obtain:

$$\ln\frac{p^0_{(L)}}{p^0_{(S)}}\bigg|_{T=T^c} - \ln\frac{p^0_{(L)}}{p^0_{(S)}}\bigg|_{T=T^0} = \frac{q}{RT^c T^0}(T^0 - T^c) \tag{9.4}$$

At the bulk triple point ($T = T^0$) we have $p^0_{(S)} = p^0_{(L)}$ and hence the second term in the left-hand side of Equation 9.4 vanishes. It is essential to emphasize here that $p^0_{(L)}$ and $p^0_{(S)}$ in the first term are the pressures of vapor being in the equilibrium with *the bulk liquid and solid phases* at the temperature T^c, corresponding to the triple point for the *confined system*. Thus, to determine the T^c on the basis of Equation 9.4, one needs to find the vapor pressures over the bulk solid and liquid phases just at that temperature, for which the freezing/melting is observed in the confined system. Having in mind that the triple point for the confined system is characterized by the equality of vapor pressures, $p^c_{(S)} = p^c_{(L)}$, for vapors in equilibrium with the confined solid and the confined liquid, respectively, the problem is reduced to establishing the correspondence at a given temperature between the equilibrium vapor pressures for the bulk state and the confined state of both the solid and the liquid.

9.2
Surface Phase Transitions at the Plane Interface

We consider first an analysis of the case of a confinement due to the plane interface. This might be a boundary between a solid and a vapor, between a liquid and a vapor, and between a liquid and a wall, where the wall can be either solid or liquid. Following the classical studies of Guggenheim [18], Butler [19] and Zhukhovitskii [20], we will consider a surface layer at the interface as a boundary phase. The external borders of this boundary phase separate the non-uniform (in the general case) boundary phase from the uniform bulk phases I and II. According to the Gibbs integral equation, the excess free energy, U^E, per unit of interfacial area of this phase can be written as follows:

$$U^E = TS^E + \gamma + \Gamma\mu \tag{9.5}$$

9.2 Surface Phase Transitions at the Plane Interface

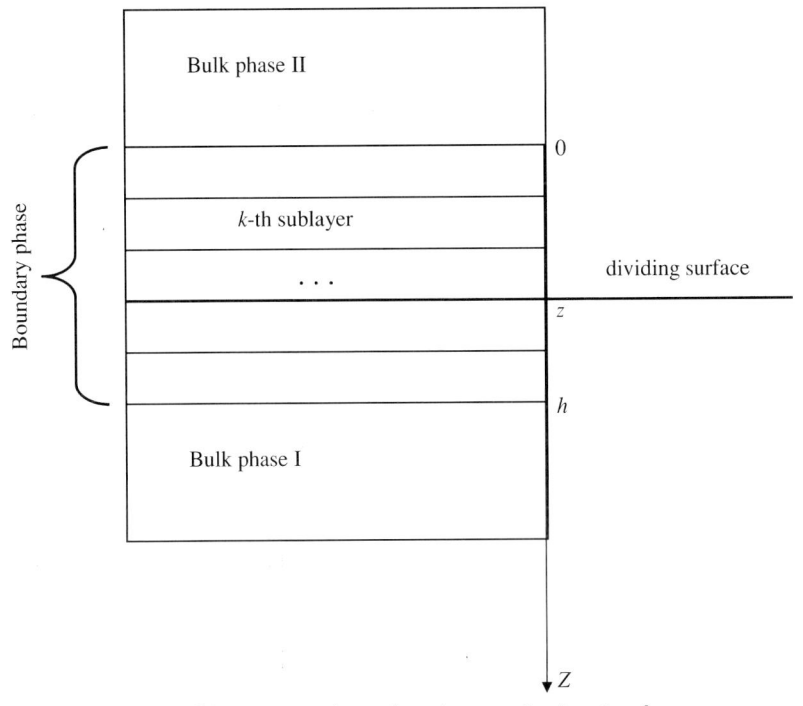

Figure 9.1 Scheme of the system under study in the case of a plane interface.

where S^E is the excess entropy per unit of interfacial area, Γ is the surface excess and μ is the molar chemical potential, γ is the energy of formation of unit of interfacial area. All excesses are to be defined with respect to a definite position of dividing surface. We will choose this position within the boundary phase (Figure 9.1). Note that for the case of the curved interface, which we will consider in the next section, it is reasonable to choose the dividing surface coinciding with the surface of tension, for which the variation of interfacial tension associated with pure mathematical shift of the dividing surface (i.e. provided that all the physical parameters inside the system as well as the external conditions remain constant under this shift) turns out to be zero [21].

In choosing the dividing surface indicated in Figure 9.1 the following identities can be stated:

$$U^E = U_\alpha - U_{II}\frac{z}{h} - U_I\frac{h-z}{h} \tag{9.6}$$

$$S^E = S_\alpha - S_{II}\frac{z}{h} - S_I\frac{h-z}{h} \tag{9.7}$$

$$\Gamma = n_\alpha - n_{II}\frac{z}{h} - n_I\frac{h-z}{h} \tag{9.8}$$

where U_α, S_α and n_α are the internal energy, entropy and number of moles of component per unit of area of the boundary phase, U_I, U_{II}, S_I, S_{II} and n_I, n_{II} are the corresponding quantities for the equivalent volume of bulk phases I and II, h is the thickness of boundary phase and z is the distance between the dividing surface and the border, separating the boundary phase from the bulk phase II.

Taking into account that G_I and G_{II}, the Gibbs free energies of the equivalent volumes of bulk phases per unit of interfacial area, might be expressed as:

$$\frac{h-z}{h}G_I = \frac{h-z}{h}n_I\mu = \frac{h-z}{h}(U_I - TS_I) + P(h-z) \tag{9.9}$$

$$\frac{z}{h}G_{II} = \frac{z}{h}n_{II}\mu = \frac{z}{h}(U_{II} - TS_{II}) + Pz \tag{9.10}$$

and combining Equations 9.5–9.10 we get:

$$\gamma = U_\alpha - TS_\alpha + Ph - n_\alpha\mu \tag{9.11}$$

Equation 9.11 for γ, as would be expected, does not depend on the choice of the position z of the dividing surface, and in particular coincides with the equation derived in Ref. [22], where the dividing surface was chosen at one of the external borders of phase α.

The Gibbs free energy of boundary phase per unit of interfacial area is:

$$G_\alpha = n_\alpha\mu \tag{9.12}$$

Notably, constancy of the chemical potentials of the component in different regions of the compositionally and structurally inhomogeneous boundary phase α is maintained due to the presence of a non-uniform external field (here, this role is played by the macroscopic field of surface forces). This field is uniform in sections parallel to the phase boundaries and decays inside each of confining phases along the normal to the interface.

If one virtually released the boundary phase from the action of surface forces field but preserved the composition and structure of this phase, then for the Gibbs free energy (G'_α) of such a virtual phase the following relation would be valid:

$$G'_\alpha = U_\alpha - TS_\alpha + Ph = \int_h \rho_\alpha(z)\lambda(z)dz \tag{9.13}$$

where the integration is performed over the whole thickness, h, of the boundary phase, $\rho_\alpha(z)$ is the molar density of this phase, and $\lambda(z)$ is a positionally dependent (owing to structural non-uniformity) molar chemical potential of the component in the boundary phase released from the field of surface forces. For brevity, hereinafter we refer to this phase as the reference phase.

Dividing the reference phase into sublayers, within each of which $\lambda(z)$ might be assumed as being constant, one can rewrite Equation 9.13 as:

$$G'_\alpha = \sum_k n_k\lambda_k \tag{9.14}$$

where k counts the sublayers and n_k is the number of moles of component per unit area of the k-th sublayer.

With allowance for Equation 9.14, Equation 9.11 takes the form:

$$\gamma = \sum_k n_k(\lambda_k - \mu) = \sum_k \gamma_k \tag{9.15}$$

where γ_k is the contribution of the corresponding sublayer of the boundary phase to the energy of formation of the unit area of the interface. Note that the generalization of Equation 9.15 for the case of multicomponent systems is straightforward and was presented in Ref. [22], leading to:

$$\gamma = \sum_{i,k} n_{ik}(\lambda_{ik} - \mu_i) = \sum_k \gamma_k \tag{9.16}$$

where index i characterizes the quantities for i-th component of the mixture.

Evidently, owing to the fast decay of the surface forces field, $\lambda_{ik} \to \mu_i$ for large k and a limited number of sublayers contribute to the total surface free energy. Note that because of the non-equilibrium state of the reference phase one always has $\lambda_{ik} > \mu_i$ and all of the summands in Equations 9.15 and 9.16 is non-negative.

For further analysis of phase transitions in different sublayers of the boundary phase it is necessary to mention the following circumstance. Because the reference phase was chosen in such a manner that its composition and structure in each layer at all temperatures is the same as the composition and structure of the corresponding layer in the real boundary phase, the transition temperature for the k-th sublayer of the reference phase and for the corresponding sublayer of the phase α are the same. This is a consequence of an unambiguous relation between the phase state of a substance and its temperature and density at a given pressure. Thus, on the one hand, to study a phase transition in the k-th sublayer of the real boundary phase it is sufficient to study the phase transition in the corresponding sublayer of the reference phase, considering the latter as a region of a certain bulk phase. On the other hand, the inhomogeneous structure of the boundary phase leads to a sequence of phase transitions as the temperature of the system approaches the bulk transition temperature and the structure of the boundary sublayers approaches the structure of the bulk phase.

Let us consider now the pressures, p^0, of vapor, in equilibrium with the condensed phase I, on the one hand, and p_k^α, in equilibrium with the reference bulk phase, having the same structure as the k-th sublayer of boundary phase α, on the other hand. Assuming the vapor phase as an ideal gas, one can write:

$$\mu = \bar{\mu} + RT \ln p^0 \tag{9.17}$$

$$\lambda = \bar{\mu} + RT \ln p_k^\alpha \tag{9.18}$$

where $\bar{\mu}$ is the temperature-dependent standard chemical potential independent of the state of the component under study [18]. Equations 9.15, 9.17 and 9.18 lead to the relation:

$$p_k^\alpha = p^0 \exp\frac{\lambda_k - \mu}{RT} = p^0 \exp\frac{\gamma_k}{n_k RT} \tag{9.19}$$

Thus, as follows from Equation 9.19 and by virtue of the non-negative values of γ_k, the equilibrium vapor pressure above the bulk phase, having the structure of the k-th sublayer of boundary phase, should not be less than, and for sublayers with $\gamma_k > 0$ should be greater than, the vapor pressure above the parent phase I. Equation 9.19 is in fact the equation of state for the k-th surface sublayer and can be written for both the liquid (L) and the solid (S) states of this sublayer:

$$p_{k(L)}^\alpha = p_{(L)}^0 \exp \frac{\gamma'_{k(L)}}{n_{k(L)} RT} \tag{9.20}$$

$$p_{k(S)}^\alpha = p_{(S)}^0 \exp \frac{\gamma'_{k(S)}}{n_{k(S)} RT} \tag{9.21}$$

Notably, $\gamma'_{k(S)}$ (or $\gamma'_{k(L)}$) have the sense of the contributions to the surface energy given by the k-th sublayer for the solid (liquid) state of both bulk and boundary phase and should not be confused with the case of a frozen layer on the top of the bulk liquid characterized by $\gamma_{(L)}$ [or melted layer on the top of the bulk solid characterized by $\gamma_{(S)}$]. Lines of state in P–T plane described by Equations 9.20 and 9.21 intersect at the point $P = p_L^\alpha = p_S^\alpha$, $T = T_k^c$, corresponding to the triple point for the k-th sublayer of boundary phase. Now we can deduce from Equations 9.20 and 9.21 the value of:

$$\ln \frac{p_{(L)}^0}{p_{(S)}^0} \bigg|_{T=T^c}$$

which needs to be substituted into Equation 9.4 to calculate the shift of the triple point for the boundary phase:

$$\Delta T_k = T^0 - T_k^c = \frac{T^0}{q} \left[\frac{\gamma'_{k(S)}}{n_{k(S)}} - \frac{\gamma'_{k(L)}}{n_{k(L)}} \right]_{T=T_k^c} \tag{9.22}$$

Taking into account Equation 9.15, one can write:

$$\Delta T_k = \frac{T^0}{q} \left[\frac{\gamma'_{(S)}}{n_{k(S)}} - \frac{\gamma'_{(L)}}{n_{k(L)}} - \sum_{j \neq k} \frac{\gamma'_{j(S)}}{n_{k(S)}} - \frac{\gamma'_{j(L)}}{n_{k(L)}} \right]_{T=T_k^c} \tag{9.23}$$

where $\gamma'_{(S)}$ and $\gamma'_{(L)}$ are the surface energies at $T = T_k^c$ for the solid (liquid) state of both bulk and boundary phase. Equations 9.22 and 9.23 might be applied to describe the shift of the triple point for the boundary phases adjacent to vapor, or to a solid or liquid wall. In each case one should substitute in place of $\gamma'_{(S)}$ ($\gamma'_{(L)}$) the appropriate value of the energy of interface formation.

From Equations 9.22 and 9.23 it follows that in general case for the boundary phase at the plane surface one might expect either premelting or prefreezing according to the relation between $\gamma'_{(S)}/n_S$ and $\gamma'_{(L)}/n_L$. Moreover, these equations predict that the shift of the triple point temperature might be experimentally detectable either for the topmost surface monolayer or for the sequence of sublayers, depending on the character of decaying of the surface forces field, which affects the extent of the boundary phase. Indeed, the numerous experimental data collected to date provide examples of various systems that exhibit a decreased or increased triple point temperature for either single monolayer or several sequentially freezing (melting) sublayers at the plane interface.

For example, the (1 1 0) faces of Pb and Al demonstrate premelting [23–25], where the thickness of melted layer diverges as the bulk melting point is approached on heating. At the same time the (1 1 1) faces of these metals exhibit nonmelting at temperatures above the bulk melting point (i.e. prefreezing) [26, 27]. Long-chain alkanes, confined by air [28, 29], graphite [30] or surfactant solution on top of a water substrate [31, 32] also display prefreezing.

Prefreezing in the boundary layers of alcohols adjacent to graphon was discovered in the 1970s by Findenegg [33, 34] and reconfirmed in more recent experiments with alcohols confined by highly oriented pyrolytic graphite [35]. In these experiments up to three sequential phase transitions were reliably detected at temperatures above the bulk triple point.

To illustrate the applicability of Equation 9.22 for quantitative analysis we begin with the simplest case, when the surface forces field fades into the bulk phase very sharply, and the thickness of the boundary phase is restricted within the height of one monolayer. Experimental data show that such a case is realized for many different systems [28, 29, 36, 37]. For a monolayer boundary phase Equation 9.22 takes the form:

$$\Delta T_1 = \frac{T^0}{q} \left[\frac{\gamma'_{(S)}}{n_{\alpha(S)}} - \frac{\gamma'_{(L)}}{n_{\alpha(L)}} \right]_{T=T_1^c} \tag{9.24}$$

It follows from Equation 9.24 that the prefreezing phenomenon, namely, the shift of the triple point for the monolayer towards the higher temperatures, will certainly take place under the conditions:

$$\gamma'_S < \gamma'_L \qquad n_{\alpha(S)} > n_{\alpha(L)} \tag{9.25}$$

The latter inequality in Equation 9.25 means that the area per molecule in the liquid monolayer is higher than that in the solid one. This situation is realized, for instance, for the interface between normal alkanes and air. If we suppose that at the bulk freezing temperature the surface energy of the solid bulk phase is equal to the surface energy of the liquid phase covered by a frozen monolayer, then, as follows from Ref. [28], for alkanes with the number of carbon atoms per molecule (n) in the range $16 \leq n \leq 50$ the condition $\gamma'_S < \gamma'_L$ is fulfilled. It is also known that the melting of bulk alkanes is accompanied by increasing molecular volume [38]. Further, SFG spectroscopy data for normal alkanes and alcohols [39] indicate the similarity of structure of liquid and solid surface monolayers. Although in both the solid and the liquid state of a monolayer the alkyl chains of molecules tend to be oriented normally to the interface, the liquid monolayer is characterized by a larger fraction of gauche conformations, while in the solid monolayer the all-trans conformations of molecules dominate. All these facts strongly suggest that the molar volume of an alkane in the surface monolayer increases upon melting. According to data from Ref. [28], we took

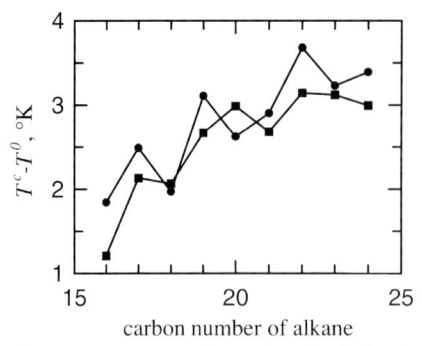

Figure 9.2 Experimental (■) [28] and calculated (●), on the basis of Equation 9.24, shifts of surface freezing temperature for alkanes with different numbers of carbon atoms per molecule.

the molecular area for alkanes in the rotator phase, which is characteristic for a frozen surface monolayer, to be $0.195 \, \text{nm}^2$ per molecule, and the density of the liquid monolayer to be 20% higher than in bulk liquid. Taking into account the relation between molar volume, v, monolayer thickness, h, and the number, n, of molecules per unit of interfacial area in the form $n = h/v$, and using the molar volume data for liquid bulk alkanes [38, 40, 41], we evaluated the shift of the triple point for surface monolayers on the basis of Equation 9.24. Data for the thicknesses of surface monolayer, which were assumed to be independent of the phase state of the latter, were taken from Ref. [28]. The results of calculations are presented in Figure 9.2 along with the experimental data of Ocko *et al.* [28].

The accordance between experimental and calculated data in Figure 9.2 is rather satisfactory. At the same time, Figure 9.2 clearly shows the tendency of ΔT_1 values, calculated on the basis of Equation 9.24, to overestimate the experimentally measured shifts of surface freezing temperature. Evidently, this results from using the values of n, evaluated for the bulk solid (rotator phase) and liquid states, instead of the unavailable corresponding characteristics for surface monolayers.

Another case of unambiguous prediction of the shift of the triple point for a surface monolayer is realized under the simultaneous fulfillment of conditions:

$$\gamma'_S > \gamma'_L \qquad n_{\alpha(S)} < n_{\alpha(L)} \tag{9.26}$$

In systems satisfying the inequalities (Equation 9.26), the triple point for the surface monolayer will shift towards lower temperatures, that is surface melting will take place. The interface between a germanium (1 1 1) face and air is an example of such systems. The melting of bulk Ge is accompanied by a 5% decrease in molar volume [42], with $v_S = 1.36 \times 10^{-5} \, \text{m}^3 \, \text{mol}^{-1}$ and $v_L = 1.296 \times 10^{-5} \, \text{m}^3 \, \text{mol}^{-1}$. Taking into account the distance between (1 1 1) crystallographic planes, $d_{111} = 3.27 \times 10^{-10} \, \text{m}$, it is straightforward to obtain the number of atoms in the solid (1 1 1) monolayer per unit area, $n_{S(111)} = 1.447 \times 10^{19} \, \text{m}^{-2}$. To evaluate the number of atoms per unit of area in the liquid state of a monolayer, the following arguments were taken into account. It is known that liquids in the vicinity of confining walls exhibit a tendency to smectic-like ordering, or layering [43]. Such layering has been observed experimentally for many nonmetal and some metal systems [44–46]. One of the characteristics of layering in the boundary layers of liquids is the oscillating character of interaction forces between solid surfaces through the liquid interlayers (solvation forces) [47]. Numerous experimental studies of the oscillating forces show that, for atomic or molecular liquids with nearly spherical molecules, the period of force oscillations is close to the atomic or molecular diameter. It is therefore not unreasonable to suppose that the thickness of a surface monolayer in liquid Ge is equal to double its atomic radius, 3.04×10^{-10} m. From this, we estimate the number of Ge atoms per unit of area in the liquid state of monolayer to be $n_L = 1.412 \times 10^{-19}$ m^{-2}.

To compute the shift of the melting temperature for the surface monolayer according to Equation 9.24, it is necessary to know also the surface energies for liquid and solid states of bulk Ge at the temperature of the surface phase transition. We were unable to find precise literature data on γ_S for the (1 1 1) face of Ge. However, it might be evaluated on the basis of the Young equation, $\gamma_{SV} = \gamma_{LV} \cos \theta + \gamma_{SL}$, by using $\gamma_{LV} = 0.63$ J m^{-2} [48] and $\gamma_{SL} = 0.24$ J m^{-2} [37], values for liquid/vapor and solid/liquid interfaces, respectively, and the contact angle, $\theta = 30°$ [44], for a drop of liquid Ge on Ge(1 1 1). Note that the value of γ_{SV} (0.79 J m^{-2}) obtained on this basis should be considered as an underscore of γ_S, since γ_{SV} corresponds to the equilibrium surface energy of the Ge/vapor interface covered by a wetting film of liquid Ge, whereas γ_S is related to the dry Ge surface. With the bulk melting temperature of Ge, $T^0 = 1210$ K, and the latent heat of fusion, $q = 36.9$ kJ mol^{-1} [49], from Equation 9.24 we obtain $\Delta T_1 = 190$ K. This gives the melting temperature of a surface monolayer as $T_1^c = 1020$ K, which is in a reasonable agreement with the experimentally observed temperature, $T \sim 1050$ K, of surface melting on a Ge(1 1 1) face [37].

The water/air interface is another well known example of systems for which the fulfillment of the inequalities (Equation 9.26) [49, 50] should provide the premelting behavior. Indeed, numerous experimental methods reveal the existence of a liquid water layer on the surface of ice well below 0 °C [51–53]. However, for the ice the formation of a polymolecular boundary liquid phase is characteristic. Convincing evidence of the structural non-uniformity of the boundary layer, decaying continuously from the ice surface into the bulk phase, was provided by different experimental techniques (see Ref. [53] and references therein) as well as by molecular dynamic studies [54]. Therefore, for a quantitative description of the shift in temperature of the phase transitions in boundary sublayers Equation 9.23 should be applied. Quantitative comparison of the theoretical calculations with the experimental results is hindered by lack of knowledge about $\gamma_{k(S)}$, $\gamma_{k(L)}$, on the one hand, and by the fact that the experiments performed in different research groups strongly differ from each other in terms of the temperature dependence of melted layer thickness, on the other hand. Nevertheless, qualitative analysis, based on the Equation 9.23, predicts dissimilar melting behavior of boundary phases adjacent to different crystal phases of the ice, in agreement with the results of glancing angle X-ray scattering experiments [55] demonstrating different

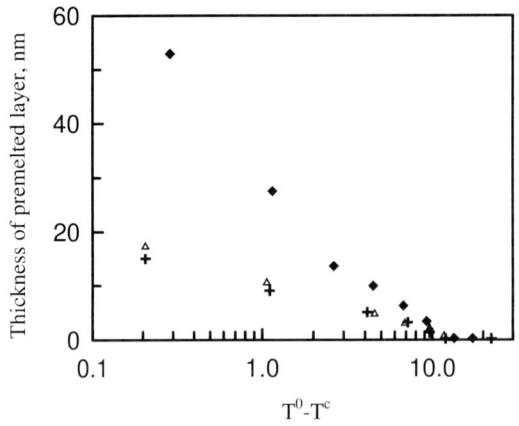

Figure 9.3 Thickness of premelted surface layer versus $T^0 - T^c$ for (♦) basal (00.1), (Δ) (11.0) and (+) (10.0) faces of ice I_h [55].

temperature dependences of the layer thickness on the (00.1), (10.0) and (11.0) surfaces of an ice I_h single crystal (Figure 9.3).

9.3
Confinement by Curved Interfaces

Now let us consider curved interfaces. As mentioned in Section 9.1, to calculate the shift of the triple point in a system containing phases confined by curved interfaces, compared to the triple point for the bulk phases, on the basis of Equation 9.4 one needs to establish a relation between the pressure of vapors in equilibrium with bulk condensed phases and the corresponding phases confined by curved interfaces. To determine the vapor pressure p^c in equilibrium with the condensed phase under the curved interface, it is necessary to take into account two physical reasons for the deviation of this pressure from the pressure above a plane interface (p^0). These are the increasing (decreasing) pressure inside a condensed phase under a convex (concave) meniscus and the surface forces-induced non-uniformity of structure of the condensed media near the interface. We will analyze the influence of above factors separately and show that they additively contribute to the shift of the triple point of boundary layers.

As a first step we consider the part of the condensed media confined by the curved interface that is far from the interface and can be considered as uniform.

Since at the triple point we can always consider the boundary between the vapor phase and each of the condensed phases (solid and liquid), we will use the Laplace equation to relate the pressure in vapor phase P_V and the pressure in liquid phase P_L:

$$P_V - P_L = P_{capil} = \gamma_{LV} K \qquad (9.27)$$

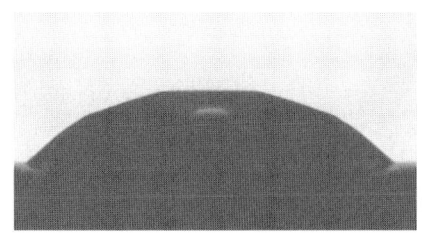

Figure 9.4 A crystal of $C_{12}H_{25}OH$ obtained by solidification of a sessile drop on a mica substrate.

where P_{capil} is the capillary pressure, γ_{LV} is the surface energy of liquid, $K = \pm (1/r_1) \pm (1/r_2)$ is the curvature of the interface, where r_1 and r_2 are the two principal radii of curvature; the positive sign is taken for the case when the center of curvature lies in the vapor phase and negative when this center is in the liquid phase.

As with the relation between the pressure in vapor phase and the pressure in solid polycrystalline phase P_S, it was proven in Ref. [9] that the equation similar to Equation 9.27 is valid for that case, with the surface energy of solid γ_{SV} substituted in place of γ_{LV}. Finally, in the case of a faceted single crystal (Figure 9.4) the difference in pressures in two phases is determined by [9]:

$$P_V - P_S = 2\gamma_{SV}^{(i)}/r^{(i)} = \gamma_{SV}^{(i)} K^{(i)} \tag{9.28}$$

where $\gamma_{SV}^{(i)}$ is the surface energy of i-th face of the crystal and $r^{(i)}$ is the distance between that face and the point O, chosen in such a way that for all faces of the crystal the Wulff's relation:

$$\gamma_{SV}^{(1)}/r^{(1)} = \gamma_{SV}^{(2)}/r^{(2)} \ldots = \gamma_{SV}^{(i)}/r^{(i)} = \text{const} \tag{9.29}$$

is fulfilled.

Then, for the pressure of vapor in equilibrium with a condensed phase confined by a curved interface the Kelvin equation is applicable:

$$\begin{aligned} p_{(L)}^c &= p_{(L)}^0 \exp\left(\frac{-v_L \gamma_{LV} K}{RT}\right) \quad \text{for liquid} \\ p_{(S)}^c &= p_{(S)}^0 \exp\left(\frac{-v_S \gamma_{SV} K}{RT}\right) \quad \text{for polycrystalline solid} \\ p_{(S)}^c &= p_{(S)}^0 \exp\left(\frac{-v_S \gamma_{SV}^{(i)} K^{(i)}}{RT}\right) \quad \text{for faceted solid} \end{aligned} \tag{9.30}$$

where v_L and v_S are the molar volumes of liquid and solid phases, respectively. The fundamental reasons for the similarity of vapor behavior in equilibrium with liquids and solids is thoroughly discussed in the classical monograph of Defay et al. [9], where the significance of the concept of the surface energy of a crystal face and the precise meaning of crystal curvature are also considered. Notably, the logic of derivation of Equation 9.30 does not infer any limitation on the detailed structure of interface and the size of confined phase, provided (i) we are still within the range of validity of the thermodynamics and (ii) the confined phase contains the uniform part exhibiting the bulk properties. That is to say Equation 9.30 is equally applicable for small particles if one replaces γ by curvature-dependent $\gamma(K)$. Moreover, when at any temperature the curved interface contains the premelted (prefreezed) sublayers of boundary phase, it is necessary to distinguish the surface energy of such an interface, γ, from that, γ', for the case when at the same temperature both the bulk and the boundary phases are solid (liquid).

Now, if at any temperature T_c the solid phase confined by the interface with the effective curvature K_S is in the equilibrium with the liquid phase confined by the interface with the curvature K_L, the equality $p^c_{(S)} = p^c_{(L)}$ should be held. By equating the right-hand sides of Equation 9.30 we obtain:

$$\left.\frac{p^0_{(L)}}{p^0_{(S)}}\right|_{T=T^c} = \exp\frac{1}{RT^c}(v_L\gamma_{LV}K_{LV} - v_S\gamma_{SV}K_{SV}) \qquad (9.31)$$

The next step is related to the consideration of the equilibrium vapor pressure $p^{\alpha,c}_k$ above the reference bulk phase, having the structure of k-th sublayer of solid or liquid boundary phase under the curved interface. Following the way outlined in Section 9.2 for derivation of Equation 9.19, it is easy to obtain:

$$p^{\alpha,c}_k = p^c \exp\frac{\lambda^c_k - \mu^c}{RT} = p^c \exp\frac{\gamma'^c_k}{n^c_k RT} \qquad (9.32)$$

where μ^c, λ^c_k correspond to the chemical potentials of components in the uniform part of condensed media confined by curved interface and in the k-th sublayer of boundary phase, released from the field of surface forces, respectively, γ'^c_k takes into account the influence of curvature on the surface energy and sensibly differs from the corresponding value of γ'_k if the radius of curvature approaches 1 nm.

Equations 9.30 and 9.32 provide the relation between the equilibrium vapor pressures above the bulk phase with plane interface and the bulk phase, having the structure of the k-th sublayer of curved boundary phase for both the liquid and the solid states:

$$p^{\alpha,c}_{k(L)} = p^0_{(L)}\exp\frac{1}{RT}\left(\frac{\gamma'^c_{k(L)}}{n^c_{k(L)}} - v_L\gamma_{LV}K_{LV}\right) \qquad (9.33)$$

$$p^{\alpha,c}_{k(S)} = p^0_{(S)}\exp\frac{1}{RT}\left(\frac{\gamma'^c_{k(S)}}{n^c_{k(S)}} - v_S\gamma_{SV}K_{SV}\right) \qquad (9.34)$$

Now we can deduce from Equations 9.33 and 9.34 the value of $\left.\frac{p^0_{(L)}}{p^0_{(S)}}\right|_{T=T^{\alpha,c}}$

which after substitution into Equation 9.4 gives the shift of the triple point for the curved boundary phase:

$$T^0 - T^{\alpha,c}_k = \frac{T^0}{q}\left[\left(\frac{\gamma'^c_{k(S)}}{n^c_{k(S)}} - \frac{\gamma'^c_{k(L)}}{n^c_{k(L)}}\right) + (v_L\gamma_{LV}K_{LV} - v_S\gamma_{SV}K_{SV})\right]_{T=T^{\alpha,c}_k} \qquad (9.35)$$

whereas from Equations 9.31 and 9.4 we obtain:

$$T^0 - T^c = \frac{T^0}{q}[(\nu_L \gamma_{LV} K_{LV} - \nu_S \gamma_{SV} K_{SV})]_{T=T^c} \quad (9.36)$$

for the shift of the triple point for the uniform phase confined by a curved interface.

Before proceeding to further analysis of various systems containing curved interfaces, it is worth emphasizing the following. Firstly, as follows from the above analysis, all the quantities in Equations 9.35 and 9.36 except q should be taken at the corresponding triple point temperature. Secondly, the derivation based on the Clausius–Clapeyron equation considers the latent heat of fusion q as independent of the temperature. This assumption is not very important when the shift of the triple points is of the order of a few degrees. However, these shifts might reach as much as tens or even hundreds of degrees [56]. At the same time the variation of the latent heat of fusion with temperature is essential. For example, the heat of water freezing at $-50\,^\circ$C is 40% lower than at the bulk triple point [57]. To take into account such a decrease of q, Defay et al. [9] proposing the use of the value corresponding to the average temperature in the interval between T^0 and T^c.

Analysis of Equations 9.35 and 9.36 indicates that the shift of the triple point for different confinements of the same material should be very sensitive to the magnitude and the character of the interface curvature. In fact, wires and spherical particles of the same radius will demonstrate different melting points. At first glance it seems that for wires the shift of the triple point should be twice lower that for spheres due to the twice lower curvature. But this estimate is quite rough. On one hand, it is related to the fact that we do not take into account the above-mentioned difference of the average heat of fusion for different depression/elevation of triple point temperature. On the other hand, the temperature variations of real solid/vapor and liquid/vapor interfacial tensions are also usually missing in such estimations, based on the accounting of the change in the curvature only. The results of molecular dynamics simulation of premelting of Pd [58] and Pb [59] nanoparticles and nanowires indeed show that the ratio $(T^0 - T^{\text{particle}})/(T^0 - T^{\text{wire}})$ deviates from 2.

We can now apply the general relations Equations 9.35 and 9.36 to analysis of particular classes of systems that contain curved interfaces.

9.3.1
Phase Transitions at the Surface and in the Interior of Small Particles

For small particles one can neglect the influence of gravity on their equilibrium shapes and consider the particles as being spherical for liquid and spherical or faceted when solid. Note that in this section we consider particles that contain at their core a uniform part that is not influenced by surface forces. According to the convention for the sign of curvature (see previous section) for a spherical particle $K = -2/r$, where r is the particle radius. Then Equation 9.36 for the shift of the triple point for the equilibrium between the vapor, liquid and solid spherical particles transforms into:

$$\Delta T^c = T^0 - T^c = \frac{2T^0}{q} \left(\frac{v_S \gamma_{SV}}{r_{SV}} - \frac{v_L \gamma_{LV}}{r_{LV}} \right)_{T=T^c} \quad (9.37)$$

and Equation 9.35 for the k-th sublayer of the boundary phase on spherical particle into:

$$\Delta T_k^{\alpha,c} = T^0 - T_k^{\alpha,c} = \frac{T^0}{q} \left[\left(\frac{\gamma_{k}'^c(SV)}{n_k^c(SV)} - \frac{\gamma_{k}'^c(LV)}{n_k^c(LV)} \right) + \left(\frac{2v_S \gamma_{SV}}{r_{SV}} - \frac{2v_L \gamma_{LV}}{r_{LV}} \right) \right]_{T=T_k^{\alpha,c}} \quad (9.38)$$

For the case of a faceted single crystal, Wulff's shape should be taken into account and γ_{SV}/r_{SV} in Equations 9.37 and 9.38 will be replaced with $\gamma_{SV}^{(i)}/r_{SV}^{(i)}$.

If one requires the masses of solid and liquid particles to be equal, then from Equation 9.37 follows Pawlow's formula [8], associated with the homogeneous melting model:

$$T^0 - T^c = \frac{2T^0}{q r_{SV} \rho_S} \left[\gamma_{SV} - \gamma_{LV} \left(\frac{\rho_S}{\rho_L} \right)^{2/3} \right] \quad (9.39)$$

where ρ_S and ρ_L are the molar densities of solid and liquid, respectively.

One of the most important points to be discussed in this section is the mutual influence of the bulk and the boundary part of the medium confined by the curved interface on their melting behavior. From Equation 9.38 it follows that the shift of the triple point temperature for the sublayers of a curved boundary phase will differ from that for the case of plane interface (as described by Equation 9.22), first of all due to the effect of curvature described by the terms in the second set of brackets (within []). This shift, in turn, affects the values of $\gamma_k'^c$ and n_k^c (because of their temperature dependence), which should be substituted into the first set of brackets and the magnitude of the effective latent heat of fusion (see the discussion after Equation 9.36). As a rule, the terms in the first and in the second sets of brackets act in the same direction (while it is not necessary for the general case). This synergetic action makes the sequential phase transitions in boundary sublayers more pronounced and more separated on the temperature scale compared to the case of plane interfaces. Therefore, the phase transitions in deeper boundary sublayers become experimentally detectable. This effect is revealed in most experimental methods as an apparent increase in the thickness of a "skin" layer (melted layer in the case of premelting) for the curved interfaces [11, 60].

Another aspect of the mutual influence of the bulk and the boundary part of the medium confined by the curved interface on their melting behavior consists in the effect of the existence of premelted (prefrozen) sublayers on the shift of the triple point for the rest of the particle in comparison to T^0. Equations 9.37 and 9.39 already account for this effect if one correctly uses the value of γ_{SV} for solid particle having melted sublayers in the case of premelting or γ_{LV} for liquid particle having

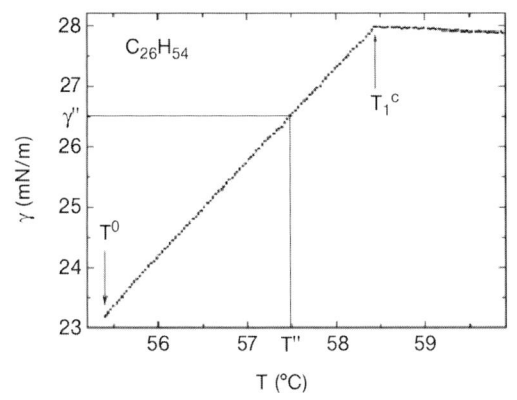

Figure 9.5 Temperature dependence of the surface tension of hexacosane [32], exhibiting prefreezing of the surface monolayer at T_1^c. T^0 indicates the bulk freezing point. Temperatures in the interval between T^0 and T_1^c correspond to the solid monolayer on the top of bulk liquid phase. If, for example, the freezing of liquid droplet occurs at some temperature T'', then the value of γ'' should be substituted in Equation 9.37 as γ_{SV}.

frozen sublayers in the case of prefreezing (see Figure 9.5 for an example). The approach developed here is valid when a melted (frozen) layer at the interface is a few molecular diameters thick. The consideration of such a melted layer as a liquid film bearing the properties of bulk liquid and characterized by γ_{LV} and γ_{SL} on two sides of the film, as it is assumed in the liquid shell model [11, 61] is problematic. In contrast, the concept of the non-uniform boundary phase allows one to describe the shift of triple point for small particles on the basis of equations similar to those derived within the framework of the homogeneous melting and growth model [8, 10], and, in addition, naturally includes the effects related to the phase transitions in boundary sublayers.

Importantly, Equations 9.37 and 9.39 are usually considered as predicting the linear dependence of the shift of the melting point on the reciprocal particle radius. Indeed, experimental data for several systems (see, for example, Refs. [62, 63]) demonstrate such a linear trend. At the same time the deviation of experimental data from linear dependence is generally treated as indicative of a failure of the homogeneous melting and growth model and of a necessity to invoke the liquid shell model. However, as is clear from the previous analysis, to describe quantitatively the increase of the absolute value of ΔT^c accompanying the decreasing of particle size one should take into account the temperature variation of both the effective latent heat of fusion and the surface energies. Besides, very small particles cannot be considered as having a uniform bulk core. With all this in mind, it is necessary to note that the linearity of ΔT^c versus $1/r$ for a wide range of temperatures is a lucky coincidence, due to mutual compensation of above-mentioned variations, rather than a rule.

To conclude this section, it is necessary to emphasize that Equations 9.37 and 9.38 allow to estimate reliably the order of magnitude for ΔT^c or $\Delta T_k^{\alpha,c}$ and predict for the

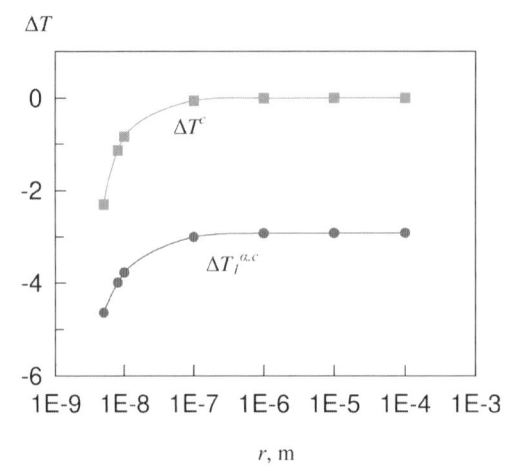

Figure 9.6 Shift of triple point temperatures ΔT^c and $\Delta T_1^{\alpha,c}$ for the core part (■) and the boundary phase (●) of spherical particles of eicosan aerosol as a function of particle radius, r. Calculations performed on the basis of Equations 9.38 and 9.39.

triple point temperature in the boundary and core part of spherical particles the possibility of both the depression, as observed for most metals (see Ref. [2] for a review and references therein), and the elevation, for example, for aerosols of hydrocarbons (Figure 9.6). Interestingly, the triple point temperature for the pre-freezed layer on a surface of aerosol particle starts to increase sensibly only for a particle radius of less than 100 nm, when the essential increasing of the triple point temperature for the core part of the particle is predicted.

9.3.2
Phase Transitions at the Interfaces and in the Interior of the Substance Condensed in a Porous Matrix

As in the previous section, we again neglect the influence of gravity on the equilibrium profiles of a meniscus at the interface between the phases condensed in pore and the vapor. We assume also that pores are large enough to provide the existence of a uniform part, not influenced by surface forces field, in the core of substance embedded in pores. Then the pressure in the condensed phase confined by pore walls and the meniscus will be unambiguously determined by the curvature of the meniscus. Accordingly, the shift of the triple points in the boundary layers and in the interior of the substance condensed in pores might be described by Equations 9.35 and 9.36, with substitution of the proper curvature of meniscus characteristic for different types of pores. We demonstrate below the analysis for the case of cylindrical pores, whereas for slits the algebra is similar, with just a twice lower curvature of meniscus.

Thus, for cylindrical pores with concave spherical meniscus, according to the convention for the sign of curvature $K = 2(\cos\theta)/r$, where r is the pore radius and θ is the contact angle between the meniscus (solid or liquid) and the pore walls. Then

Equation 9.36 for the shift of the triple point for the equilibrium between the vapor, liquid and solid confined in pores transforms into:

$$\Delta T^c = T^0 - T^c = \frac{2T^0}{qr}(v_L \gamma_{LV} \cos\theta_L - v_S \gamma_{SV} \cos\theta_S)_{T=T^c} \tag{9.40}$$

Applying the Young equations:

$$\gamma_{LV}\cos\theta_L = \gamma_{WV} - \gamma_{LW}; \quad \gamma_{SV}\cos\theta_S = \gamma_{WV} - \gamma_{SW} \tag{9.41}$$

where the index W stands for the pore wall, and neglecting the difference between molar volumes of solid and liquid phases one can rewrite Equation 9.40 in the form:

$$\Delta T^c = T^0 - T^c \approx \frac{2T^0 v_L}{qr}(\gamma_{SW} - \gamma_{LW})_{T=T^c} \tag{9.42}$$

Both Equations 9.40 and 9.42 can be used for the analysis of the shift of triple point temperature in a substance confined in pores. The choice between the two depends on the information available about the system, and on the applicability of the assumption $v_S \approx v_L$. From our point of view, Equation 9.40 is more convenient when analyzing the melting behavior of one and the same substance in pores of various matrices, since the only parameters that will vary from system to system are the experimentally accessible values of contact angles and pore sizes. Data on θ_L and θ_S formed by liquid or solid meniscus with the pore wall can be obtained in the plane geometry, since the contact angles are not very sensitive to the curvature of the interfaces for moderate curvatures. Moreover, it is clearly seen from Equation 9.40 that the character of interaction between the substance in the pore and the pore walls, obviously reflected by the magnitude of contact angle, essentially affects the shift of the triple point temperature.

The influence of the premelting/prefreezing at the substance/wall and substance/vapor interfaces on the shift of triple point for the interior of the substance in pores is naturally accounted for in Equations 9.40 and 9.42 through the values of corresponding surface energies in the same way as it was described above for the phase transitions in the small particles. For instance, in Equation 9.40 the value of γ_{SV} is sensitive to the state of the boundary phase at solid/vapor interface and is affected by the melted sublayer if present. Similarly, in Equation 9.42 the magnitude of γ_{SW} immediately reflects the appearance and the number of premelted sublayers at the interface between confined solid and the pore wall. We stress, once again, that the radii involved in Equations 9.40 and 9.42 are the total pore radii and no additional correction for the presence of premelted boundary sublayers should be invoked.

The analysis shows that, depending on the character of interaction between the substance and the pore walls (which is reflected either by values of contact angles in Equation 9.40 or ratio between γ_{SW} and γ_{LW} in Equation 9.42), these equations might predict for the triple point temperature both depression, which is typically detected for many substances condensed in mesoporous silica [64–66], and elevation, as found for a range of materials confined within slit-like pores of activated carbon [4, 67, 68].

Let us now consider the shift of the triple point in the boundary phases. Since in the pore the substance has two confining phases (the wall and the vapor), Equation 9.35 should be applied to both substance/vapor and substance/wall interfaces. Thus, for the shift of triple point temperature for the k-th sublayer at the substance/vapor interface:

$$\Delta T_k^{\alpha,c} = \frac{T^0}{q}\left[\left(\frac{\gamma'^c_{k(SV)}}{n^c_{k(SV)}} - \frac{\gamma'^c_{k(LV)}}{n^c_{k(LV)}}\right) + \frac{2}{r}(\nu_L\gamma_{LV}\cos\theta_L - \nu_S\gamma_{SV}\cos\theta_S)\right]_{T=T_k^{\alpha,c}} \quad (9.43)$$

and for the k-th sublayer at the substance/wall interface:

$$\Delta T_k^{\alpha,c} = \frac{T^0}{q}\left[\left(\frac{\gamma'^c_{k(SW)}}{n^c_{k(SW)}} - \frac{\gamma'^c_{k(LW)}}{n^c_{k(LW)}}\right) + \frac{2}{r}(\nu_L\gamma_{LV}\cos\theta_L - \nu_S\gamma_{SV}\cos\theta_S)\right]_{T=T_k^{\alpha,c}} \quad (9.44)$$

Note that in both Equations 9.43 and 9.44 the terms reflecting the influence of curvature-initiated excess pressure in a confined medium, that is, the terms in the second set of round brackets, might be expressed through the parameters of solid/wall and liquid/wall interfaces (cf. Equations 9.40 and 9.42). Further simplification of the analysis may then be achieved by neglecting the difference between $\nu_{k(L)}$ and $\nu_{k(S)}$ where ν_k is the molar volume in the k-th sublayer of liquid or solid boundary phase. From Equation 9.44 we will obtain:

$$\Delta T_k^{\alpha,c} = \frac{T^0\nu_L}{qr}\left[\frac{r}{h}\frac{\nu_{k(L)}}{\nu_L}(\gamma'^c_{k(SW)} - \gamma'^c_{k(LW)}) + 2(\gamma_{SW} - \gamma_{LW})\right]_{T=T_k^{\alpha,c}} \quad (9.45)$$

Clearly, from Equation 9.45, depending on the character of decay of the surface forces field, which affects the extent of the boundary phase, the influence of the second term is revealed at different pore sizes. Thus, if the surface forces field fades into the bulk phase very sharply, and the thickness of boundary phase is restricted within the height of one monolayer, then $(\gamma'^c_{1(SW)} - \gamma'^c_{1(LW)}) \approx (\gamma_{SW} - \gamma_{LW})$ and due to the factor r/h the curvature term, $2(\gamma_{SW} - \gamma_{LW})$, will sensibly increase the shift of triple point only for pores with r of the same order as h, whereas for wider pores $\Delta T_k^{\alpha,c}$ will be close to that at plane interfaces. In contrast, when the interaction of a confined substance with pore walls is strong enough to induce the multilayer boundary phase (such as takes place for water in hydroxylated mesoporous silica [64, 69]), then $(\gamma'^c_{k(SW)} - \gamma'^c_{k(LW)}) \ll (\gamma_{SW} - \gamma_{LW})$ even for small k, and the shift of the triple point for boundary layers becomes curvature dependent already at moderate pore sizes.

Returning to the dependence of the shift of triple point temperature for the interior of the substance in pores on the reciprocal pore radius, we stress again that the temperature variation of both the effective latent heat of fusion and the surface energies, especially important for the large temperature shifts (and hence, for the nanosize pores), causes the deviation of this dependence from linearity, expected at the first glance from the Equations 9.40 or 9.42. This conclusion is well illustrated by data on freezing of cyclohexane, benzene and water in mesoporous silica [65, 66], demonstrating nearly linear behavior at small melting point depressions and strong nonlinearity for $\Delta T^c > 30\,°K$.

9.4
Concluding Remarks

In this chapter we have proposed a unified thermodynamic approach to the analysis of melting/freezing phenomena in confined systems. The approach is based on the Clausius–Clapeyron relations for coexistence of solid and liquid phases jointly with equations relating the vapor pressures above bulk (with plane interfaces) and confined states of a substance. For illustration we have applied our analysis to three types of confinement: plane interfaces, small particles and pores. The analysis for other types of confinement like free, wetting and adsorption films, emulsions, and so on is straightforward. Notably, the logic of derivation allows us to better understand the influence and physical content of the parameters involved in the key equations.

Although the equations discussed in this chapter allow us to calculate the magnitudes and the signs of the shift of the triple point temperature for both the boundary layers and the core part of confined substance, their main advantage consists not in the predictive power but in the ability to solve the inverse problem. Namely, on the basis of reliably measured shifts of melting/freezing temperatures one can extract basic information about the structure and properties of boundary phases, surface energies as well as their temperature dependences.

References

1 Alcoutlabi, M. and McKenna, G.B. (2005) *Journal of Physics: Condensed Matter*, **17**, R461.

2 Mei, Q.S. and Lu, K. (2007) *Progress in Materials Science*, **52**, 1175.

3 Christensen, H.K. (2001) *Journal of Physics: Condensed Matter*, **13**, R95.

4 Radhakrishnan, R., Gubbins, K.E. and Sliwinska-Bartkowiak, M. (2002) *Journal of Chemical Physics*, **116**, 1147.

5 Tkachenko, A.V. and Rabin, I. (1997) *Physical Review E*, **55**, 778.

6 Mikheev, L.V. and Chernov, A.A. (1987) *Soviet Physics JETP*, **65**, 971.

7 Wilen, L.A., Wettlaufer, J.S., Elbaum, M. and Schick, M. (1995) *Physical Review B-Condensed Matter*, **52**, 12426.

8 Pawlow, P. (1909) *Zeitschrift fur Physikalische Chemie*, **65**, 1; Pawlow, P. (1909) *Zeitschrift fur Physikalische Chemie*, **65**, 545.

9 Defay, R., Prigogine, I., Bellemans, A. and Everett, D.H. (1966) *Surface Tension and Adsorption*, Green & Co, Longmans.

10 Buffat, P. and Borel, J.-P. (1976) *Physical Review A*, **13**, 2287.

11 Kofman, R., Cheyssac, P., Aouaj, A. *et al.* (1994) *Surface Science*, **303**, 231.

12 Cahn, J.W., Dash, J.G. and Fu, H. (1992) *Journal of Crystal Growth*, **123**, 101.
13 Couchman, P.R. and Jesser, W.A. (1977) *Nature*, **269**, 481.
14 Vanfleet, R.R. and Mochel, J.M. (1995) *Surface Science*, **341**, 40.
15 Skripov, V., Koverda, V. and Skokov, V. (1981) *Physica Status Solidi*, **66**, 109.
16 Zhang, Z., Li, J.C. and Jiang, Q. (2000) *Journal of Physics D-Applied Physics*, **33**, 2653.
17 Batchelor, R.W. and Foster, A.G. (1944) *Transactions of the Faraday Society*, **40**, 300.
18 Guggenheim, E.A. (1933) *Modern Thermodynamics by the Methods of Willard Gibbs*, Methuen and Co, London.
19 Butler, J.A.V. (1932) *Proceedings of the Royal Society A: Mathematical, Physical & Engineering Sciences*, **135**, 348.
20 Zhukhovitskii, A.A. (1944) *Acta Physica et Chemica USSR*, **18**, 176.
21 Kondo, S. (1956) *Journal of Chemical Physics*, **25**, 662.
22 Boinovich, L. and Emelyanenko, A. (2007) *Colloids Surfaces A*, **300**, 321.
23 Frenken, J.W.M., Maree, P.M. and van der Veen, J.F. (1986) *Physical Review B-Condensed Matter*, **34**, 7506.
24 van der Gon, A.W.D., Smith, R.J., Gay, J.M. et al. (1990) *Surface Science*, **227**, 143.
25 Frenken, J.W.M. and van Pinxteren, H.M. (1994) *Surface Science*, **307–309**, 728.
26 Carnevali, P., Ercolessi, F. and Tosatti, E. (1987) *Physical Review B-Condensed Matter*, **36**, 6701.
27 Herman, J.W. and Elsayed-Ali, H.E. (1992) *Physical Review Letters*, **69**, 1228.
28 Ocko, B.M., Wu, X.Z., Sirota, E.B. et al. (1997) *Physical Review E*, **55**, 3164.
29 Earnshaw, J.C. and Hughes, C.J. (1992) *Physical Review A*, **46**, R4494.
30 Castro, M.A., Clarke, S.M., Inaba, A. and Thomas, R.K. (1997) *The Journal of Physical Chemistry B*, **101**, 8878.
31 Lei, Q. and Bain, C.D. (2004) *Physical Review Letters*, **92**, 176103.
32 Sloutskin, E., Bain, C.D., Ocko, B.M. and Deutsch, M. (2005) *Faraday Discussions*, **129**, 339.
33 Findenegg, G.H. (1972) *Journal of the Chemical Society-Faraday Transactions I*, **68**, 1799.
34 Findenegg, G.H. (1973) *Journal of the Chemical Society-Faraday Transactions I*, **69**, 1069.
35 Messé, L., Perdigon, A., Clarke, S.M. et al. (2003) *Journal of Colloid and Interface Science*, **266**, 19.
36 Maeda, N. and Yaminsky, V.V. (2001) *International Journal of Modern Physics B*, **15**, 3055.
37 van der Gon, A.W.D., Gay, J.M., Frenken, J.W.M. and van der Veen, J.F. (1991) *Surface Science*, **241**, 335.
38 Ubbelohde, A.R. (1965) *Melting and Crystal Structure*, Clarendon Press, Oxford.
39 Miranda, P.B. and Shen, Y.R. (1999) *The Journal of Physical Chemistry*, **103**, 3292.
40 Sackmann, H. and Venker, P. (1952) *Zeitschrift für Physikalische Chemie*, **199**, 100.
41 Tatevsky, V.M.(ed.) (1960) *Physicochemical Properties of Neat Hydrocarbons (in Russian)*, Gostoptekhizdat, Moscow.
42 Stishow, S.M. (1968) *Soviet Physics Uspekhi*, **96**, 467.
43 Mikheev, L.V. and Chernov, A.A. (1987) *Soviet Physics JETP*, **65**, 971.
44 Naidich, Y.V., Grigorenko, N.F. and Perevertailo, V.M. (1981) *Journal of Crystal Growth*, **53**, 261.
45 Regan, M.J., Pershan, P.S., Magnussen, O.M. et al. (1997) *Physical Review B-Condensed Matter*, **55**, 15874.
46 Nakada, T., Miyashita, S., Sazaki, G. et al. (1996) *Japanese Journal of Applied Physics*, **35**, 52.
47 Boinovich, L.B. and Emelyanenko, A.M. (1999) *Progress in Colloid and Polymer Science*, **112**, 64.
48 Weast, R.S. and Astle, M.J.(eds) (1981) *Handbook of Chemistry and Physics*, 62th edn, CRC Press, Boca Raton.
49 *Handbook of Chemistry (in Russian)*, Volume 2, 2nd edn, Khimia, Moscow-Leningrad (1964).
50 Fletcher, N.H. (1971) *Reports on Progress in Physics*, **34**, 913.

51 Barer, S.S., Churaev, N.V., Derjaguin, B.V. et al. (1980) *Journal of Colloid and Interface Science*, **74**, 173.
52 Golecki, I. and Jaccard, C. (1978) *Journal of Physics C-Solid State Physics*, **11**, 4229.
53 Li, Y. and Somorjai, G.A. (2007) *Journal of Physical Chemistry C*, **111**, 9631.
54 Ikeda-Fukuzawa, T. and Kawamura, K. (2004) *Journal of Chemical Physics*, **120**, 1395.
55 Lied, A., Dosch, H. and Bilgram, J.H. (1994) *Physical Review Letters*, **72**, 3554.
56 Sambles, J.R. (1971) *Proceedings of the Royal Society A: Mathematical, Physical & Engineering Sciences*, **324**, 339.
57 Efimov, S.S. (1985) *Inzhenerno-Fizicheskii Zhurn*, **49**, 658.
58 Miao, L., Bhethanabotla, V.R. and Joseph, B. (2005) *Physical Review B-Condensed Matter*, **72**, 134109.
59 Tartaglino, U., Zykova-Timan, T., Ercolessi, F. and Tosatti, E. (2005) *Physics Reports-Review Section of Physics Letters*, **411**, 291.
60 Peters, K.F., Cohen, J.B. and Chung, Y.W. (1998) *Physical Review B-Condensed Matter*, **57**, 13430.
61 Reiss, H. and Wilson, I.B. (1948) *Journal of Colloid Science*, **3**, 551.
62 Zhang, M., Efremov, M.Y., Schiettekatte, F. et al. (2000) *Physical Review B-Condensed Matter*, **62**, 10548.
63 Allen, G.L., Gile, W.W. and Jesser, W.A. (1980) *Acta Metallurgica*, **28**, 1695.
64 Morishige, K. and Kawano, K. (1999) *Journal of Chemical Physics*, **110**, 4867.
65 Dosseh, G., Xia, Y. and Alba-Simionesco, C. (2003) *The Journal of Physical Chemistry B*, **107**, 6445.
66 Buntkowsky, G., Breitzke, H., Adamczyk, A. et al. (2007) *Physical Chemistry Chemical Physics*, **9**, 4843.
67 Sliwinska-Bartkowiak, M., Dudziak, G., Sikorski, R. et al. (2001) *Physical Chemistry Chemical Physics*, **3**, 1179.
68 Watanabe, A. and Kaneko, K. (1999) *Chemical Physics Letters*, **305**, 71.
69 Kittaka, S., Ishimaru, S., Kuranishi, M. et al. (2006) *Physical Chemistry Chemical Physics*, **8**, 3223.

10
Manipulation of DNA by Surfactants

Björn Lindman, Rita S. Dias, M. Graça Miguel, M. Carmen Morán, and Diana Costa

10.1
Introduction

Double-stranded DNA (Figure 10.1) is a highly charged polyelectrolyte and, therefore, associates strongly with any oppositely charged cosolute, from simple ions to polymers, proteins, surfactants, lipids and particles. As expected, the association increases strongly with the number of charges of the cosolute as well as its charge density; on the other hand, the association is weakened in the presence of a screening electrolyte. Another important property of ds-DNA is its rigidity – the persistence length is normally considered to be of the order of 500 Å [1, 2]. The stiffness of a polyelectrolyte will be unfavorable for the association but this will very much depend on the properties of the cosolute.

For the formation of the double-helix structure we best consider DNA as an amphiphilic polymer with the hydrophobic interactions driving association and the hydrophilic interactions – mainly electrostatic due to phosphate groups but also hydration plays a role – opposing association. Other interactions, like hydrogen-bonding and base stacking, while not driving association, control the three-dimensional structure formed.

A simple and important illustration of the delicate balance between driving and opposing forces is the fact that without electrolyte the electrostatic repulsion between the strands becomes strong enough to lead to dissociation, denaturation or "melting" of native DNA; similarly a modest temperature increase also drives dissociation, with the electrostatic forces (essentially temperature-independent) complemented by an increased conformational entropy.

ss-DNA and ds-DNA differ in three important respects: ss-DNA is more flexible (persistence length of 0.75–8.5 nm, depending on the salt concentration and technique used to evaluate it [1–4]) and is, because of the exposed bases, more hydrophobic, while ds-DNA has twice as high a linear charge density. In comparing the association of DNA with different cationic cosolutes we can deduce information on the relative significance of the different factors.

Highlights in Colloid Science. Edited by Dimo Platikanov and Dotchi Exerowa
Copyright © 2009 WILEY-VCH Verlag GmbH & Co. KGaA, Weinheim
ISBN: 978-3-527-32037-0

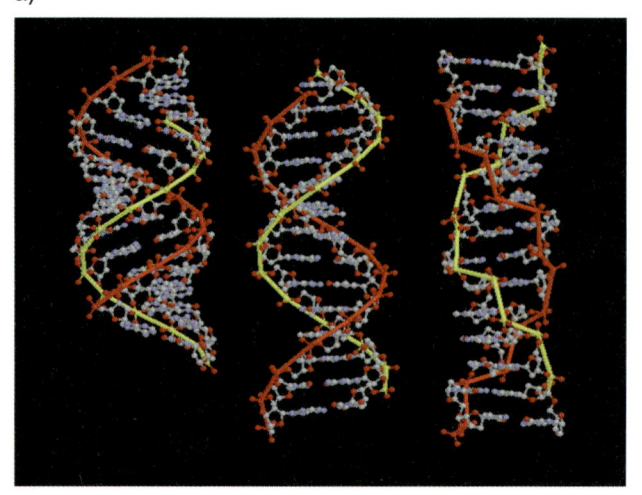

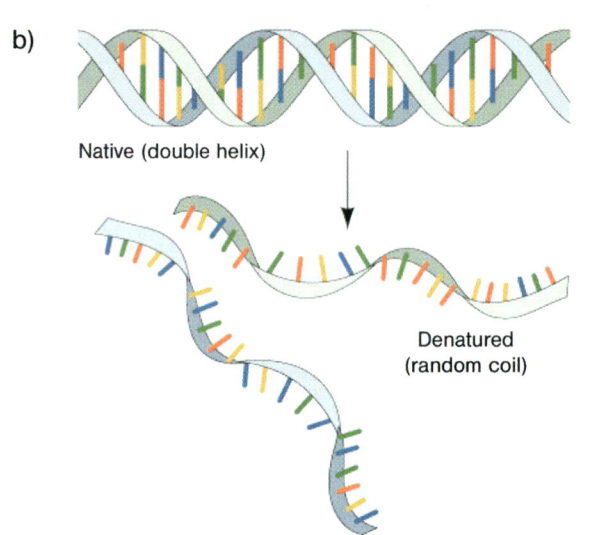

Figure 10.1 (a) Secondary structures of, from left to right, A-, B- and Z-forms of DNA. "Ball and stick" representation with the phosphate backbones highlighted (taken from www.biochemistry.ucla.edu/biochem/Faculty/Martinson/Chime/abz_dna/abz_master.html, last accessed 28/07/2008) and (b) transition from double-stranded DNA (ds-DNA) to single-stranded random coil DNA (ss-DNA). Terms that describe the change from ds-DNA to ss-DNA are: melting, denaturation, strand separation. Terms that describe the change from ss-DNA to ds-DNA are annealing, renaturation, and, in certain contexts, hybridization. (Taken from www.mun.ca/biochem/courses/3107/Topics/DNA_properties.html, last accessed 28/07/2008.)

There is a quite extensive literature relating to the association of DNA with cationic cosolutes. Recently we have, together with several leading contributors to the field, attempted to provide a broad picture of these interactions as well as some of the consequences for complex systems and applications [5]. While we refer the reader to this treatise for a more complete account and for extensive literature references, in this chapter we focus our discussion on the differences between ss- and ds-DNA, the role of the amphiphilic nature of DNA and some very recent findings in our own studies relating to these issues. Thus, while the biological consequences and biotechnological applications of the interactions between DNA and cationic species are numerous – ranging from the folding of DNA into the cell nucleus to transfection and protection – we focus here on the more general aspects; several of the consequences and applications are described elsewhere [5–7].

10.2
Surfactants Bind to ds-DNA and Induce Compaction

If a cationic surfactant is progressively added to a DNA solution, one will observe that initially there is no significant interaction. However, at a quite well-defined concentration the surfactant activity, equaling the concentration of free surfactant, becomes rather independent of the surfactant concentration, signifying an association of surfactant to DNA [8–10]. The binding isotherm that is obtained is illustrated in Figure 10.2.

As can be seen, there is a strongly cooperative binding of cationic surfactant to DNA. This type of behavior is observed also for all other mixed solutions of a polyelectrolyte and an oppositely charged surfactant, as well as for many other polymer–surfactant systems [11–13]. While in earlier literature this was usually discussed in terms of a cooperative binding it is now found more elucidating to refer to a surfactant self-assembly process, which is facilitated by the polymer. The surfactant in an aqueous solution starts to form micelles at a certain well-defined concentration, termed the critical micelle concentration, cmc. For an ionic surfactant the cmc is orders of magnitude higher than for a non-ionic one. This is because the formation of highly charged aggregates attracts a large number of counterions; the concomitant loss of counterion translational entropy is strongly unfavorable for micelle formation. In the presence of an oppositely charged polyelectrolyte this effect is largely eliminated and the cmc is much lowered, often by orders of magnitude. As expected from a surfactant self-assembly picture, the cmc – in the presence of a polymer also often referred to as cac (critical association concentration) – does not vary significantly with either the DNA concentration or its chain length [14].

Interestingly, the cac is much lower when DNA is denatured, and in the single-stranded conformation, than for the double-helix DNA, as illustrated by binding isotherms (curve 1 vs. curve 2 in Figure 10.2) [10] and in a recent study that deduced the cac from conductivity data [14]. This is a simple example of a stronger DNA–cationic surfactant interaction for ss- than for ds-DNA.

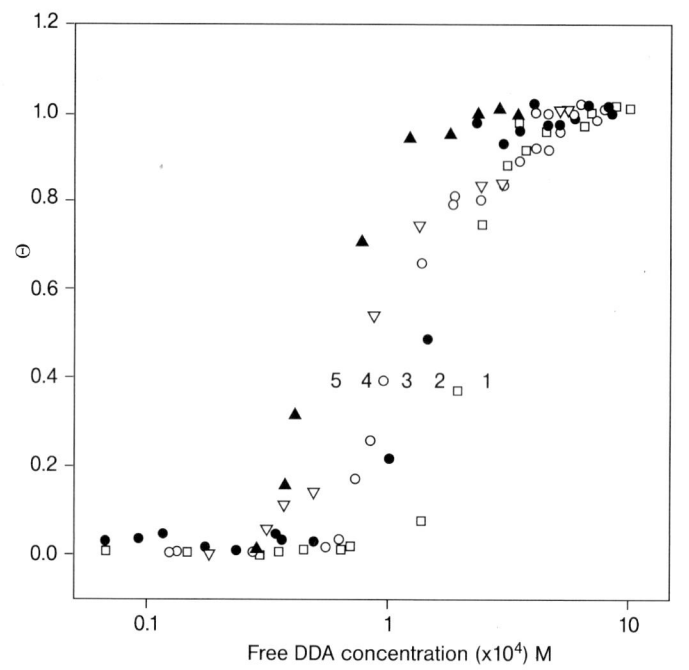

Figure 10.2 Binding isotherms of DDA with native (1) and denatured (2) DNA, with single-stranded polyribonucleotides poly(U) (3), poly(A) (5), as well as with double helix poly(U)_poly(A) (4). (From Ref. [10].)

For lower concentrations of DNA, the conductivity measurements presented a novel "anomalous" feature, namely, a second inflection point for surfactant concentrations below the cac; this feature was not displayed at higher concentrations of DNA. This was assigned to a mixture of ss- and ds-DNA molecules due to the low ionic strength of the solutions in these experiments.

The binding of cationic surfactant to DNA reduces its charge density, ultimately to a very low value. This has many consequences and manifestations. One effect is that the conformational state is affected to the extent that the stiffness is related to electrostatic repulsions – a polyelectrolyte is extended because of an osmotic swelling effect associated with the counterion entropy. A second effect is that the solubility is decreased since the solubility of polyelectrolytes is often determined by the counterion entropy. Below, we deal with phase separation in DNA solutions induced by the addition of cationic surfactant.

A macroscopic phase separation only occurs if the concentration of DNA is high enough. For dilute solutions another phenomenon takes place, namely the compaction or condensation of DNA. This can to some extent be considered as a phase separation for a single DNA molecule: surfactant molecules bind to DNA not as individual ions but as self-assembly aggregates. These have a high charge and act effectively as multivalent counterions. As surfactant is progressively bound to DNA

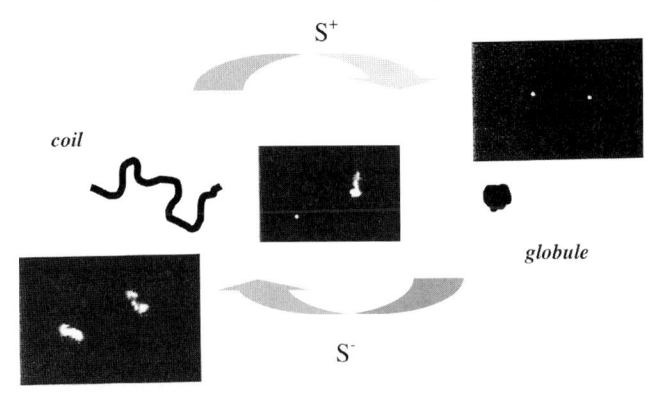

10.2 Surfactants Bind to ds-DNA and Induce Compaction

Figure 10.3 Schematic diagram of the compaction and decompaction of DNA by cationic and anionic surfactants. Fluorescence microscopy allows the visualization of DNA molecules in compacted conformation after the addition of a positively charged surfactant. With addition of an anionic surfactant, the DNA molecule is released into solution as a coil. For intermediate concentrations of surfactants both DNA conformations coexist in solution. (From Ref. [91].)

the different parts of the DNA molecule start to attract each other by electrostatic correlation effects [15]. For high molecular weight ds-DNA, the compaction process can be easily followed by fluorescence microscopy, adding a fluorescent probe that binds to DNA. Figure 10.3 gives an illustration.

Fluorescence microscopy has been used extensively to monitor the compaction of DNA in the presence of several cationic species [16–23] as well as some other cosolutes [24]. For cationic surfactants, compaction starts at concentrations only slightly above the onset of binding, as characterized by the cac [19] and shows the same dependence of surfactant alkyl chain length as other self-assembly processes [25].

While this approach is extremely useful in following the interaction of DNA with cosolutes, recent work has indicated a richer conformational behavior than inferred in earlier work [26–28]. Thus, earlier, the problem of DNA secondary structure was discussed in a simple two-state model with either fully extended or compacted DNA chains (and with coexistence between the two in a wide range of conditions). However, recent work by Yoshikawa's group as well as in our own has demonstrated firstly that there may be intermediate states and secondly that the compacted state is different for different cosolutes. Ongoing work focuses on this aspect as well as elucidating differences in compaction between different DNAs, for example the effect of molecular weight.

Several examples demonstrate that DNA compaction is reversible [29]. For surfactants this can be easily illustrated. Thus, addition of an anionic surfactant will "extract" cationic surfactant from DNA and thus convert compacted DNA into an extended coil state. This is also illustrated in Figure 10.3 while Figure 10.4 shows the typical chain length effect expected for a self-assembly process.

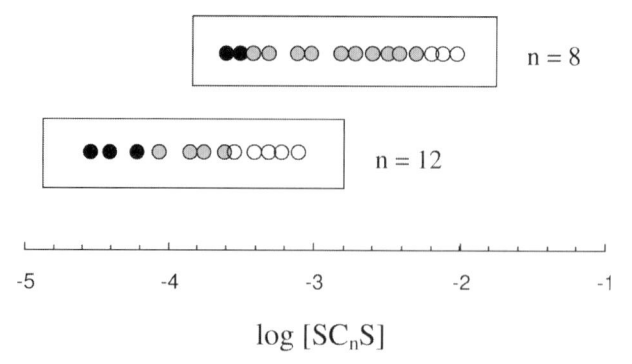

Figure 10.4 Dependence of the conformational behavior of single T4DNA molecules, 0.5 mM in aqueous buffer solution and a constant dodecyltrimethylammonium bromide concentration of 3.16×10^{-4} M, on the stepwise addition of sodium dodecyl sulfate and sodium octyl sulfate. Filled circles correspond to the globular DNA conformation, and shaded circles to the coexistence between elongated coils and compacted DNA molecules, whereas open circles correspond to the extended conformation of DNA. $T = 25\,°C$. (From Ref. [29].)

10.3
Surfactant Addition Can Lead to Phase Separation of DNA

A common phenomenon observed for mixed solutions of a pair of oppositely charged colloids, surfactant–surfactant, surfactant–polymer, polymer–polymer, is associative phase separation, namely, the two oppositely charged colloids concentrate in one phase in equilibrium with a very dilute solution [30].

Such behavior has been extensively studied for polyelectrolyte–oppositely charged surfactant systems. The extent of the phase separation increases with increasing alkyl chain length of the surfactant and with the polyelectrolyte charge density and molecular weight [31–33]. The nature of the concentrated phase varies with polymer and surfactant and with conditions; it can be a concentrated solution or gel, a liquid crystalline phase (lamellar, hexagonal, cubic) or a solid crystalline phase.

DNA shows a particularly strong associative phase behavior with cationic surfactants. While we have described in detail the behavior in previous publications [34, 35] we focus our discussion here on the effect of DNA secondary structure. As a basis for our discussion we give in Figure 10.5a a schematic observed phase diagram and in Figure 10.6 the main DNA–surfactant phase structures identified.

We exemplify here the precipitation behavior of DNA with the addition of dodecyltrimethylammonium bromide (DTAB) [36]. This is a surfactant with a relatively short alkyl chain, which allows for a rather fast equilibration of the samples; also this surfactant forms a micellar solution up to relatively high surfactant concentrations [37]. Figure 10.5a shows a schematic representation of the ternary phase diagram. As expected, the aqueous mixture of DNA and cationic surfactant phase separates associatively into one dilute phase and one phase concentrated in both polyelectrolyte and surfactant, a precipitate. The electrostatic interactions

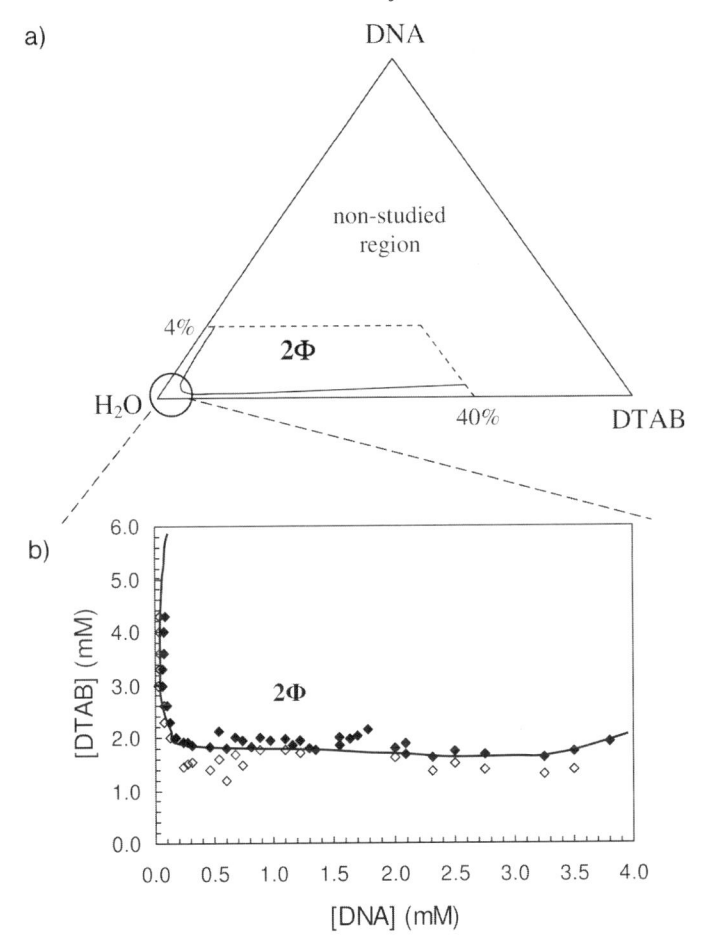

Figure 10.5 (a) Schematic representation of the isothermal pseudo-ternary phase map for the system DNA/DTAB/water. There is a phase separation in almost the entire considered region; (b) expanded view of the water corner of the system. Open symbols correspond to clear one-phase solutions and filled symbols to two-phase samples. $T = 25\,°C$. (Redrawn from Ref. [36].)

between the components are obviously strong and lead to a strong association. Surfactant aggregates induced by the polymer act as its counter-ions, thereby reducing the charge of the complex and the entropic driving force for mixing as well as the interpolymer repulsions [38].

However, contrary to other polyelectrolyte–surfactant systems [39–42], the precipitate does not redissolve with an excess of surfactant, at least in the examined, very broad, interval of concentrations. The difficulty of the redissolution of complexes composed of very highly charged polymers has also been observed in some other studies [41, 43, 44].

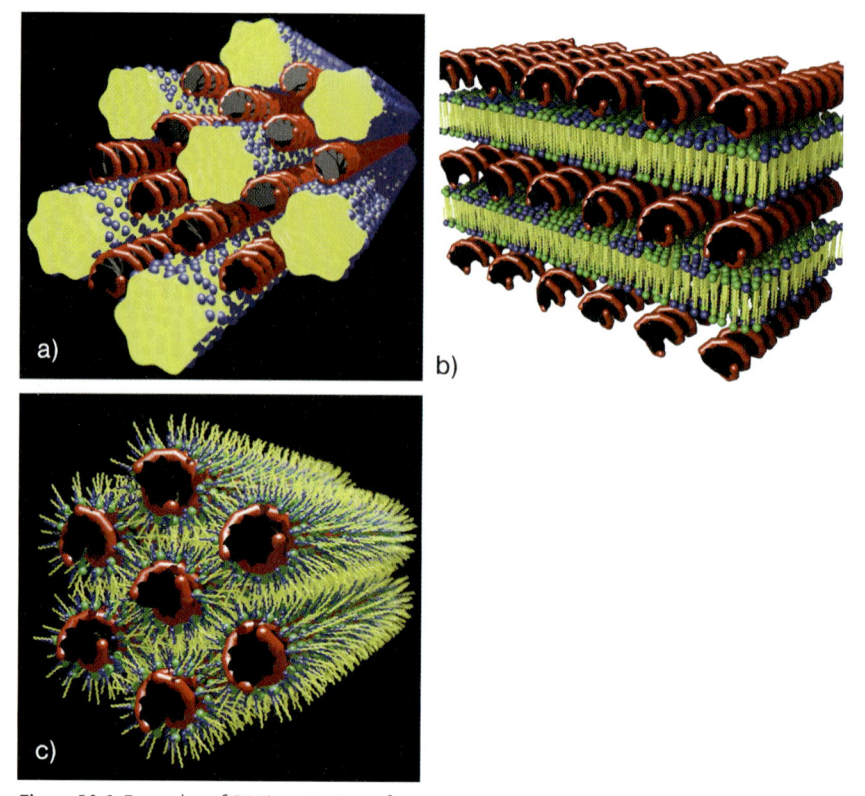

Figure 10.6 Examples of DNA–cationic surfactant structures: (a) normal hexagonal; (b) lamellar; and (c) reversed hexagonal phases. Illustrations rendered by Daniel Topgaard and Cecilia Leal using POVRAY©.

Further information drawn from the phase map is that the precipitate is formed at very low amounts of DNA and low surfactant concentrations, far below the surfactant cmc. This is a logical observation, since, as mentioned above, the polyelectrolyte–oppositely charged surfactant systems have a cac lower than the cmc of free surfactants, often by orders of magnitude.

A quite rigorous study of the two-phase border for these systems was performed. For this, turbidity studies were conducted and the results are presented in Figure 10.5b as an expansion of the dilute region of the pseudo-ternary phase map. This solubility diagram is presented in a simplified two-dimensional representation. Since the amount of water in these systems is very high, this type of representation provides a better visualization.

One interesting point is that the precipitate region shows an asymmetry with respect to the surfactant and DNA axes (note the difference in the scale of the axes). This was observed for other systems as well (e.g. Refs. [45–47]) and means that an addition of small amounts of DNA to a surfactant solution will cause precipitation,

10.3 Surfactant Addition Can Lead to Phase Separation of DNA

while a larger amount of surfactant will be required to cause phase separation of a more concentrated polyelectrolyte solution.

Here we are interested in the differences in behavior between single- and double-stranded DNA when interacting with cationic surfactants. We have performed then two different studies within this system: salt dependence and temperature dependence.

10.3.1
Effect of Salt

As mentioned above, the melting of DNA is dependent on the salt concentration. For solutions with a low concentration of DNA and no addition of simple salt, the molecules present a single-stranded conformation, whereas an addition as low as 1 mM of NaBr is sufficient to keep the DNA molecules in their native double-helix state.

We performed precipitation studies of DNA solutions by the addition of DTAB in the absence and presence (1 and 100 mM) of NaBr. Figure 10.7 presents the results. We can see that for the ssDNA solutions a lower amount of surfactant is necessary to induce phase separation than for the dsDNA solutions. It can be argued that the

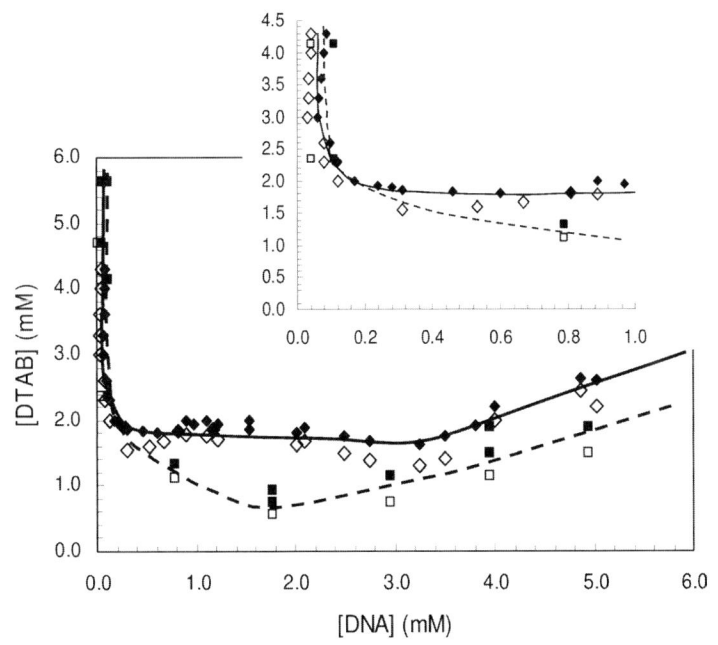

Figure 10.7 Precipitation map for the system DNA–DTAB–water at different salt concentrations: 1 and 100 mM NaBr (solid line) and 0 mM NaBr (dashed line). As in Figure 10.5, open symbols correspond to clear one-phase solutions and filled symbols to two-phase samples. $T = 25\,°C$. (From Ref. [36].)

behavior is a general feature for polyelectrolyte–oppositely charged surfactant systems. Thus, it is commonly accepted that the cac of polyelectrolyte–oppositely charged surfactant systems increases on addition of salt [13]. This is due to a weakened interaction between the polymer and surfactant induced by the stabilization of (free) micelles and a screening of the electrostatic interactions. This delays the precipitation of the system, and decreases the two-phase region. It should be stressed, however, that in our case we observe a crossing of the two precipitation lines, which is novel and not expected from previous work; furthermore, there are no noticeable differences in the phase separation of the system in the presence of 1 or 100 mM, suggesting that the system is not very sensitive to the salt concentration changes for these intermediate concentrations.

10.3.2
Effect of Temperature

It appears then that these differences in behavior arise from the fact that we have different DNA conformations. To make this point clear we studied the temperature dependence of the same system. For the temperature studies we used a fixed concentration of NaBr, 10^{-5} M, and performed the studies at 4, 25 and 50 °C. Under these conditions we have at 4 °C double-helix DNA solutions and for the two higher temperatures single-stranded DNA molecules. The precipitation diagram is presented in Figure 10.8.

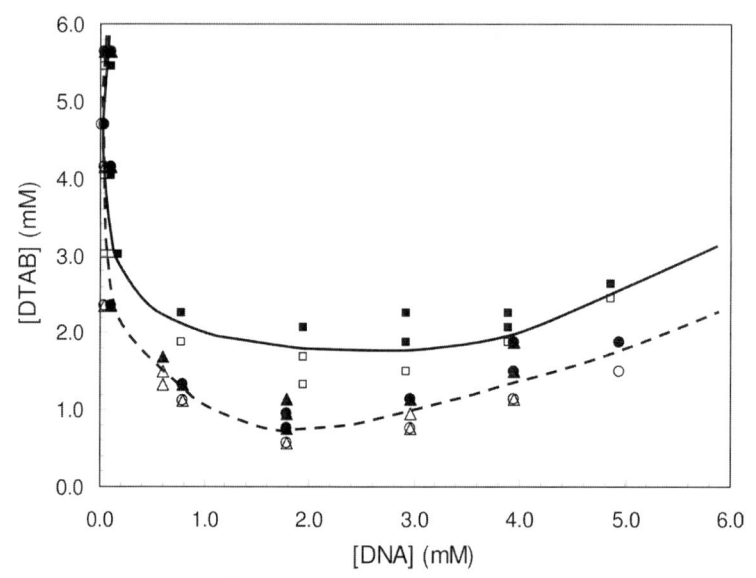

Figure 10.8 Phase map for the systems DNA–DTAB–water at different temperatures: 4 °C (solid line); 25 and 50 °C (dashed line). As above, open symbols correspond to clear one-phase solutions and filled symbols to two-phase samples. (From Ref. [36].)

We can see the same features as in the precipitation map (Figure 10.7). For the two higher temperatures, the precipitation regions overlap, less surfactant is required to induce the phase separation of the DNA–DTA complexes. For the samples mixed and kept at 4 °C, higher concentrations of DTAB are necessary for precipitation.

We can conclude that the precipitation of the DNA–cationic surfactant systems does depend on whether the DNA is in the single- or double-stranded conformation. The fact that the ss-DNA molecules do precipitate for lower concentrations of surfactant again shows a stronger association than for ds-DNA.

The fact that the two conformational states show a different behavior brings interesting perspectives from a separation and purification point of view, as it seems that, with an appropriate choice of surfactant concentration, it is possible to separate single- from double-stranded DNA molecules in solution.

10.4
DNA is an Amphiphilic Polyelectrolyte

From the binding isotherms and the phase diagrams we can infer a considerably stronger association to cationic surfactants for ss-DNA than for ds-DNA; further illustrations of this are given below. From a simple electrostatic picture ds-DNA with twice as high a linear charge density would be expected to interact significantly stronger. We have noted above that the two conformational states differ also in two other important respects, flexibility and amphiphilicity. While Linse, in Monte Carlo simulations [48], has demonstrated that the strength of association between a linear poly-ion and an oppositely charged spherical macro-ion increases with poly-ion flexibility, thus offering a possible explanation, we here consider the role of amphiphilicity.

As is well known, and as already alluded to above, the association of two DNA strands into the double helix is driven by the hydrophobic interactions between the bases. Polar interactions, associated with the phosphate and carbohydrate groups, counteract the association. Hydrogen bonding and specific packing of the bases control the details of the double helix structure.

While the electrostatic interactions of DNA have been analyzed in detail, the hydrophobic interactions have been much less discussed. In particular the balance between the polar and nonpolar interactions have a deep impact on how DNA interacts with cosolutes, including electrolytes, nonpolar molecules, surfactants, lipids and macromolecules.

In the context of DNA–surfactant interactions we will briefly comment on the amphiphilic nature of DNA and its consequences for the solution behavior. In discussing the self-assembly behavior of DNA, we begin by broadly discussing other amphiphilic compounds and their self-assembly. Amphiphilic compounds – those that have distinct hydrophilic and lipophilic parts – range from low molecular weight molecules, like surfactants and lipids, to macromolecules, consisting of synthetic graft and block copolymers, and biomacromolecules, like proteins, lipopolysaccharides and nucleic acids.

Surfactant and lipid self-assembly can lead to a diverse range of aggregate structures, the type of aggregate formed being mainly determined by the chemical structure and the relative strength of the hydrophilic and lipophilic parts [49].

Amphiphilic/associating water-soluble polymers, in particular block copolymers and hydrophobically modified water-soluble polymers, have been studied extensively during the last decade and are well understood [50, 51]. For graft copolymers that are hydrophobically modified water-soluble polymers, which are common as thickeners and dispersants, the self-assembly is very different than for block copolymers. For such graft copolymers there is a strong opposing force due to the hydrophilic polymer backbone. In particular, due to the entropic penalty in folding the polymer chain, only small discrete hydrophobic micro-domains ("micelles") are formed.

DNA is clearly different from both these cases, but closer to the graft copolymer situation. However, the segregation between hydrophilic and lipophilic parts is less pronounced and the force opposing self-assembly is stronger due to a high charge density and a large persistence length.

We have already noted that for DNA the balance between the hydrophobic force driving self-assembly and the opposing force is very subtle. Two consequences are: firstly, the stability of the double helix (ds-DNA) is critically dependent on the electrolyte concentration. In the absence of electrolyte the opposing force dominates and dissociation occurs. Small amounts of electrolyte or essentially any cationic cosolute overcomes the electrostatic repulsion and stabilizes ds-DNA. Secondly, if the driving force is changed, for example by changing the base composition, there is a significant change in the stability of the double helix.

A simple illustration of the balance between forces is given by calorimetry studies of DNA solutions (Figure 10.9). As can be seen the peak characterizing the melting of ds-DNA into ss-DNA disappears at low salt concentrations, which can be referred to the instability of ds-DNA in the absence of screening electrolyte. Strikingly, for a higher DNA concentration elimination of the electrolyte does not lead to denaturation; this is due to a self-screening effect: the counterions dissociating from DNA provide an electrolyte atmosphere [34, 52, 53].

Other illustrations of the significance of hydrophobic interactions are the solubilization of hydrophobic molecules in the ds-DNA (e.g. [54, 55]); the adsorption of DNA on hydrophobic surfaces [56–58]; the effect of hydrophobic cosolutes on DNA melting [59–61]; DNA–protein interactions; and DNA chemical and physical gels [62–64].

It was, for example, observed by ellipsometry that, whereas both ds- and ss-DNA molecules adsorb on hydrophobic surfaces, ss-DNA adsorbs in general more strongly than ds-DNA [57, 58]. Also, while ds-DNA molecules form a very thick and diffuse layer on the surface, the ss-DNA molecules adsorb in a thin layer of about 20 Å, indicating that the molecules are parallel to the surface [58]. This is naturally due to the larger hydrophobicity of the ss-DNA, as each base will serve as an attachment point to the surface, which overcomes the entropy loss of the adsorption; ss-DNA is much more flexible than ds-DNA.

The interactions between DNA and alkyltrimethylammonium bromide salts with short hydrophobic chains and the influence of the chain length on the melting have

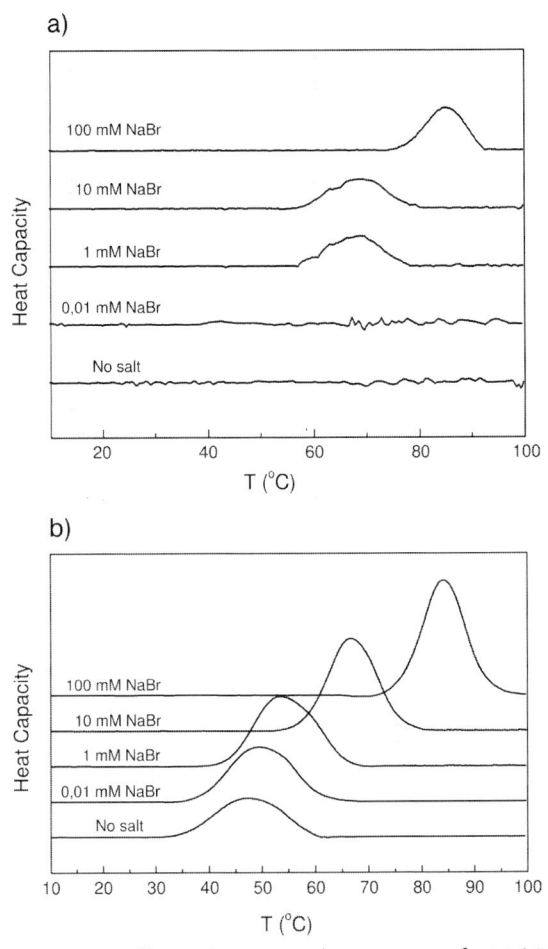

Figure 10.9 Differential scanning calorimetry traces for (a) 0.5 and (b) 6.0 mM DNA solutions at different concentrations of salt (as indicated). (From Ref [34].)

been studied previously [59]. It was observed that the melting temperature of DNA decreases with the increase of the hydrophobic group in a linear fashion up to the pentyl substitution.

Short-chain alcohols showed the same behavior. The melting temperature of DNA decreases in water/methanol solutions [65]. Furthermore, the midpoint of the solvent denaturation decreased in the order methanol, ethanol and propanol; that is, the secondary structure stability was lowered as the length of the aliphatic chain was increased [65]. Following the same line, in another contribution it was realized that an increase in the number or size of alkyl substituents on amides, ureas, carbamates and alcohols increased the "denaturating effectiveness" towards DNA [61]. The contribution of non-specific hydrophobic interactions to DNA denaturation was then brought up, and in fact it is not surprising that these small hydrophobic molecules destabilize the double-helix of DNA since the hydrophobic interactions between the two species are very favorable.

One other indication that points to the importance of the hydrophobic moieties of DNA on the interaction with co-solutes is the mentioned difference in interactions of ss- and ds-DNA with cationic surfactants.

The interaction of DNA with proteins is a good example of systems where hydrophobic interactions are of great importance. Essential genetic functions such as transcription, replication, cleavage and recombination rely on the ability of proteins to recognize and interact with specific sequences of DNA. Extensive studies have thus been performed to understand the underlying mechanism of the binding of proteins to DNA (e.g. for reviews see Refs. [66–68]).

Hydrophobic interactions between nonpolar amino-acid residues and the bases of DNA contribute to the function of DNA binding proteins through conformational effects and direct interactions with the DNA nucleotides. Direct evidence of the role of hydrophobic interactions in DNA–protein complexes has been found for T7 RNA polymerase [69, 70], Epstein–Barr virus replication factor EBNA1 [71], and bovine papillomavirus helicase E1 [72, 73], for example.

The fact that the melting temperature of DNA depends on its base sequence is a good indication of the importance of the nearest neighbors for the stability of the helix [74], and it is normally attributed to the stacking interactions between adjacent bases.

As mentioned above, different bases have different hydrophobicities. The purine bases, with two aromatic rings instead of one for the pyrimidine bases, are more hydrophobic, and so the adsorption of the bases on graphite surfaces was found to increase in the series guanine > adenine > thymine > cytosine [75].

10.5
Phase Separation Phenomena Underlie the Preparation of Novel Particles

As described above, interactions between oppositely charged surfactants and polyelectrolytes in aqueous solutions can lead to associative phase separation, where the concentrated phase assumes the form of a viscous liquid, gel, liquid crystal or precipitate. This behavior has been exploited to form gel particles, which have been prepared by drop-wise addition of cellulose-based polycation solution (chitosan or-, N,N,N-trimethylammonium derivatized hydroxyethyl-cellulose) to anionic (sodium dodecyl sulfate, sodium perfluorooctanoate) and cationic (cetyltrimethylammonium bromide/sodium perfluorooctanoate) surfactant solutions [76–80].

Using this approach, we have recently developed a way to form DNA gel particles at water/water emulsion type interfaces by mixing DNA [either single- (ssDNA) or double-stranded (dsDNA)] with the cationic surfactant cetyltrimethylammonium bromide (CTAB) or the protein lysozyme [64]. The originality of this work consists in forming DNA reservoir gel particles without adding any kind of crosslinker or organic solvent. Under optimal conditions, droplets from DNA solutions instantaneously gelled into discrete particles upon contact with the surfactant solution. The

Figure 10.10 Representative morphology of CTAB–dsDNA particles. (From Ref. [64].)

size of the resulting particle reflects the size of the parent drop and varies between 1 and 2 mm (Figure 10.10).

The formation of these DNA gel particles constitutes an example of strong associative phase separation. An indication of the strength of this interaction is the formation of a stronger film (or skin) constituted by the polyelectrolyte–surfactant complex. Preliminary results of SAXS measurements have supported the existence of an ordered structure formed on the hydrated skin of the obtained particles. SEM images of the cross-section of the particles have given evidence for the existence of a shell structure, its formation being more pronounced in the case of ss-DNA. The capsule shells obtained may be considered as physical networks in which surfactant micelles form polycationic-multianionic electrostatic complexes as crosslink points.

Determinations of the degree of DNA entrapment show loading efficiency (LE) values of CTAB–DNA particles higher than 99%, confirming the effectiveness of DNA entrapment in the surfactant solution. DNA loads of up to 2% were achieved. The binding of CTAB to DNA depends of the secondary structure of the polyelectrolyte. Quantification of the surfactant–DNA complexes in the particles shows significative differences (52% versus 79% in particles formed with native ds-DNA and denatured ss-DNA, respectively).

When the DNA gel particles are inserted in a medium different responses are encountered: swelling or deswelling, dissolution, and release of DNA [81]. We modified the contribution of electrostatic interactions in CTAB–DNA binding by monitoring the effect of the medium ionic strength on the weight of the particles. The data were then transformed into the relative weight loss. Figure 10.11 depicts the effect of salt on the relative weight ratio of CTAB–dsDNA and CTAB–ssDNA particles, respectively.

The stability of the gel particles is given mainly by the electrostatic attraction between DNA and the oppositely charge surfactant. A weakening of this association is expected to result in a partial or complete dissolution. The electrostatic attraction is obviously expected to be weaker in the presence of electrolyte. Particles placed in either deionized water or pH 7.6 10 mM Tris HCl buffer show water uptake from the medium and swelling could be observed. The swelling continues during the entire time interval studied. Only in the case of particles placed in the buffer is there a return to the original particle weight.

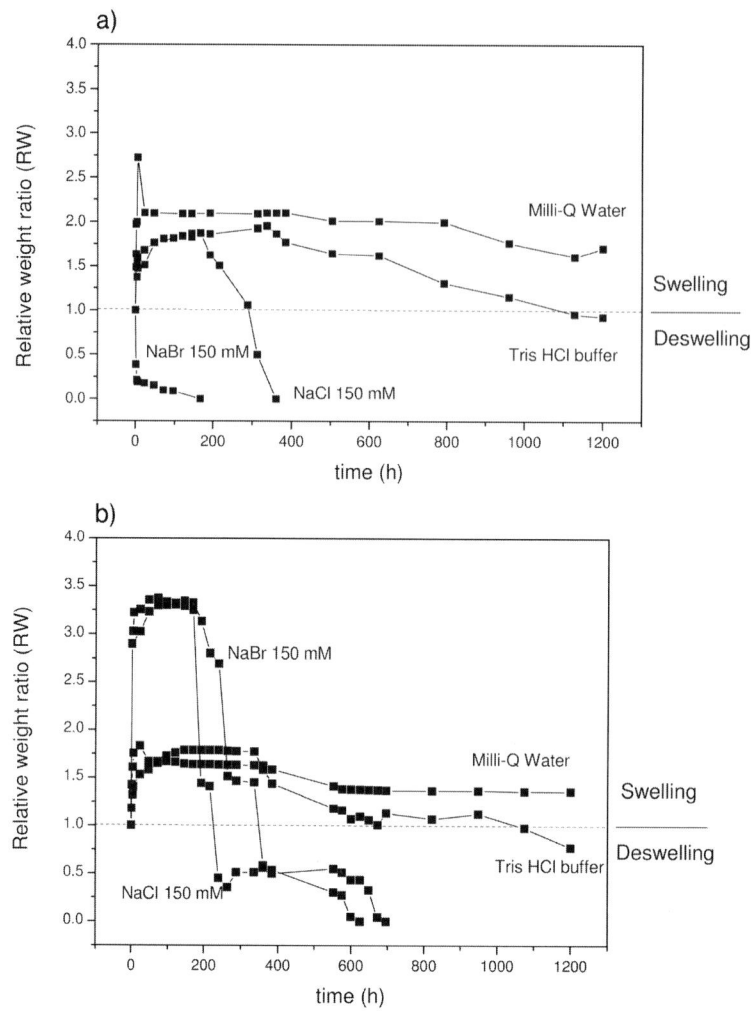

Figure 10.11 Relative weight ratio measurements performed on (a) CTAB–dsDNA and (b) CTAB–ssDNA particles after exposure to different solutions. (From Ref. [81].)

Differences in CTAB–DNA interactions between the secondary structures of DNA are displayed during the swelling behavior experiments in the presence of high salt content. While CTAB–dsDNA particles placed in 150 mM NaBr monotonously dissolve with time, particles formed with denatured DNA show an initial swelling and dissolve only after 600 h. The observed response is related to the capacity to form stronger DNA–surfactant complexes in the latter system, to which both higher flexibility and amphiphilic character contribute.

The effect of ionic strength on the release of DNA from the different particles has also been explored. Figure 10.12 shows the observed cumulative DNA release.

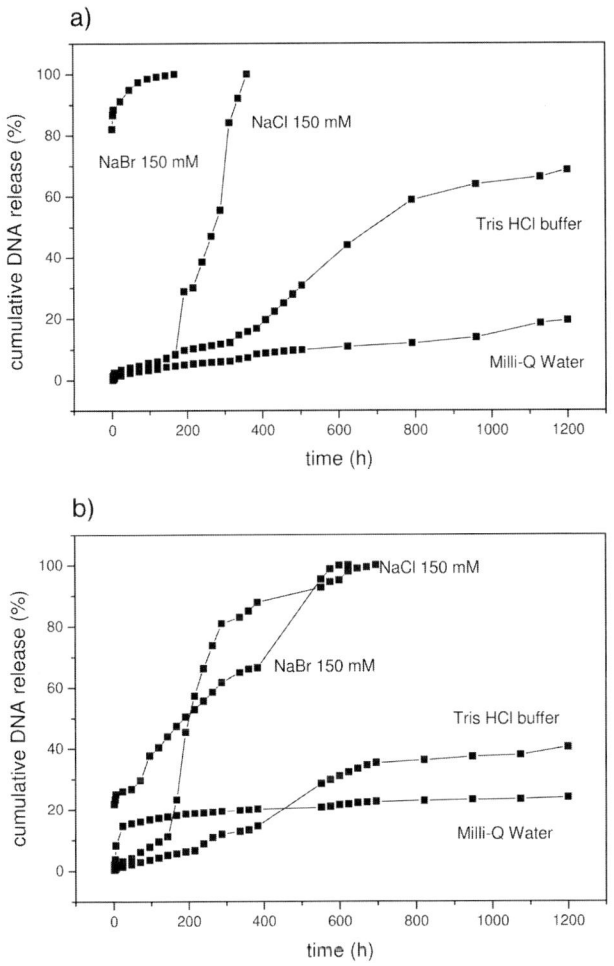

Figure 10.12 Release of DNA from (a) CTAB–dsDNA and (b) CTAB–ssDNA particles in different solutions. (From Ref. [81].)

Generally, the release pattern resembles that observed in the swelling/dissolution profiles (Figure 10.11).

Information about the secondary structure of the DNA molecules in the gels can be obtained by fluorescence microscopy using acridine orange staining. Based on the observation of green or red fluorescence [82] acridine orange is used to differentiate between native, double-stranded DNA and denatured, single-stranded DNA in the CTAB–DNA particles (Figure 10.13). These results are consistent with the secondary structure of the DNA used for the particle preparation.

Interestingly, DNA molecules are released from particles for a long period of time under *in vitro* conditions, which may provide an intracellular sustained release of DNA *in vivo*. Current studies are focused on characterizing the structure of these

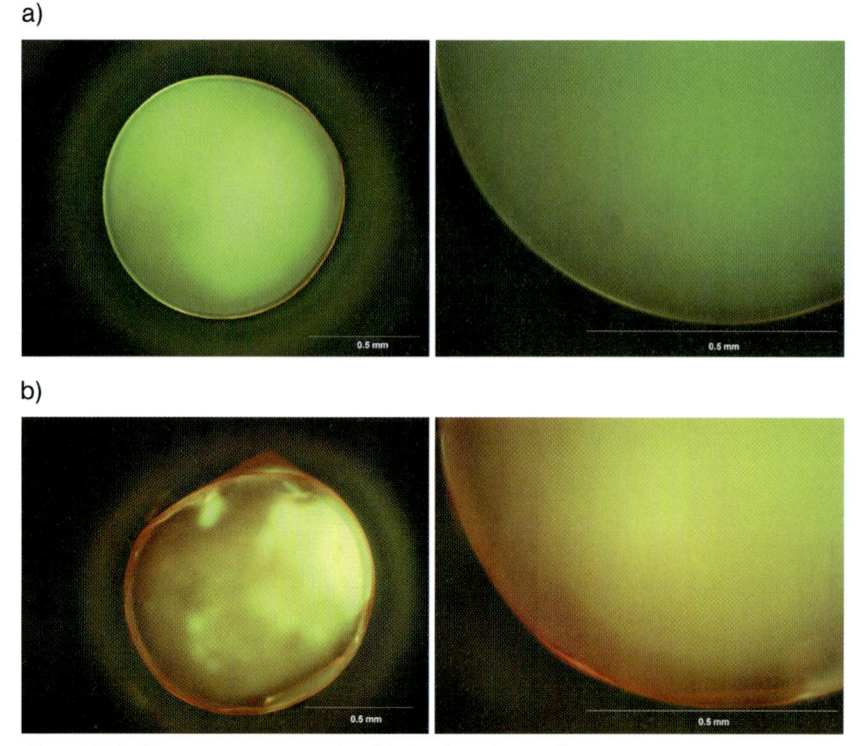

Figure 10.13 Fluorescence micrographs of CTAB–dsDNA (a) and CTAB–ssDNA (b) particles: individual particles (left) and the same particles at high magnification (right). (From Ref. [81].)

DNA gel particles and modifying the experimental conditions for further applications in gene transfection. Importantly, these particles represent a conceptual step in the design and development of new non-viral vectors for the delivery of therapeutic DNA.

10.6
DNA Can be Crosslinked into Gels

Crosslinking of polyelectrolytes like DNA gives gels, which are osmotically highly swollen but contract on addition of electrolytes and, in particular, of oppositely charged cosolutes that associate with the polyelectrolyte chain. The deswelling behavior of the DNA networks, prepared by crosslinking double-stranded DNA with ethylene glycol diglycidyl ether (EGDE), thus reports on DNA–cosolute interactions and gives a basis for the development of responsive DNA formulations. Both single- and double-stranded DNA gels have interesting applications and, furthermore, a comparison between them gives a basis for understanding mechanisms. The swelling behavior on addition of different cosolutes, such as metal ions, polyamines, charged proteins and surfactants, has been investigated for different DNA gel

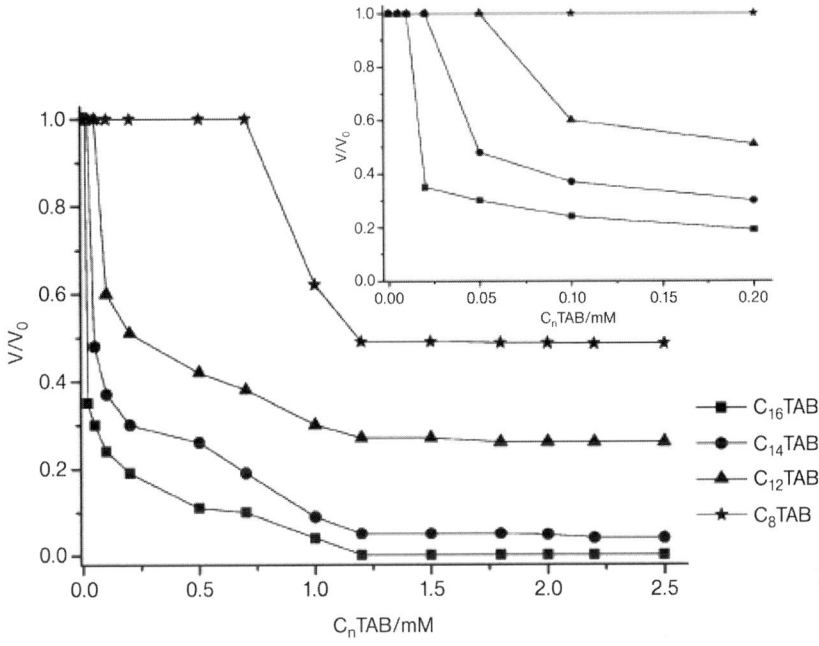

Figure 10.14 Swelling isotherm (V/V_0) for DNA gels (1% crosslinker) immersed in solutions of the cationic surfactants $C_{16}TAB$, $C_{14}TAB$, $C_{12}TAB$ and C_8TAB. The concentration range 0–0.2 mM is also represented in detail (see inset). Temperature 25 °C, pH 9. (From Ref. [83].)

samples, which include long and short ds-DNA and long and short ss-DNA [63]. The deswelling on addition of metal ions occurs at lower concentrations with increasing valency of the counterion. Striking features were found in the deswelling of DNA gels by chitosan, spermine, spermidine, lysozyme, poly-L-lysine and poly-L-arginine. Chitosan is the most efficient cosolute of those investigated with respect to DNA gel collapse [83].

Cationic surfactants at low concentrations do not affect the gel volume, but effectively collapsed the gel from a critical aggregation concentration (cac), which decreases with increasing length of the hydrophobic tail (Figure 10.14). This is because at the cac the surfactants start to form micelles, which act as multivalent counterions, and thus cause a very important osmotic deswelling.

The collapse of the gels is reversible as can be inferred from Figure 10.15, which presents the volume change on addition of an anionic surfactant to a DNA gel, which was collapsed by a cationic surfactant; the anionic surfactant interacts strongly with the cationic surfactant, forming different mixed aggregates, and effectively "extracts" the cationic surfactant from the gel.

Interestingly, there is a significant difference in the collapse between gels based on ds- and ss-DNA. In general, single stranded DNA gels exhibit a larger collapse, in the presence of cations, than double stranded ones do. As Linse has demonstrated, the

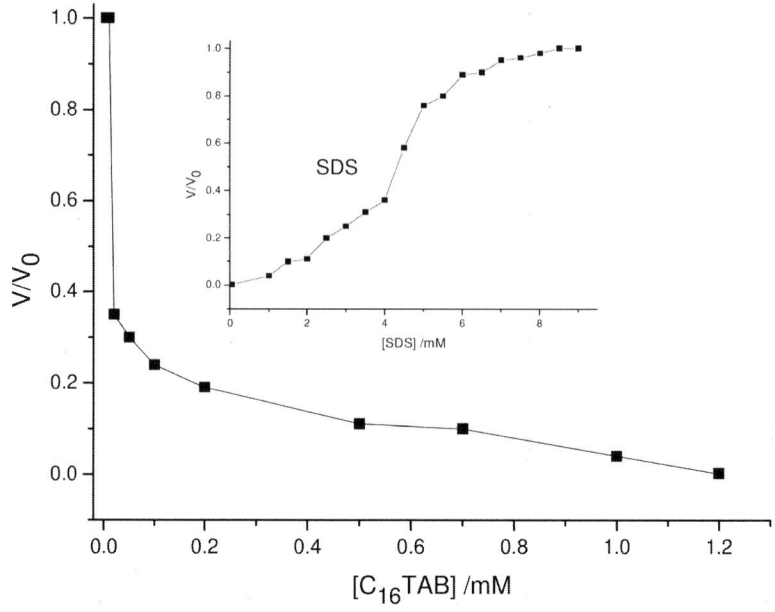

Figure 10.15 Swelling isotherm (V/V_0) for DNA gels (1% crosslinker) pre-collapsed first in solutions of the cationic surfactant C_{16}TAB and then immersed in solutions of the anionic surfactant sodium dodecyl sulfate. Temperature 25 °C, pH 9. (From Ref. [83].)

collapse of a covalent gel depends on several factors, in particular the gel charge density and the polymer chain flexibility [84]. The larger deswelling for ss-DNA gels would indicate that the effect of flexibility is very important. However, the difference is more pronounced with surfactants than with the other cosolutes investigated, which can only be understood if hydrophobic interactions between the gel network and the surfactants are significant. This deduction is supported by the concentration dependences, since the onset of deswelling occurs at lower concentrations for ss- than for ds-DNA; this is also in agreement with the behavior displayed by DNA solutions as described above.

Surfactant-induced deswelling of the DNA gels under some conditions appears to be quite homogeneous, while in others there is a separation into a collapsed region in the outer parts of the gel sample and an inside swollen part. Such a "skin" formation is quite different for ss- and ds-DNA, with ss-DNA giving a more pronounced skin formation over a wider range of binding ratios, β. For example, no macroscopic separation into collapsed and swollen regions is observed at intermediate degrees of binding for ds-DNA gels whereas a dense surfactant-rich surface phase (skin) is found to co-exist with a swollen core network for ss-DNA gels with $\beta > 0.5$. One explanation for this difference is the large deformation energy required for the compression of the very stiff ds-DNA chains.

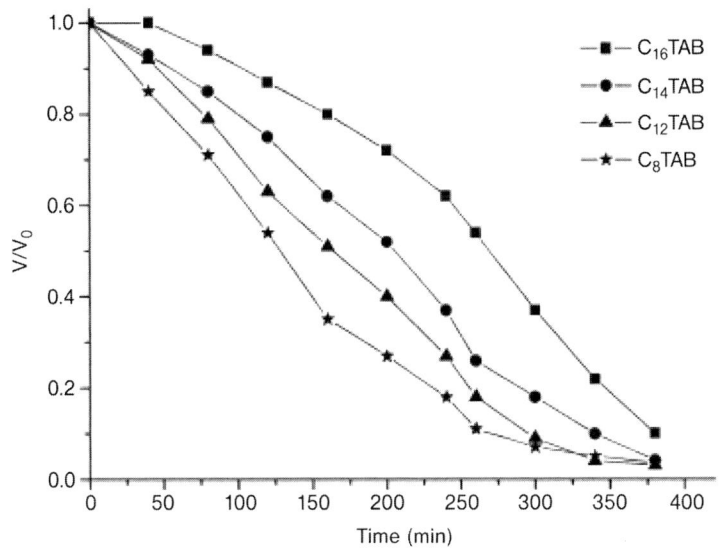

Figure 10.16 Transient volume changes (V/V_0) of a DNA gel (1% crosslinker) immersed in solutions of 1 mM of the cationic surfactants C_{16}TAB, C_{14}TAB, C_{12}TAB and C_8TAB. Temperature 25 °C, pH 9. (From Ref. [83].)

Covalent DNA gels clearly respond sensitively to interactions with surfactants and are useful for characterizing surfactant association to DNA. The dependence on the secondary structure parallels that found in other types of studies. Regarding kinetics it can be seen that in some cases equilibrium appears to be reached while in others there is formation of trapped non-equilibrium states. The time dependence for the former type of systems is illustrated in Figure 10.16. As can be seen a long-chain surfactant, while giving rise to a larger deswelling, is characterized by significantly slower kinetics. This can easily be understood since the determining factor would be the gradient in the surfactant activity; the surfactant activity equals the free surfactant concentration, that is the cac.

10.7
Perspectives

As can be seen, a general understanding of DNA–surfactant interactions, and in particular the phase behavior, gives a basis for developing novel DNA-based materials, including particles, gels and membranes [85]. The systems can be used for delivery of ss- or ds-DNA or they can be made responsive, triggering for changes in temperature, salinity or for different specific interactions. Cationic surfactants offer particularly efficient control of the properties of DNA-based particles, gels and membranes but suffer from being typically toxic. While toxicity certainly applies for most classical surfactants, we are engaged in current work focusing on novel surfactants with much improved biocompatibility. These include surfactants with the cationic functionality based on an amino-acid [86–88], quaternary ammonium surfactants with hydrophilic substituents [14, 89] and so-called temporary surfactants [90].

References

1 Tinland, B., Pluen, A., Sturm, J. and Weill, G. (1997) *Macromolecules*, **30**, 5763–5765.
2 Murphy, M.C., Rasnik, I., Cheng, W. et al. (2004) *Biophysical Journal*, **86**, 2530–2537.
3 Mills, J.B., Vacano, E. and Hagerman, P.J. (1999) *Journal of Molecular Biology*, **285**, 245–257.
4 Smith, S.B., Cui, Y.J. and Bustamante, C. (1996) *Science*, **271**, 795–799.
5 Dias, R.S. and Lindman, B.(eds) *DNA Interactions with Polymers and Surfactants*, Wiley-Blackwell, New Jersey, 2008.
6 Mahato, R.I.(ed.) (2005) *Biomaterials For Delivery and Targeting of Proteins and Nucleic Acids*, CRC Press.
7 Schleef, M.(ed.) (2005) *DNA-Pharmaceuticals: Formulation and Delivery in Gene Therapy, DNA Vaccination and Immunotherapy*, Wiley-VCH, Verlag Gmbh.
8 Hayakawa, K., Santerre, J.P. and Kwak, J.C.T. (1983) *Biophysical Chemistry*, **17**, 175–181.
9 Shirahama, K., Takashima, K. and Takisawa, N. (1987) *Bulletin of the Chemical Society of Japan*, **60**, 43–47.
10 Petrov, A.I., Khalil, D.N., Kazaryan, R.L. et al. (2002) *Bioelectrochemistry*, **58**, 75–85.
11 Kwak, J.C.T.(ed.) (1998) *Polymer-Surfactant Systems*, Marcel Dekker, New York.
12 Holmberg, K., Jönsson, B., Kronberg, B. and Lindman, B. (2003) *Surfactants and Polymers in Aqueous Solution*, 2nd edn, John Wiley & Sons, Ltd, West Sussex.
13 Lindman, B. and Thalberg, T. (1993) in *Interactions of Surfactants with Polymers and Proteins* (eds E. Goddard and K. Ananthapadmanabhan), CRC Press, Boca Raton, pp. 203–276.
14 Dias, R.S., Magno, L.M., Valente, A.J.M. et al. *Journal of Chemical Physics*. Accepted for publication.
15 Guldbrand, L., Jönsson, B., Wennerström, H. and Linse, P. (1984) *Journal of Chemical Physics*, **80**, 2221–2228.
16 Minagawa, K., Matsuzawa, Y., Yoshikawa, K. et al. (1991) *FEBS Letters*, **295**, 67–69.
17 Takahashi, M., Yoshikawa, K., Vasilevskaya, V.V. and Khokhlov, A.R. (1997) *Journal of Physical Chemistry B*, **101**, 9396–9401.
18 Khan, M.O., Mel'nikov, S.M. and Jönsson, B. (1999) *Macromolecules*, **32**, 8836–8840.
19 Melnikov, S.M., Sergeyev, V.G. and Yoshikawa, K. (1995) *Journal of the American Chemical Society*, **117**, 9951–9956.
20 Melnikov, S.M., Sergeyev, V.G. and Yoshikawa, K. (1995) *Journal of the American Chemical Society*, **117**, 2401–2408.
21 Mel'nikov, S.M., Dias, R., Mel'nikova, Y.S. et al. (1999) *FEBS Letters*, **453**, 113–118.
22 Yamasaki, Y. and Yoshikawa, K. (1997) *Journal of the American Chemical Society*, **119**, 10573–10578.
23 Yoshikawa, K., Kidoaki, S., Takahashi, M. et al. (1996) *Berichte Der Bunsen-Gesellschaft-Physical Chemistry Chemical Physics*, **100**, 876–880.
24 Mel'nikov, S.M., Khan, M.O., Lindman, B. and Jönsson, B. (1999) *Journal of the American Chemical Society*, **121**, 1130–1136.
25 Dias, R., Mel'nikov, S., Lindman, B. and Miguel, M.G. (2000) *Langmuir*, **16**, 9577–9583.
26 Takagi, S., Tsumoto, K. and Yoshikawa, K. (2001) *Journal of Chemical Physics*, **114**, 6942–6949.

27 Zinchenko, A., Sakaue, T., Araki, S. *et al.* (2007) *The Journal of Physiology*, **111**, 3019–3031.
28 Morán, M.C., Pais, A.A.C.C., Gaweda, S. *et al.* (2008) *Journal of Colloid and Interface Science*, **323**, 75–83.
29 Dias, R.S., Lindman, B. and Miguel, M.G. (2002) *Journal of Physical Chemistry B*, **106**, 12608–12612.
30 Piculell, L. and Lindman, B. (1992) *Advances in Colloid and Interface Science*, **41**, 149–178.
31 Thalberg, K., Lindman, B. and Karlström, G. (1991) *The Journal of Physical Chemistry*, **95**, 3370–3376.
32 Thalberg, K., Lindman, B. and Bergfeldt, K. (1991) *Langmuir*, **7**, 2893–2898.
33 Norrman, J., Lynch, I. and Piculell, L. (2007) *Journal of Physical Chemistry B*, **111**, 8402–8410.
34 Rosa, M., Dias, R., Miguel, M.G. and Lindman, B. (2005) *Biomacromolecules*, **6**, 2164–2171.
35 Dias, R., Mel'nikov, S.M., Lindman, B. and Miguel, M.G. (2000) *Langmuir*, **16**, 9577–9583.
36 Dias, R., Rosa, M., Pais, A.C. *et al.* (2004) *Journal of the Chinese Chemical Society*, **51**, 447–469.
37 McGrath, K.M. (1995) *Langmuir*, **11**, 1835–1839.
38 Goddard, E. and Ananthapadmanabhan, K. (1993) *Interactions of Surfactants with Polymers and Proteins*, CRC Press, Boca Raton.
39 Thalberg, K. and Lindman, B. (1989) *The Journal of Physical Chemistry*, **93**, 1478–1483.
40 Carnali, J. (1993) *Langmuir*, **9**, 2933–2941.
41 Goddard, E. and Hannan, R. (1977) *Journal of the American Oil Chemists Society*, **54**, 561–566.
42 Ohbu, K., Hiraishi, O. and Kashiwa, I. (1982) *Journal of the American Oil Chemists Society*, **59**, 108–112.
43 Chen, L., Yu, S., Kagami, Y. *et al.* (1998) *Macromolecules*, **31**, 787–794.
44 Kim, B., Ishizawa, M., Gong, J. and Osada, Y. (1999) *Journal of Polymer Science Part A-Polymer Chemistry*, **37**, 635–644.
45 Thalberg, K., Lindman, B. and Karlström, G. (1990) *The Journal of Physical Chemistry*, **94**, 4289–4295.
46 Hansson, P. and Almgrem, M. (1994) *Langmuir*, **10**, 2115–2124.
47 Guillemet, F. and Piculell, L. (1995) *The Journal of Physical Chemistry*, **99**, 9201–9209.
48 Akinchina, A. and Linse, P. (2002) *Macromolecules*, **35**, 5183–5193.
49 Evans, D.F. and Wennerström, H. (1999) *The Colloidal Domain-Where Physics, Chemistry, Biology and Technology Meet*, 2nd edn, Wiley-VCH Verlag Gmbh, New York.
50 Alexandridis, P. and Lindman, B. (2000) *Amphiphilic Block Copolymers, Self-Assembly and Applications*, Elsevier, Amsterdam.
51 Piculell, L., Thuresson, K. and Lindman, B. (2001) *Polymers for Advanced Technologies*, **12**, 44–69.
52 Inman, R.B. and Jordan, D.O. (1960) *Biochimica et Biophysica Acta*, **42**, 421–426.
53 Korolev, N.I., Vlasov, A.P. and Kuznetsov, I.A. (1994) *Biopolymers*, **34**, 1275–1290.
54 Gaugain, B., Barbet, J., Capelle, N. *et al.* (1978) *Biochemistry*, **17**, 5078–5088.
55 Maheswari, P.U., Rajendiran, V., Palaniandavar, M. *et al.* (2006) *Inorganica Chimica Acta*, **359**, 4601–4612.
56 Eskilsson, K., Leal, C., Lindman, B. *et al.* (2001) *Langmuir*, **17**, 1666–1669.
57 Cárdenas, M., Braem, A., Nylaner, T. and Lindman, B. (2003) *Langmuir*, **19**, 7712–7718.
58 Cárdenas, M., Terán, J.C., Nylander, T. and Lindman, B. (2004) *Langmuir*, **20**, 8597–8603.
59 Orosz, J.M. and Wetmur, J.G. (1977) *Biopolymers*, **16**, 1183–1199.
60 Geiduschek, E.P. and Herskovits, T.T. (1961) *Archives of Biochemistry and Biophysics*, **95**, 114–129.
61 Levine, L., Jencks, W.P. and Gordon, J.A. (1963) *Biochemistry*, **2**, 168–175.
62 Costa, D., dos Santos, S., Antunes, F.E. *et al.* (2006) *Arkivoc*, 161–172.

63 Costa, D., Miguel, M.G. and Lindman, B. (2007) *Journal of Physical Chemistry B*, **111**, 10886–10896.
64 Morán, M.C., Miguel, M.G. and Lindman, B. (2007) *Langmuir*, **23**, 6478–6481.
65 Geiduschek, E.P. and Herskovits, T.T. (1961) *Archives of Biochemistry and Biophysics*, **95**, 114–129.
66 Hard, T. and Lundback, T. (1996) *Biophysical Chemistry*, **62**, 121–139.
67 Jen-Jacobson, L., Engler, L.E. and Jacobson, L.A. (2000) *Structure*, **8**, 1015–1023.
68 Garvie, C.W. and Wolberger, C. (2001) *Molecular Cell*, **8**, 937–946.
69 Cheetham, G.M.T., Jeruzalmi, D. and Steitz, T.A. (1999) *Nature*, **400**, 89–189.
70 Cheetham, G.M.T., Jeruzalmi, D. and Steitz, T.A. (1999) *Nature*, **399**, 80–83.
71 Ceccarelli, D.F.J. and Frappier, L. (2000) *Journal of Virology*, **74**, 4939–4948.
72 West, M., Flanery, D., Woytek, K. *et al.* (2001) *Journal of Virology*, **75**, 11948–11960.
73 West, M. and Wilson, V.G. (2002) *Journal of Virology*, **296**, 52–61.
74 Borer, P.N., Dengler, B., Tinoco, I. and Uhlenbeck, O.C. (1974) *Journal of Molecular Biology*, **86**, 843–853.
75 Sowerby, S., Cohn, C.A., Heckl, W.M. and Holm, N.G. (2001) *Proceedings of the National Academy of Sciences of the United States of America*, **98**, 820–822.
76 Babak, V.G., Merkovich, E.A., Galbraikh, L.S. *et al.* (2000) *Mendeleev Communications*, **3**, 94–95.
77 Julia Ferres, M.R., Erra Serrabasa, P., Muñoz Liron, I. and Ayats Llorens, A. (1998) Procedure for preparing capsules and for encapsulation of substances (Patent No ES2112150) Spain.
78 Lapitsky, Y. and Kaler, E.W. (2004) *Colloids and Surfaces A*, **250**, 179–187.
79 Lapitsky, Y., Eskuchen, W.J. and Kaler, E.W. (2006) *Langmuir*, **22**, 6375–6379.
80 Lapitsky, Y. and Kaler, E.W. (2006) *Colloids Surfaces A-Physicochemical and Engineering Aspects*, **282–283**, 118–128.
81 Morán, M.C., Miguel, M.G. and Lindman, B. (2007) *Biomacromolecules*, **8**, 3886–3892.
82 Peacocke, A.R. (1973) The interaction of acridines with nucleic acids, in *Acridines* (ed. R.M. Acheson), Interscience Publishers, New York, pp. 723–775
83 Costa, D., Miguel, M.G. and Lindman, B. (2007) *Journal of Physical Chemistry B*, **111**, 8444–8452.
84 Schneider, S. and Linse, P. (2003) *Journal of Physical Chemistry B*, **107**, 8030–8040.
85 Papancea, A., Valente, A.J.M., Patachia, S. *et al.* (2008) *Langmuir*, **24**, 273–279.
86 Rosa, M., Moran, M.D., Miguel, M.G. and Lindman, B. (2007) *Colloids and Surfaces A-Physicochemical and Engineering Aspects*, **301**, 361–375.
87 Rosa, M., Pereira, N.P., Simões, S. *et al.* (2007) *Molecular Membrane Biology*, 1–12.
88 Jadhav, V., Maiti, S., Dasgupta, A. *et al.* (2008) *Biomacromolecules*, **8**, 1852–1859.
89 Dasgupta, A., Das, P.K., Dias, R.S. *et al.* (2007) *Journal of Physical Chemistry B*, **111**, 8502–8508.
90 Lundberg, D., Stjerndahl, M. and Holmberg, K. (2005) *Langmuir*, **21**, 8658–8663.
91 Miguel, M., Pais, A.A.C.C., Dias, R.S. *et al.* (2003) *Colloids and Surfaces A-Physicochemical and Engineering Aspects*, **228**, 43–55.

11
Deposition of Colloid Particles at Heterogeneous Surfaces
Zbigniew Adamczyk, Jakub Barbasz, and Małgorzata Nattich

11.1
Introduction

Deposition (irreversible attachment) of colloid and bioparticles on solid/liquid interfaces is of major significance for predicting the efficiency and kinetics of many processes such as self-assembly, filtration, separation by affinity chromatography, immobilization at interfaces, removal of pathological cells, immunological assays, biofouling of transplants and artificial organs, and so on. Accordingly, particle deposition on homogeneous surfaces has been studied extensively both theoretically [1–6] and experimentally [7–26].

From a practical point of view, however, especially interesting seems the problem of particle deposition at heterogeneous surfaces, covered by sites such as polyelectrolyte chains [27–30], proteins [31, 32] or colloid nanoparticles [33–36].

Recently, much interest has focused on heterogeneous surfaces bearing patterned surfaces features of a regular shape like circles and dots [37, 38], squares [39], long rectangles (stripes) [40, 41] and others [42]. These features are usually produced on homogeneous surfaces by microcontact printing, photolithography or laser ablation. The growing interest in such surface architecture stems from their practical significance as anti-reflecting and self cleaning surfaces, biosensors (protein arrays) optical filters, masks, photonic crystals, microfluidic and microelectronic devices. This concerns, for example, the problem of particle deposition on stripes pertinent to microcircuitry.

Besides practical importance, the deposition of particles on heterogeneous surfaces, in particular surface features, has a major significance for colloid science, because interesting clues concerning the dynamic particle center interactions can be extracted by determining the maximum coverage and the topology of particle layers [43, 44].

Despite this considerable significance for basic science and practice, the problem of particle deposition at heterogeneous surfaces, especially on surface features of various shapes, has not been reviewed recently in the literature. Therefore, the main

Highlights in Colloid Science. Edited by Dimo Platikanov and Dotchi Exerowa
Copyright © 2009 WILEY-VCH Verlag GmbH & Co. KGaA, Weinheim
ISBN: 978-3-527-32037-0

goal of this chapter is to present recent developments in this field, with the emphasis on the theoretical predictions concerning the jamming limit and the structure of particle monolayers.

11.2
Theoretical Models

Depending on the ratio of the adsorbing particle dimension d_p to the surface feature characteristic dimension, denoted L_s, three main particle deposition regimes can be distinguished (Figure 11.1):

1. Quasi-continuous surface regime, appearing when $d_p/L_s \gg 1$;
2. Random site surface regime, $d_p/L_s \sim 1$;
3. Patterned surface regime, $d_p/L_s < 1$.

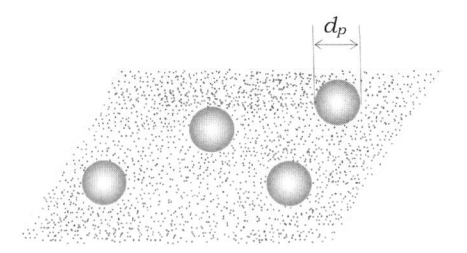

quasi – continuous surfaces $d_p/L_s \gg 1; (L_s - \text{site size})$

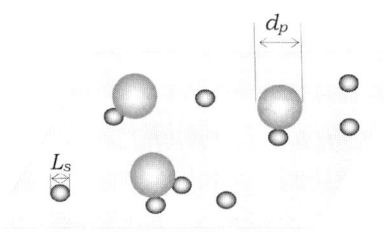

random site surfaces $d_p/L_s \sim 1; (L_s - \text{site size})$

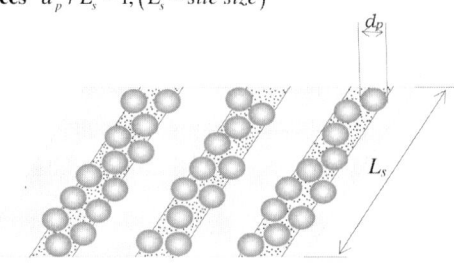

patterned surfaces $d_p/L_s < 1; (L_s - \text{surface feature size})$

Figure 11.1 Particle deposition regimes.

Interestingly, in case 3 two different situations can be distinguished, namely when the distance between surface features is much larger than particle dimensions, and the second limiting situation when the distance between surface features is comparable with particle size. In this work we focus solely on the former situation, referred to as the isolated surface feature regime.

11.2.1
Random Sequential Adsorption Approach

Deposition at quasi-continuous surfaces is often pertinent to larger colloid particles at uniform surfaces like mica, quartz, glass, polymeric surfaces, bearing sites, usually surface charges or ionic species of the nanometer size range [10, 13, 14, 17–22]. Because of its practical significance and the accessibility of a large quantity of experimental data, obtained by direct experimental techniques such as optical microscopy, this deposition regime has been the subject of extensive theoretical studies. Most of the theoretical results have been derived using various modifications of the random sequential adsorption (RSA) approach based on the Monte-Carlo type simulations [1–5, 13, 45–48]. The basic rules of the RSA simulation scheme are:

1. An adsorbing (virtual) particle is created whose position and orientation is selected at random within prescribed limits defining the simulation domain.
2. If the virtual particle fulfils prescribed criteria, it is adsorbed with unit probability and its position remains unchanged during the entire simulation process (localized and irreversible deposition postulate).
3. If the criteria are violated, for example, if the particles overlap, a new deposition attempt is made that is fully uncorrelated with previous attempts.

The dimensionless coverage of particles is defined as [48]:

$$\Theta_p = (S_g/\Delta S) N_p \qquad (11.1)$$

where S_g is the characteristic cross-section of the particle and ΔS is the surface area of the adsorbing surface (feature).

For spherical particles one can define S_g unequivocally as πa^2, so $\Theta_p = \pi a_p^2 N_p/\Delta S$. For non-spherical (anisotropic) particles the choice of the characteristic cross-section is not unique, however.

The basic RSA simulation scheme is applicable not only for the quasi-continuous deposition regime but also for deposition at surface features of various shapes and dimensions.

In contrast, deposition at surfaces bearing isolated sites of the size comparable with particles (Figure 11.1) is modeled in terms of the modified RSA simulation scheme whose first step consists in the deposition of sites according to the classical RSA of desired surface density N_s and the coverage $\Theta_s = \pi a_s^2 N_s/\Delta S$ (where a_s is the site radius). Then, modeling of the RSS process for the site covered surface is carried out according to the algorithm whose main steps are [49, 50]:

1. An adsorbing (virtual) particle is generated at random within the simulation area ΔS with periodic boundary conditions on its perimeter; if it does not touch any of sites, the particle is rejected and another virtual particle is generated (the number of attempts N_{at} is increased by one).

2. Otherwise, if the particle touches any of the sites, the overlapping test is performed according to the usual RSA rules, that is, it is checked if there is any previously adsorbed particle within the exclusion area; if so, the simulation loop is repeated (the number of attempts was increased by one).

3. If there is no overlapping, the virtual particle is assumed to be adsorbed irreversibly at a given position.

Simulations of random site deposition can be carried out either for point-like sites distributed perfectly random over a homogeneous surface (Jin et al. [51]) or for sites having finite dimensions and the shape of hard disks of diameter $2a_s$ incorporated into the substrate (Figure 11.2a) [49] or hard spheres attached to the surface, (Figure 11.2b) [50].

From an experimental point of view, of more interest is the deposition model exploiting spherically shaped sites. The basic assumption of this model is that the colloid particle can only be adsorbed upon touching the site (Figure 11.2b). Otherwise, at bare interface, the particle will not adsorb. Physically, this corresponds to the situation where the particles are irreversibly bound to the sites due to short-range attractive interactions of an electrostatic or chemical nature. Interestingly, particle deposition at heterogeneous surfaces covered by spherically shaped sites is a three-dimensional process, contrary to deposition at disk-shaped sites.

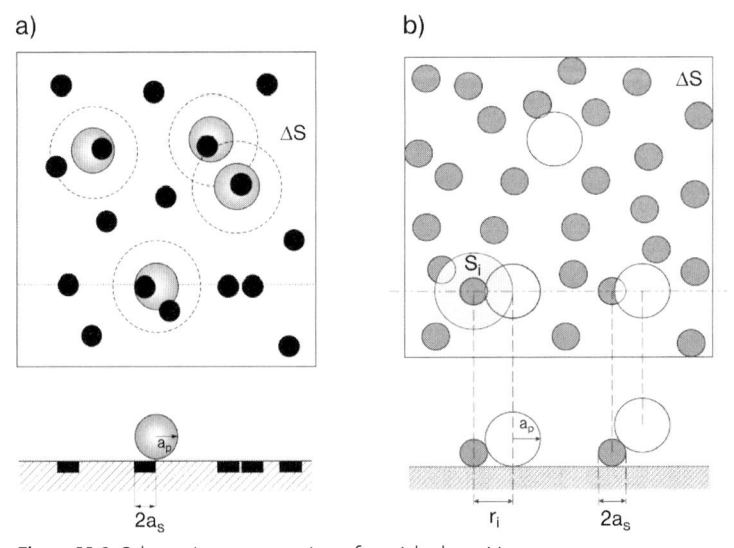

Figure 11.2 Schematic representation of particle deposition on heterogeneous surfaces bearing disk-shaped adsorption sites (a) and spherically-shaped sites (b).

11.3
Illustrative Theoretical Results

11.3.1
Deposition at Quasi-Continuous Surfaces

Because of its major significance, deposition at quasi-continuous surfaces has been investigated extensively in terms of the RSA model. Most results concern hard spherical particle deposition at planar interfaces of infinite extension [1, 5, 44–48]. However, there exist also results for polydisperse spherical particles [52] and for anisotropic hard particles of a convex shape like squares [53], rectangles (cylinders), spherocylinders (disk rectangles) and ellipses (spheroids) [2]. Results are also available for particles interacting via the short-range repulsive potential stemming from the electric double layers [5, 13, 43, 44, 48].

The most important parameter determined in these simulations was the jamming coverage Θ_∞, which characterizes the interface capacity for accommodating particles. For spherical particles $\Theta_\infty = 0.547$ [1, 46, 48], which is markedly smaller than the maximum hexagonal packing of spheres in 2D, which is equal to $\pi/2\sqrt{3} = 0.9069$, or the regular packing, which is $\pi/4 = 0.7854$ [48]. For squares, $\Theta_\infty = 0.53$ and for an ellipse of axis ratio 2 : 1, $\Theta_\infty = 0.583$ [2]. Values of Θ_∞ for particles of other shapes are given in Ref. [48]. As discussed there, the variation in Θ_∞ among various particle shapes is rather insignificant, with the exception of very elongated objects. Thus, for example, Θ_∞ for a cylinder with the axis ratio 15 : 1 the jamming coverage is 0.445. These values are pertinent to the side-on deposition of particles. However, for very elongated particles, their deposition at quasi-continuous surfaces can also occur at arbitrary orientations, which can be described in terms of the unoriented deposition regime, analyzed in detail in Refs. [3, 48, 54, 55].

It was shown [48] that the above results obtained for hard particles can also be used in interpreting the experimental results for particles interacting via the short-range Yukawa potential upon defining the effective interaction range h^* [48]. Using this concept one can calculate the jamming coverage for interacting particles (referred to as the maximum coverage) from the simple relationship valid for both spherical and anisotropic particles [48]:

$$\Theta_{mx} = \Theta_\infty \frac{1}{(1+h^*)^2} \tag{11.2}$$

The above theoretical data obtained for quasi-continuous surfaces can be used as reference states for the interpretation of particle deposition at random site surfaces, as discussed below.

11.3.2
Deposition at Random Site Surfaces

The random site surface (RSS) regime is pertinent to nanoparticle (colloids, proteins, polyelectrolytes) deposition at surfaces bearing isolated centers such as ions, smaller particles or macromolecules able to bind the solute particles irreversibly.

The most important parameters characterizing the RSS process are the initial deposition probability governing the kinetics, expressed usually in terms of the site coverage Θ_s, and the jamming coverage of particles. For disk shaped sites, one can easily deduce that the initial deposition probability of particles (in the limit when their coverage remains very low) equals [49]:

$$p_o = \Theta_s \tag{11.3}$$

This is so because a particle can only adsorb if the projection of its center lies within the disk area (Figure 11.2). Interestingly, Equation 11.3 is valid for arbitrary coverage of sites and their distribution.

In contrast, for spherical sites, particles can be adsorbed upon touching the surface of the site, so the effective interaction area in this case is $S_i = 4\pi a_s a_p$ (Figure 11.2b). Therefore, the initial deposition probability for low site coverage is given by:

$$p_o = 4\lambda' \Theta_s \tag{11.4}$$

As can be noticed, in this case, p_o increases proportionally to the particle-to-site size ratio λ', exceeding significantly the adsorption probability on disk shaped sites, characterized by the same coverage.

However, for higher coverage of spherical sites, p_o increases at a smaller rate with their coverage as Equation 11.4 predicts, because the effective interaction areas start to overlap. Therefore, for higher coverage Θ_s, p_o can be well approximated by the Poisson distribution [50]:

$$p_o = 1 - e^{-S_i N_s} = 1 - e^{-h\lambda' \Theta_s} \tag{11.5}$$

More accurate expressions for p_o can be derived in the limit of higher Θ_s by exploiting the scaled particle theory (SPT) of a 2D hard particle fluid, which predicts the following expression [50]:

$$p_o = 1 - (1-\Theta_s) e^{-\frac{(4\lambda'-1)\Theta_s}{1-\Theta_s} - \left[\frac{(2\lambda'^{1/2}-1)\Theta_s}{1-\Theta_s}\right]^2} \tag{11.6}$$

The p_o function is of a primary interest because it represents the averaged probability of adsorbing a particle at surfaces covered by a given number of sites. Hence, by knowing p_o one can calculate the initial flux of solute (particles) to heterogeneous surfaces. Figure 11.3 shows the dependence of p_o on Θ_s calculated numerically for spherical sites using the RSA simulation scheme, for $\lambda' = 2$, 5 and 10. As can be noticed, the p_o of particles increases abruptly with Θ_s, especially for larger λ' values. For $\lambda' = 10$, p_o reaches unity (the value pertinent to homogeneous surfaces) for Θ_s as low as 0.1. This behavior is well reflected by Equation 11.6, being in a quantitative agreement with the numerical data for the entire range of Θ_s and λ' studied.

In addition, p_o at spherical sites becomes considerably larger than for disk-shaped sites (dashed line in Figure 11.3). For example, at $\lambda' = 2$, p_o for spherical sites increases proportionally to $8\Theta_s$ (for $\Theta_s < 0.1$) whereas for the disk sites it increases proportionally to Θ_s. For $\lambda' = 10$, this difference becomes even more pronounced because for spherical sites $p_o \sim 400\,\Theta_s$ initially, whereas for disks $p_o \sim \Theta_s$. This

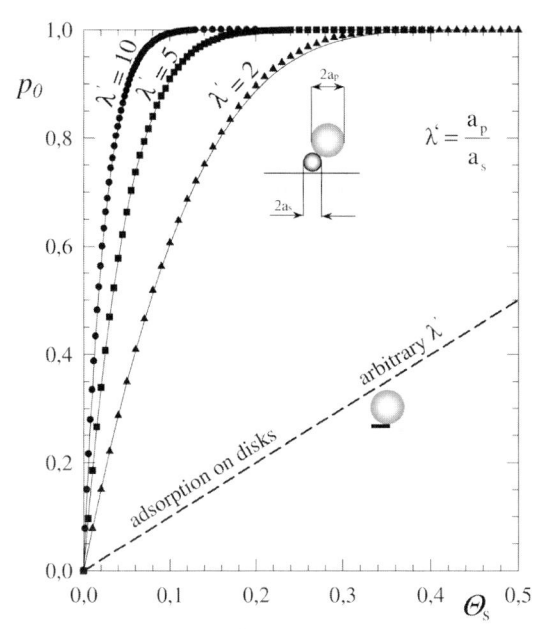

Figure 11.3 Dependence of the initial deposition probability p_0 on site coverage Θ_s. The points denote the numerical simulations performed for spherically shaped sites and $\lambda' =$ 10, 5 and 2; the solid lines represent the analytical results calculated from Equation 11.6 and the dashed line shows the analytical predictions for disk-shaped sites, that is, $p_0 = \Theta_s$.

observation has practical implications, showing that the geometry of adsorption sites plays a more decisive role than their surface concentration.

The results shown in Figure 11.3 describe the particle deposition rate at heterogeneous surfaces in the limit when their accumulation is negligible, that is, for $\Theta_p \to 0$ only. If Θ_p becomes finite, the probability of particle deposition decreases as a result of volume exclusion effects, which are analyzed in detail in Refs. [48–50].

It was shown [51] that the jamming coverage of particles in the RSS process involving point-like sites can be predicted from the interpolating function:

$$\Theta_p^\infty(\Theta_s) = 1 - \frac{1 + 0.314\Theta_s^2 \lambda'^4 + 0.450\Theta_s^3 \lambda'^6}{1 + 1.83\Theta_s \lambda'^2 + 0.660\Theta_s^3 \lambda'^6 + \Theta_s^{7/2} \lambda'^7} \tag{11.7}$$

For not too large Θ_s, Equation 11.7 can well be interpolated by a much simpler function:

$$\Theta_p^\infty(\Theta_s) = \Theta_\infty \frac{\Theta_s \lambda'^2}{\Theta_\infty + \Theta_s \lambda'^2} \tag{11.8}$$

In the limit of $\Theta_s \lambda'^2 \ll 1$ this simplifies to:

$$\Theta_p^\infty = \Theta_s \lambda'^2 \tag{11.9}$$

The analytical results expressed by Equations 11.7 are strictly valid in the limit of negligible site dimension compared to adsorbing particle size, that is, for $\lambda' \gg 1$. Extensive simulations for the more realistic situation of finite site dimension were performed in Ref. [49] by applying the RSS algorithm described above. Figure 11.4a shows the dependence of the jamming coverage derived from these simulations on the site coverage. As can be seen, the low coverage asymptotic formula, Equation 11.9, describes well the exact results for $\Theta_p^\infty < 0.2$ only. Interestingly, the jamming coverage Θ_p^∞ shown in Figure 11.4a in the case of disk-shaped sites never exceeds the jamming limit in the RSA process for homogeneous surfaces, that is, $\Theta_\infty = 0.547$.

Theoretical results were also derived in the case of random site surfaces bearing spherical sites [50]. As already noticed in Figure 11.3, the initial deposition probability is much higher in this case than for the disk sites. Also, in contrast to disks, there appears a possibility of multiple coordination of one site for $\lambda' < 4$, which influences profoundly the jamming coverage and the topology of particle monolayers. Unfortunately, for spherical sites, because of the overlapping of the interaction areas, no simple analytical relationships describing the jamming limit, can be derived. These data are accessible from numerical simulations only, obtained by applying the RSS algorithm [35, 36]. In Figure 11.4b the dependence of the jamming coverage of particles Θ_p^∞ on the site coverage Θ_s derived from these simulations is plotted on a logarithmic scale to cover a broad range of site coverage. The jamming coverage dependence on Θ_s can well be approximated for $\lambda' > 2$ by the interpolating function [35, 36]:

$$\Theta_p^\infty = \Theta_\infty \left(1 - e^{-\frac{\lambda'^2 \Theta_s}{\Theta_\infty}}\right) \tag{11.10}$$

Equation 11.10 gives a satisfactory accuracy for the entire range of Θ_s studied. However, it breaks down for $\lambda' = 2$ and 1 when a maximum on the Θ_p versus Θ_s dependence is observed at Θ_s equal to 0.25. This maximum jamming coverage for $\lambda' = 2$ attained the value of 0.57, which is slightly larger than $\Theta_\infty = 0.547$. For $\lambda' = 1$ the maximum attained the value of 0.66. The maxima appear because the area accessible for adsorption becomes larger than the geometrical interface area. This is so because particles can touch the sites at various distances from the interface. When Θ_s increases further above this critical value, the accessible area again becomes very close to the geometrical interface area. This hypothesis is confirmed by the fact that in the case of disk-shaped sites, where all particles are adsorbed at the interface only in one plane, no such maximum appeared (Figure 11.4a).

It is also interesting to observe, by comparing Figure 11.4a, b, that for $\lambda' < 4$ the jamming coverage in the case of spherical adsorption sites increases with Θ_s more abruptly than for disk-shaped sites. This is so because for $\lambda' < 4$ one site can coordinate more than one particle. The number of particles coordinated by one site, referred to as the site coordination number, can be determined most directly by plotting the numerically determined dependence of the jamming coverage Θ_p^∞ on the parameter $\lambda'^2 \Theta_s$. The slope of this dependence in the limit of Θ_s tending to zero, when the volume exclusion effects between adsorbed particles remain negligible,

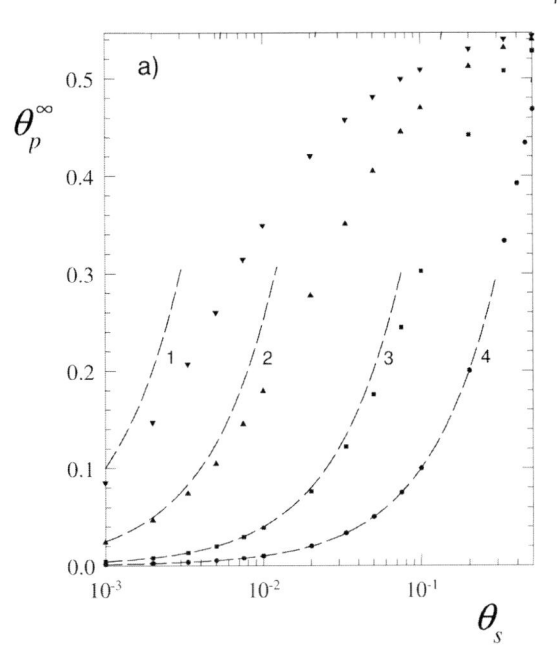

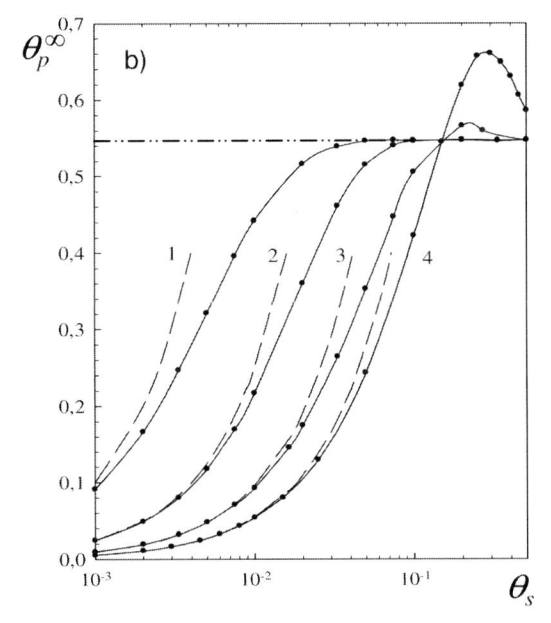

Figure 11.4 (a) Dependence of the jamming coverage of particles Θ_p^∞ on the coverage of the disk-shaped adsorption sites Θ_s; the points denote the results of numerical simulations, performed for $\lambda' =$ (1) 10, (2) 5, (3) 2 and (4) 1; dashed lines show the results derived from the equation $\Theta_p^\infty = \lambda'^2 \Theta_s$. (From Ref. [51].) (b) Dependence of jamming coverage of particles Θ_p^∞ on the coverage of the spherically-shaped sites Θ_s; points denote the results of numerical simulations, performed for $\lambda' =$ (1) 10, (2) 5, (3) 2 and (4) 1; solid lines represent the results calculated from the fitting function given by Equation 11.10 (for $\lambda' < 2$) and dashed lines show the results derived from Equation 11.12., $\Theta_p^\infty = l_s \lambda'^2 \Theta_s$. (From Ref. [36].)

gives directly the site coordination number l_s. The dependence of l_s on λ' can be well fitted by the simple interpolating function [35]:

$$l_s = 5.967/\lambda' - 0.517 \tag{11.11}$$

Obviously, Equation 11.11 is valid for $1 \leq \lambda' < 4$, otherwise for $\lambda' \geq 4$, $l_s = 1$.

As can be predicted from Equation 11.11, l_s increases abruptly when the size of the particle approaches the site dimension, that is for $\lambda' \to 1$, where it becomes equal to 5.45. This means that one center can coordinate on average more than five adsorbing particles. In contrast, for $\lambda' = 2$, $l_s = 2.45$. This suggests that surface clusters composed of a desired number of particles can be produced by appropriately adjusting the particle-to-site size ratio λ'.

Using Equation 11.11 one can also deduce that the jamming coverage can well be approximated for small values of $\lambda'^2 \Theta_s$ by the interpolating function:

$$\Theta_p^\infty = \lambda'^2 l_s \Theta_s = \left(\frac{5.967}{\lambda'} - 0.517\right)\lambda'^2 \Theta_s \tag{11.12}$$

As can be seen in Figure 11.4b, the limiting analytical results predicted from Equation 11.12 reflect well the exact numerical data if $\Theta_p^\infty < 0.2$.

11.3.3
Particle Deposition at Surface Features

In the case of particle deposition at isolated surface features, often referred to for brevity as collectors, one can apply a modified version of the RSA simulation scheme described above. Instead of the periodic boundary conditions at the perimeter of the simulation domain, one postulates that the particles can only adsorb if the projection of their centers lies within the surface area of the feature (Figure 11.5) [56, 57].

Interpretation of numerical results obtained in this case is facilitated by the fact that a few analytical solutions exist in the limit of line segments and infinitely long lines (thin stripes). Also, useful analytical expressions can be derived for surface features whose surface area is considerably larger than the particle cross-section area [57].

For example, it was shown in Ref. [58] that, in the limit of $\bar{L} \to \infty$, the jamming coverage for infinitely line is described by the expression:

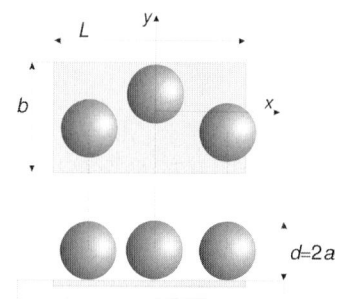

Figure 11.5 Schematic representation of particle deposition at surface features.

$$\Theta_{\infty 1D} = \int_0^\infty \exp\left[-2\int_0^{t'}\frac{1-e^{-\xi}}{\xi}d\xi\right]dt' = 0.7476 \qquad (11.13)$$

This value of $\Theta_{\infty 1D}$, referred to as the jamming limit in one dimension, was obtained originally by Renyi and others [59, 60]. Interestingly, a very similar value of the jamming limit (i.e. 0.7506) was obtained in the diffusion RSA process solved analytically in Ref. [61].

Knowing $\Theta_{\infty 1D}$ one can predict that in the limit $L/d \gg 1$ (infinitely long line), the averaged number of particles adsorbed on the collector is given by:

$$\langle N_p \rangle = 0.7476\,\bar{L} \qquad (11.14)$$

In contrast, for 2D collectors (circles, rectangles), one can predict that in the case when adsorbing particles are much smaller than the collector characteristic dimension, that is, if $L/d \gg 1$, the limiting expression for the averaged number of adsorbed particles is given by:

$$\langle N_p \rangle = \Theta_\infty S_c/S_g \qquad (11.15)$$

In particular, for circles, Equation 11.15 becomes:

$$\langle N_p \rangle = \Theta_\infty \bar{L}^2 \qquad (11.16)$$

Analogously, for rectangles, one can derive the expression:

$$\langle N_p \rangle = \frac{4}{\pi}\Theta_\infty \bar{b}\,\bar{L} \qquad (11.17)$$

where $\bar{b} = b/d$ and b is the collector width.

Therefore, for squares, where $\bar{b} = \bar{L}$, one has:

$$\langle N_p \rangle = \frac{4}{\pi}\Theta_\infty \bar{L}^2 \qquad (11.18)$$

With an intermediate range of the L/d and b/d parameters, meaningful results can be obtained by numerical simulations only, which have been performed for line segments, either rectilinear or bent [56], as well as for circles, rectangles and stripes [57].

Figure 11.6 shows the dependencies of the averaged number of particles adsorbed under the jamming state on the $\bar{L}$ parameter for line segments of various shape. As can be noticed, for straight line segments, the theoretical results for $\bar{L}>2$ can be well fitted by the function:

$$\langle N_p \rangle = 0.505 + 0.748\,\bar{L} \qquad (11.19)$$

This theoretical result suggests that the length of linear nanostructures, for example, polyelectrolyte chains, adsorbed at solid surfaces invisible under a microscope can be determined by inverting Equation 11.19, which results in the simple relationship:

$$L = d\,[1.337\langle N_p\rangle - 0.675] \qquad (11.20)$$

As can be seen in Figure 11.6, bending the collector to the form of a semicircle (curve 2) exerts a negligible effect on the value of $\langle N_p \rangle$ if $\bar{L}>2$. This observation suggests that

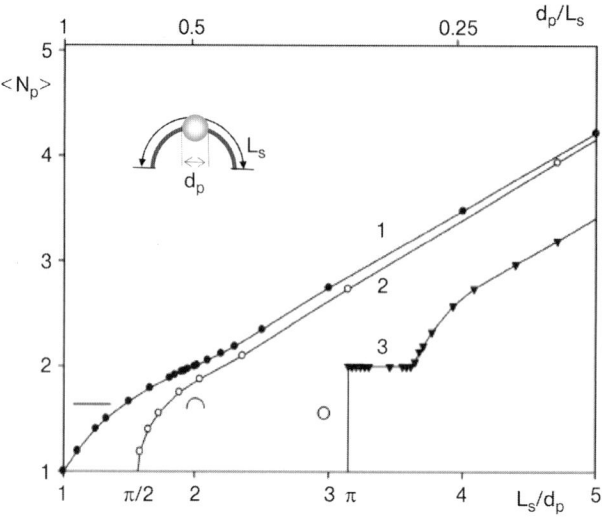

Figure 11.6 Averaged number of particles deposited at the jamming state on linear collectors of various shape $\langle N_p \rangle$ versus the L_s/d_p parameter derived from numerical simulations (points). (1) Line segments: for $1 < \bar{L} < 2$ $\langle N_p \rangle = -0.0147\bar{L}^4 + 0.402\bar{L}^3 - 2.16\bar{L}^2 + 4.90\bar{L} - 2.10$. $<N_p> = 0.505 + 0.748\ \bar{L}$, for $\bar{L} > 2$ (2) Semicircles: $\langle N_p \rangle = 1$ for $0 < \bar{L} < \pi/2$, $\langle N_p \rangle = -6.25 + 33.4\bar{L}^{-1} - 34.4\bar{L}^{-2}$ for $\pi/2 < \bar{L} < 2$, $\langle N_p \rangle = 0.747\bar{L} + 0.413$ for $\bar{L} > 2$. (3) Circles: $\langle N_p \rangle = 1$ for $0 < \bar{L} < \pi$, $\langle N_p \rangle = 2$ for $\pi < \bar{L} < 1.15\pi$, $\langle N_p \rangle = (1 - 0.302\bar{L})/(0.249 - 0.082\bar{L})$ for $1.15\pi < \bar{L} < 4$. (From Ref. [58].)

Equation 11.20 can be used effectively to evaluate the length of elongated nanoparticles having a shape that deviates significantly from a rectilinear shape. However, for circles, Equations 11.19 and 11.20 seem to be valid only for $\bar{L} > 10$.

The above theoretical results were, notably, obtained for quasi-one-dimensional collectors of negligible surface area. In this case the two-dimensional surface coverage was infinite, and the only meaningful parameter was the one-dimensional coverage Θ_{1D} (averaged number of particles per unit length of the collector). Analogous results, describing the dependence of the number of adsorbed particles on the $\bar{L}$ and b parameters, in the case of two-dimensional collectors like stripes and dots are discussed in Ref. [57].

11.4
Comparison with Experimental Results

The range of applicability of the above theoretical approaches and numerical results can be estimated by comparing them with experimental results obtained, mostly, by direct optical microscope observation and for monodisperse polystyrene latex particles [8–10, 13, 14, 19, 21, 22, 34–36]. Several sets of experimental data have also been gathered using the AFM for latex [17, 20], colloid gold [62] or dendrimer suspensions [26]. One can also apply electron microscopy to determine particle coverage [15] or indirect methods like reflectometry [12, 18, 26], ellipsometry [63], radioisotopic labeling [64–66] or streaming potential [29, 30, 67–69]. The latter method is especially sensitive and can be used for particles of arbitrary size, ranging from ions, polyelectrolytes, proteins to large colloid particles of micrometer size range.

Especially reliable results have been obtained for quasi-homogeneous surfaces, mostly mica modified by adsorption of silanes, aluminium compounds [13, 14, 19, 21, 22] or polyelectrolytes [27–32] using the diffusion [17, 20, 21, 33–35] or impinging-jet cells [8–14, 19, 21, 22]. This was so because adsorbed particles can be enumerated in real time using videomicroscopy or image analyzing systems, allowing one to measure the coordinates of thousands of particles at a prescribed time [16, 19, 21, 22, 29, 30, 33–35]. In this way one can follow not only the kinetics of particle deposition but also the distribution of particles over interfaces, the structure of their monolayers and the maximum (jamming) coverage.

Because of its significance, many systematic studies have been carried out in the literature with the aim of determining the maximum (jamming) coverage of particle as a function of the double-layer thickness $Le = (\varepsilon kT/2e^2 I)^{1/2}$, where ε is the dielectric permittivity of the dispersing medium and e is the elementary charge. Most of these studies were carried out using monodisperse latex particles [17, 20, 21], colloid gold particles [62] or dendrimers [26], under the diffusion transport conditions or forced convection transport in the impinging-jet cells. The coverage of particles was determined by optical microscopy, AFM, electron microscopy and reflectometry. These results are collected in Figure 11.7, shown in the reduced form, that is, as the dependence of $\theta_{mx}/\theta_\infty$ on the a/Le parameter (the ratio of particle size to the double layer thickness). Values of $a/Le \gg 1$ are characteristic for larger colloid particles and large ionic strength, whereas values comparable to unity are characteristic for small colloid particles (e.g. gold) or dendrimers. As can be seen in Figure 11.7, the experimental data collected under various conditions can well be reflected by theoretical predictions derived from the RSA simulations described above. Interestingly, however, a better agreement with experimental results is shown with the theoretical data obtained for finite deposition time, equal to ten in dimensionless units, rather than for infinite simulation time. This indicates that in some of these experimental studies the maximum coverage was not achieved because of the limited experimental time or too low a concentration of adsorbing particles.

Further experimental results pertinent to quasi-uniform surfaces, mostly kinetic data, have been discussed in review papers [36, 43, 44] and monographs [70, 71].

The results obtained for quasi-homogeneous surfaces can be used for the interpretation of data derived for heterogeneous surfaces.

Figure 11.8 shows typical results, obtained as the dependence of the initial deposition probability (expressed in terms of the reduced mass transfer rate k_c/k_c^0) of negatively charged polystyrene latex particles (averaged diameter 0.9 μm) on the coverage of spherically shaped sites (positively charged polystyrene latex particles 0.45 μm in diameter, deposited on a bare mica surface) [72]. As can be seen, the deposition probability (reduced initial transfer rate) increased abruptly with the site coverage and attained maximum values pertinent to quasi-continuous surfaces

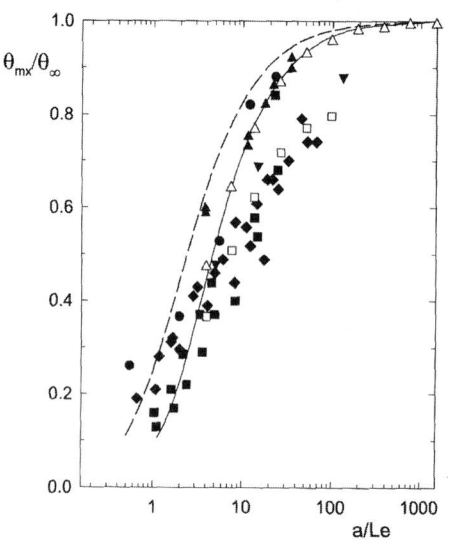

Figure 11.7 Collection of experimental data showing the dependence of the maximum coverage of colloid particles $\Theta_{mx}/\Theta_\infty$ on the a/Le parameter: (●) Johnson and Lenhoff data, AFM, natural convection [17]; (▲) Adamczyk et al. optical microscopy, gravity [19]; (▼) Adamczyk and Szyk, optical microscopy, diffusion [21]; (■) Böhmer et al., reflectometry, RIJ cell [18]; (♦) Harley et al., electron microscopy, diffusion [15]; (open triangles) theoretical RSA simulations $\tau = 10^5$; (□) theoretical RSA simulations $\tau = 10$. The solid and dashed lines represent the analytical approximation calculated from Equation 11.2 for $\phi_0 = 100$ and $10\,kT$, respectively. (From Ref. [44].)

for site coverage θ_s as low as 5%. This behavior was interpreted quantitatively in terms of the theoretical model developed in Refs. [6, 48], where the coupling between the bulk transport and the surface transport described by the deposition probability given by Equation 11.6 was considered. Physically, the coupling model accounts for the fact that a particle after a failed deposition attempt does not return immediately to the bulk, as it is postulated in the standard RSA model, but has a finite chance to adsorb nearby. According to this extended RSA model, the reduced initial deposition rate of particles can be expressed as [6, 44, 48]:

$$k_c/k_c^0 = \frac{K p_o(\theta_s)}{1+(K-1)p_o(\theta_s)} \qquad (11.21)$$

where $K = k_a/k_c^0$ is the coupling constant k_a is the deposition rate constant and $p_o(\theta_s)$ is the probability of particle deposition given by Equation 11.6. As can be seen in Figure 11.8, the theoretical predictions derived from Equation 11.21 with $K = 8$ [72] describe well the experimental data, whereas the theoretical results, which neglect the bulk transport (shown by the dashed line), proved inadequate. The results shown in Figure 11.8 have a major practical significance, indicating that, because of the coupling effect, the maximum deposition rate of particles is attained for site coverage as low as a few per cent. This is so, however, if the binding

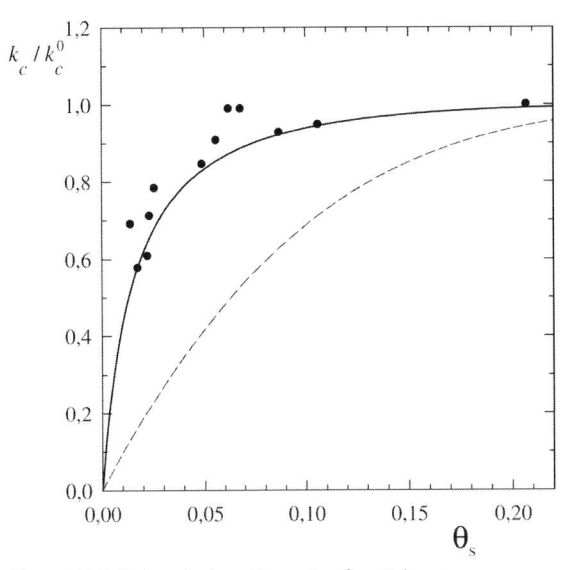

Figure 11.8 Reduced adsorption rate of particles at heterogeneous surfaces (k_c/k_c^0) versus spherical site coverage (θ_s). Points denote the experimental results. The solid line shows the theoretical results calculated from Equation 11.21 with $K = 8$ and the dashed like shows the theoretical results calculated by neglecting the coupling, that is, $K = 0$.

energy between centers and particles is large enough to ensure irreversible and localized deposition.

This assumption may no longer be valid for sites of the nanometer size range (e.g. polyelectrolyte chains), as demonstrated in Ref. [29]. In this work these sites were produced by covering bare mica with PAH (polyallyamine hydro chloride) of molecular weight of 70 000, positively charged for the pH range of these experiments. The averaged hydrodynamic radius of the polyelectrolyte as determined by dynamic light scattering was 20.7 nm [29]. The initial deposition rate of negatively charged polystyrene latex particles of 660 nm in diameter was studied by optical microscopy using the impinging-jet cell. Figure 11.9 shows the obtained dependence of the reduced deposition rate constant of particles (k_c/k_c^0) on the PAH coverage of mica. As can be seen, the theoretical results derived from Equation 11.22 with the coupling constant $K = 5.48 \times 10^3$ (corresponding to the experimental conditions) do not reflect properly the experimental data collected for the ionic strengths of 10^{-2} M and 10^{-3} M. The experimentally determined rate constant increases much more slowly than the standard coupling theory predicts (curve 1 in Figure 11.9). Thus, a "threshold" surface coverage of polyelectrolyte was needed to produce a measurable deposition rates of latex. This behavior was quantitatively accounted for by postulating that an efficient immobilization of massive latex particles was possible on centers exhibiting high local charge (positive), that is, formed by a cluster composed of a few closely adsorbed PAH molecules [29]. It was shown, by assuming the Poisson distribution of

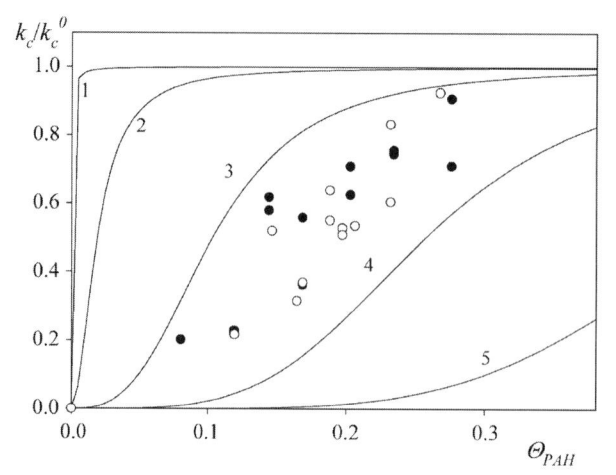

Figure 11.9 Dependence of the reduced initial deposition rate (k_c/k_c^0) on PAH coverage (θ_{PAH}) [29]. The solid lines were calculated from Equation 11.22 with the coupling constant $K = 5.48 \times 10^3$ and the number of macro-ions forming active centers: $n_s =$ (line 1) 1, (2) 2, (3) 3, (4) 4 and (5) 5. Filled and open points represent experimental data obtained for $I = 10^{-2}$ and 10^{-3} M, respectively. (From Ref. [29].)

particles in the adsorbed PAH layer, that the reduced rate constant of latex particles can be expressed in this case as:

$$k_c/k_c^0 = \frac{K \frac{\Theta_s^{n_s}}{n_s!}}{1+(K-1)\frac{\Theta_s^{n_s}}{n_s!}} \qquad (11.22)$$

where n_s is the number of PAH molecules forming the cluster (site).

As can be seen in Figure 11.9, the experimental data are well reflected by the theoretical curves calculated from Equation 11.22 for n_s between 3 and 4. This observation has a practical implication, proving that colloid particle deposition can be exploited as a sensitive tool for detecting the structure of the surface heterogeneities in the nanometer size range, in particular the charge heterogeneity.

The behavior shown in Figure 11.9 seems to be universal, because it has been observed in other polyelectrolyte/substrate systems, for example, poly(ethylene imine) (PEI)/mica [30] or pDMAEMA/silica [28]. Because of similarities in chemical composition, it can be expected that an analogous particle deposition mechanism will appear in the case of surfaces covered by protein layers.

Information on surface charge heterogeneity can be extracted easily by correlating the reduced deposition rate of particles with the averaged zeta potential rather than with the coverage of sites. The averaged zeta potential of heterogeneous surfaces can be determined directly using the streaming potential method [69, 71]. Notably, however, the zeta potential is a global quantity that characterizes the overall surface of the collector.

11.4 Comparison with Experimental Results | 219

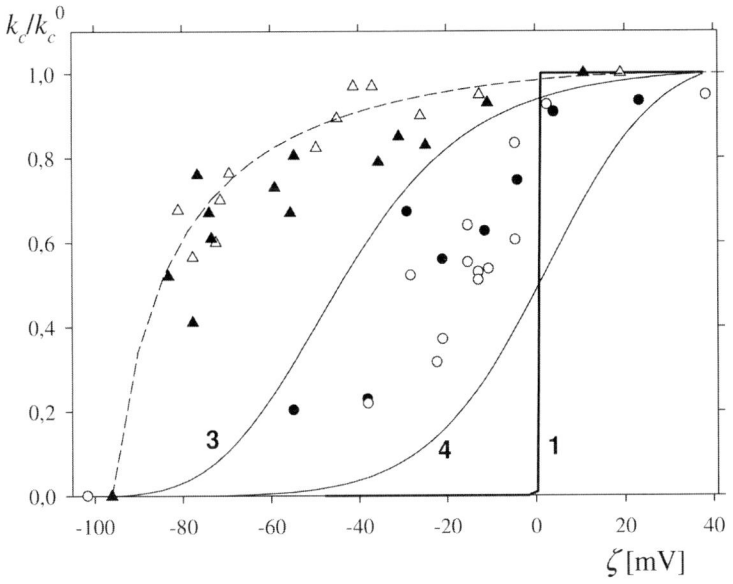

Figure 11.10 Dependence of the reduced initial deposition rate (k_c/k_c^0) on the apparent zeta potential of surfaces covered by PAH (circles) or positively charged latex particles (triangles) used as centers [29]. The solid line labeled 1 denotes the theoretical results calculated from the DLVO theory. Solid lines 3 and 4 were calculated from Equation 11.22 with the coupling constant $K = 5.48 \times 10^3$ and the number of macro-ions forming active centers $n_s = 3$ and 4, respectively. The dashed line was calculated for particle covered surfaces using Equation 11.21 with $K = 7$ and $\lambda' = 2.94$. Filled and open circles represent experimental data obtained for surfaces covered by PAH adsorbed from solutions of $I = 10^{-2}$ and 10^{-3} M, respectively. Full and empty triangles represent data obtained for Re = 2 and 8, respectively. (From Ref. [29].)

Our experimental data for latex deposition at a PAH layer, expressed in this form, are presented in Figure 11.10. The solid lines denote the theoretical predictions derived from Equation 11.22 for n_s between 3 and 4. As can be seen, there was significant deposition of particles even in the case when the apparent zeta potential of the surface was of the same sign as the particle zeta potential. Similar observations were made in the case of the deposition of silica particles onto quartz sand modified to various degrees by the silanization procedure [28]. These observations contradict theoretical predictions stemming from the classical DLVO theory depicted by the solid line 1 in Figure 11.10 [29, 71]. According to this theory, treating surface charge as uniformly smeared out over the surface, the particle flux should decrease to negligible values for zeta potentials below -1 mV.

Thus, the results shown in Figure 11.10 furnish direct support for the heterogeneity hypothesis often used to interpret colloid particle deposition kinetics under a barrier-controlled deposition regime [73]. In these experiments, particle deposition kinetics were determined as a function of salt concentration added to reduce the electrostatic energy barrier. Measured deposition rates were found by orders of magnitude to be higher than the DLVO theory prediction, analogously as for our case of particle deposition on PAH-covered surfaces.

Further support for the heterogeneity hypothesis stems from comparison of our data for PAH with previous measurements of particle deposition on sites formed by deposition of oppositely charged latex on mica [74], for a particle size ratio 2.9. The apparent zeta potential of mica as a function of site coverage was measured by the streaming potential method, analogously as in the present case. However, both the coverage of sites and the deposited particles were evaluated directly by microscope counting, which makes these data a useful reference system. As can be seen in Figure 11.10, a significant deposition of particles was observed for negative values of the apparent zeta potential of the surface, that is, of the same sign as adsorbing particles. This was so because the electrical double-layer thickness was much smaller than the site dimensions; therefore, a considerable part of the site charge remained available for adsorbing particles. As a result, particle deposition was governed by local electrostatic interactions rather than by the averaged value of the zeta potential. Interestingly, this behavior, which contradicts the classical DLVO theory, can well be accounted for by the extended RSA model, where it was postulated that only one site is sufficient to effectively immobilize the adsorbing particle (see dashed line in Figure 11.10) [52].

The results shown in Figures 11.9 and 11.10 concerned the initial deposition stage when surface blocking effects were negligible and the deposition kinetics remained linear with respect to time. For longer times, significant deviations from linearity occur [29, 30, 34–36, 44, 72] and, finally, the maximum (jamming) coverage is attained, analogously as for quasi-continuous surfaces. However, with random site surfaces, the maximum coverage of particles Θ_p^∞ depends not only on the ionic strength but also on the site coverage Θ_s. Figure 11.11 presents measurements of the jamming coverage performed for negatively charged polystyrene latex particles of diameter 0.9 μm adsorbing on a mica surface covered to a desired degree by positively charged polystyrene latex (averaged diameter 0.45 μm) [33, 75]. Hence, the λ' parameter was exactly 2 in this case. As can be seen in Figure 11.11, the experimental data obtained for ionic strengths of 10^{-2} and 10^{-3} M are adequately reflected by theoretical predictions (depicted by the solid line) derived from RSA simulations according to the algorithm described above. An interesting feature observed in Figure 11.11 is that Θ_p^∞ attains a maximum for site coverage $\Theta_s = 0.22$. The height of the maximum is 0.56, which exceeds the jamming limit predicted theoretically for homogeneous surfaces (equal to 0.547 as previously stated). This confirms that particle deposition on spherically shaped sites is a three-dimensional process with the surface area available for particles larger than the geometrical area of the substrate surface. This is so, however, if the lateral interactions between particles are described by the hard sphere potential and the particle/substrate interactions by the perfect sink model [48], which is the case for high ionic strength.

A different situation arises for lower ionic strength when the double layer thickness becomes comparable to the particle dimensions. In this case, because of the repulsion with the substrate, particles tend to adsorb preferentially on top of a site. As a result, one site can accommodate only one larger particle, whereas for high ionic

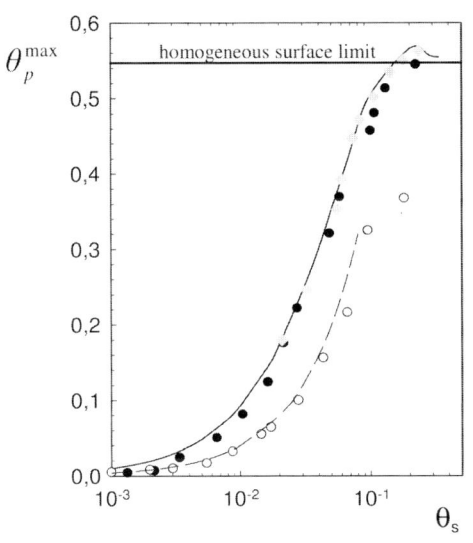

Figure 11.11 Dependence of the maximum (jamming) of latex particles Θ_p^{max} on the site (smaller latex) coverage Θ_s; the points represent the experimental results for $I =$ (●) 10^{-2}, (♦) 10^{-3} and (○) 10^{-5} M; the solid line denotes the theoretical Monte-Carlo simulations (smoothened), and the dashed line shows the analytical results derived from the dependence $\Theta_p^{max} = \lambda'^2 \Theta_s$.

strength it could accommodate on average more than two particles. Because of this effect, the surface area available for particles is reduced, which results in decreased maximum coverage. This can be seen in Figure 11.11, because for $I = 10^{-5}$ M, Θ_p^{∞} was only 0.36 for and the maximum disappeared. Note that the experimental results for $I = 10^{-5}$ M can well be reflected for a considerable coverage range by the simple analytical expression $\Theta_p^{\infty} = \lambda'^2 \Theta_s$, which indicates that the coordination number of the site was close to one.

Therefore, the results shown in Figure 11.11 support the hypothesis that one can regulate the composition and structure of surface clusters formed in deposition on random site surfaces by the ionic strength.

The site coordination number is expected to also exert a major influence on the structure of the deposition layer, quantitatively characterized in terms of the two-dimensional pair correlation function defined as [36, 48]:

$$g(r) = \frac{\pi a_p^2}{\Theta_p} \left\langle \frac{\Delta N_p}{2\pi r \Delta r} \right\rangle \quad (11.23)$$

where $\langle \rangle$ means the ensemble average and N_p is the number of particles adsorbed within the ring $2\pi r \Delta r$ drawn around a central particle. For convenience, the distance r is usually normalized using the particle radius a_p as a scaling variable.

It is possible to observe, however, for the RSS process that there are three pair correlation functions: (i) site–site, given by the quasi-continuous RSA model,

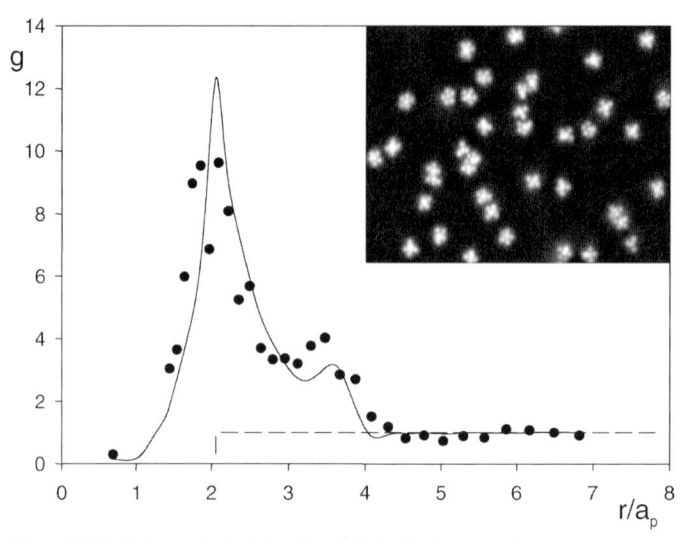

Figure 11.12 Pair correlation function $g(r/a_p)$ of polystyrene latex particles (averaged diameter 0.9 μm) adsorbed at mica pre-covered by sites, $I = 10^{-3}$ M, and $\lambda' = 0.95$, $\Theta_p = 0.06$. The solid line denotes the theoretical results from the RSA, Monte-Carlo simulations. The dashed line represents the g function for homogeneous surfaces. (From Ref. [75].) The inset shows the particle monolayer as observed under microscope.

(ii) particle–site and (iii) particle–particle. Usually, in experiments, it is feasible to determine the particle–particle correlation function, by measuring directly the coordinates of adsorbed particles. Figure 11.12 presents the pair correlation function derived experimentally for $\lambda' = 0.95$ [75]. As seen in the micrograph showing adsorbed latex particles for $\Theta_s = 0.015$ and $\Theta_p = 0.06$ ($I = 10^{-3}$ M), a significant tendency to particle clustering was observed. These surface aggregates were mainly composed of four and five particles coordinated around one site (the averaged coordination number was 4 for these experimental conditions). This confirms that particle clusters of targeted composition can be produced by particle deposition at random site surfaces. The formation of surface clusters is expressed in the pair correlation function shown in Figure 11.12. A large primary maximum (peak) is observed at r_p equal to approximately 2, that is, when two adsorbing particles are almost in contact. There also appears a minimum and a secondary maximum at $r_p = 3.5$. Interestingly, all these peculiar features of experimentally determined pair correlation function are in a full agreement with theoretical simulations (solid line in Figure 11.12) performed according to the above RSA model.

11.4.1
Deposition at Surface Features and Patterns

Because this field of research is under rapid development, few systematic experimental measurements of particle deposition on linear and curvilinear surface

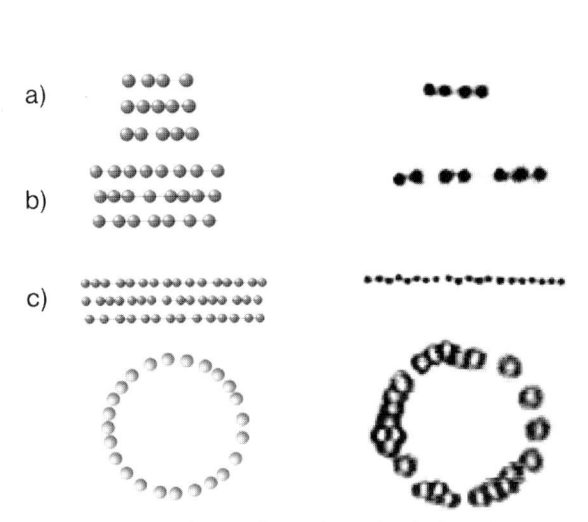

Figure 11.13 Distribution of particles under the jamming state deposited at line segments and a circle, derived from simulations and from experiments, for $\bar{L}=$ (a) 5, (b) 10 and (c) 20. The results for circles have been taken from the work of Zheng et al. [37]. (From Ref. [61].)

features have been conducted. Exceptions are the interesting experiments of Zheng et al. [37, 41]. In his work, polystyrene latex particles 1 μm in diameter were used that adsorbed at ring-shaped surface features (11 μm in diameter) produced by polymer stamping on gold-covered silicon.

In contrast, some measurements involving linear surface features (line segments) produced on mica have been reported [58]. Figure 11.13 compares the so-obtained particle distributions with theoretical predictions. There are pronounced similarities, although some aggregates can be observed in experimental "monolayers," which is probably caused by drying the sample before optical microscope examinations. Interestingly, the 1D monolayer coverage in the case of line segments was found to be 0.85 and 0.76 for $L/d = 5$ and 10, respectively, which is close to the respective theoretical predictions of 0.87 and 0.78.

As far as patterned surfaces are concerned, there exist some results for dots and stripes, which can be analyzed in terms of theoretical models discussed above. Interesting experiments of this type have been carried out by Zheng et al. [37] for circular surface features (dots). These features, having an average diameter of 11 μm, have been produced by stamping on gold-covered silicon bearing 10 bilayers of polyelectrolytes (negatively charged sulfonated polystyrene SPS and positively charged poly[(diallyldimethyl)ammonium chloride]).

A similar stamping technique was used by Chen et al. [40] to produce long stripes having an average width of 5 μm, on gold and polyelectrolyte multilayer covered silicon. In this study sulfate latex particles were used, having an average diameter of $d=0.53\,\mu m$, whose charge was regulated by surfactant addition (DTAB).

Although these experimental results agree qualitatively with theoretical predictions based on the above RSA model [57] a quantitative comparison can be performed if experimental data become available for high ionic strength when particles behave like hard spheres.

11.5
Concluding Remarks

The extended RSA model has proved efficient for analyzing theoretically particle deposition at heterogeneous surfaces, in particular the kinetics, structure and distribution of particles monolayers, jamming coverage, and so on.

Experimental data obtained mostly for model colloid suspensions using direct measurement techniques have confirmed the validity of the RSA model and these theoretical predictions. In particular, the following main features of particle deposition on heterogeneous surfaces have been confirmed:

1. The initial deposition rate of particles increased abruptly with the coverage of sites Θ_s, attaining values pertinent to continuous surfaces (maximum deposition rates) for Θ_s of the order of a few per cent. This effect, appearing because of the coupling of surface and bulk transport, was especially well pronounced for spherical sites smaller than 100 nm.

2. The maximum (jamming) coverage of particles adsorbed on spherical sites increased for low Θ_s according to the relationship $\Theta_p^\infty = l_s \lambda'^2 \Theta_s$. The site coordination number l_s, and consequently the composition and structure of surface clusters produced in this way, can be controlled by the size ratio parameter λ' and the ionic strength.

3. However, for a nanometer size range of sites, the deposition rate of particles increased with Θ_s much more slowly, because efficient immobilization of particles occurred at multiple sites only (local charge heterogeneities).

It was also suggested that by measuring microscopically the average number of adsorbed colloid particles one can determine the size (length) and shape of surface features that are invisible under an optical microscope.

The theoretical and experimental results obtained for model colloid systems can be used effectively as reference states for analyzing protein and macromolecule adsorption at heterogeneous surfaces.

Acknowledgment

This work was financially supported by the COST D43 Special Grant of MNiSzW.

References

1. Hinrichsen, E.L., Feder, J. and Jossang, T.J. (1986) *Journal of Statistical Physics*, **44**, 793–822.
2. Viot, P., Tarjus, G., Ricci, S.M. and Talbot, J. (1992) *Journal of Chemical Physics*, **97**, 5212–5218.
3. Adamczyk, Z. and Weroński, P. (1996) *Journal of Chemical Physics*, **105**, 5562–5573.
4. Oberholzer, M.R., Stankovich, J.M., Carnie, S.L. *et al.* (1997) *Journal of Colloid and Interface Science*, **194**, 138–153.
5. Weroński, P. (2005) *Advances in Colloid and Interface Science*, **118**, 1–24.
6. Adamczyk, Z., Senger, B., Voegel, J.C. and Schaaf, P. (1999) *Journal of Chemical Physics*, **110**, 3118–3138.
7. Hull, M. and Kichener, J.A. (1969) *Transactions of the Faraday Society*, **65**, 3093–3104.
8. Dąbroś, T. and van de Ven, T.G.M. (1982) *Colloid and Polymer Science*, **261**, 694–707.
9. Dąbroś, T. and van de Ven, T.G.M. (1982) *Journal of Colloid and Interface Science*, **89**, 232–244.
10. Adamczyk, Z., Siwek, B., Zembala, M. and Warszyński, P. (1989) *Journal of Colloid and Interface Science*, **130**, 578–587.
11. Albery, W.J., Fredleins, R.A., Kneebone, G.R. *et al.* (1990) *Colloids Surfaces*, **44**, 337–341.
12. Dijt, J.C., Cohent-Stuart, M.A., Hofman, J.E. and Fleer, G.J. (1990) *Colloids Surfaces*, **51**, 141–158.
13. Adamczyk, Z., Zembala, M., Siwek, B. and Warszyński, P. (1990) *Journal of Colloid and Interface Science*, **140**, 123–124.
14. Adamczyk, Z., Siwek, B. and Zembala, M. (1992) *Journal of Colloid and Interface Science*, **151**, 351–369.
15. Harley, S., Thompson, D.W. and Vincent, B. (1992) *Colloids Surfaces*, **62**, 163–176.
16. Meinders, J.M., Noordmans, J. and Busscher, H.J. (1992) *Journal of Colloid and Interface Science*, **152**, 265–279.
17. Johnson, C.A. and Lenhoff, A.M. (1996) *Journal of Colloid and Interface Science*, **179**, 587–599.
18. Böhmer, M.R., van der Zeeuw, E.A. and Koper, G.J.M. (1998) *Journal of Colloid and Interface Science*, **197**, 242–250.
19. Adamczyk, Z., Siwek, B. and Zembala, M. (1998) *Journal of Colloid and Interface Science*, **198**, 183–185.
20. Semmler, M., Mann, E.K., Ricka, J. and Borkovec, M. (1998) *Langmuir*, **14**, 5127–5132.
21. Adamczyk, Z. and Szyk, L. (2000) *Langmuir*, **16**, 5730–5737.
22. Adamczyk, Z., Siwek, B., Warszyński, P. and Musiał, E. (2001) *Journal of Colloid and Interface Science*, **242**, 14–24.
23. Goransson, A. and Tragardth, C. (2000) *Journal of Colloid and Interface Science*, **231**, 228–239.
24. Kun, R. and Fendler, J.H. (2004) *The Journal of Physical Chemistry B*, **108**, 3462–3468.
25. Pericet-Camara, R., Papastavrou, G. and Borkovec, M. (2004) *Langmuir*, **20**, 3264–3270.
26. Kleimann, J., Lecoultre, G., Papastavrou, G. *et al.* (2006) *Journal of Colloid and Interface Science*, **303**, 460–471.
27. Shin, J., Roberts, J.E. and Santore, M. (2002) *Journal of Colloid and Interface Science*, **247**, 220–230.
28. Kozlova, N. and Santore, M. (2006) *Langmuir*, **22**, 1135–1142.
29. Adamczyk, Z., Zembala, M. and Michna, A. (2006) *Journal of Colloid and Interface Science*, **303**, 353–364.
30. Adamczyk, Z., Michna, A., Szaraniec, M. *et al.* (2007) *Journal of Colloid and Interface Science*, **313**, 86–96.
31. Joscelyne, S. and Tragardh, Ch. 1997, *Journal of Colloid and Interface Science*, **192**, 294–305.
32. Garno, J.C., Amro, N.A., Wadu-Mesthrige, K. and Liu, G.-Y. (2002) *Langmuir*, **18**, 8186–8192.

33 Adamczyk, Z., Jaszczółt, K., Siwek, B. and Weroński, P. (2004) *Journal of Chemical Physics*, **120**, 1155–1162.

34 Adamczyk, Z., Jaszczółt, K., Siwek, B. and Weroński, P. (2004) *Colloids Surfaces A*, **249**, 95–98.

35 Jaszczółt, K., Adamczyk, Z. and Weroński, P. (2005) *Langmuir*, **21**, 8952–8959.

36 Adamczyk, Z., Jaszczółt, K., Michna, A. et al. (2005) *Advances in Colloid and Interface Science*, **118**, 25–42.

37 Zheng, H., Rubner, M.F. and Hammond, P.T. (2002) *Langmuir*, **18**, 4505–4510.

38 Karakurt, I., Leiderer, P. and Boneberg, J. (2006) *Langmuir*, **22**, 2415–2417.

39 Kruger, C. and Jonas, U. (2002) *Journal of Colloid and Interface Science*, **252**, 331–338.

40 Chen, K.M., Jiang, X., Kimerling, L.C. and Hammond, P.T. (2000) *Langmuir*, **16**, 7825–7834.

41 Zheng, H., Berg, M.C., Rubner, M.F. and Hammond, P.T. (2004) *Langmuir*, **20**, 7215–7222.

42 Wright, J., Ivanowa, E., Pham, D. et al. (2003) *Langmuir*, **19**, 446–452.

43 Adamczyk, Z. and Warszyński, P. (1996) *Advances in Colloid and Interface Science*, **63**, 141–149.

44 Adamczyk, Z. (2003) *Advances in Colloid and Interface Science*, **100–102**, 267–347.

45 Schaaf, P. and Talbot, J. (1989) *Journal of Chemical Physics*, **91**, 4401–4409.

46 Evans, J.W. (1993) *Reviews of Modern Physics*, **65**, 1281–1329.

47 Talbot, J., Tarjus, G., van Tassel, P.R. and Viot, P. (2000) *Colloids Surfaces A*, **165**, 287–324.

48 Adamczyk, Z. (2006) *Particles at Interfaces, Interactions, Deposition, Structure*, Academic Press, Elsevier.

49 Adamczyk, Z., Weroński, P. and Musiał, E. (2002) *Journal of Chemical Physics*, **116**, 4665–4672.

50 Adamczyk, Z., Weroński, P. and Musiał, E. (2002) *Journal of Colloid and Interface Science*, **248**, 67–75.

51 Jin, X., Wang, N.H.L., Tarjus, G. and Talbot, J. (1993) *The Journal of Physical Chemistry*, **97**, 4256–4258.

52 Adamczyk, Z., Siwek, B., Zembala, M. and Weroński, P. (1997) *Journal of Colloid and Interface Science*, **185**, 236–244.

53 Vigil, R.D. and Ziff, R.M. (1989) *Journal of Chemical Physics*, **91**, 2599–2602.

54 Adamczyk, Z. and Weroński, P. (1997) *Journal of Colloid and Interface Science*, **189**, 348–360.

55 Adamczyk, Z. and Weroński, P. (1995) *Langmuir*, **11**, 4400–4410.

56 Adamczyk, Z., Barbasz, J. and Zembala, M. (2007) *Langmuir*, **23**, 5557–5562.

57 Adamczyk, Z., Barbasz, J. and Nattich, M. *Langmuir*, **24** (2008) 1756–1762.

58 Burridge, D.J. and Mao, Y. (2004) *Physical Review E*, **69**, 037102–037111.

59 Renyi, A. (1958) *Publications of Mathematical Institute of Hungarian Academy of Sciences*, **3**, 109–127.

60 Pomeau, Y.J. (1980) *Journal of Physics A-Mathematical and General*, **13**, L193–L196.

61 Bafaluy, F.J., Choi, H.S., Senger, B. and Talbot, J. (1995) *Physical Review E*, **51**, 5985–5993.

62 Kooj, E.S., Brouwar, E.A.M., Wormaester, H. and Poelsema, B. (2002) *Langmuir*, **18**, 7677–7682.

63 Malmsten, M., Lassen, B., Westin, J. et al. (1994) *Journal of Colloid and Interface Science*, **179**, 163–172.

64 Bowen, B.D. and Epstein, N. (1979) *Journal of Colloid and Interface Science*, **72**, 81–97.

65 Aptel, J.D., Thomann, J.M., Voegel, J.C. et al. (1988) *Colloids Surfaces*, **32**, 159–171.

66 Zembala, M., Voegel, J.C. and Schorof, R. (1998) *Langmuir*, **14**, 2167–2173.

67 Norde, W. and Rouwendal, E. (1990) *Journal of Colloid and Interface Science*, **139**, 169–176.

68 Elgersma, A.V., Zsom, R.L.J., Norde, W. and Lyklema, J. (1991) *Colloids Surfaces*, **54**, 89–101.

69 Zembala, M. and Adamczyk, Z. (2000) *Langmuir*, **16**, 1593–1601.

70 Elimelech, M., Gergory, J., Jia, X. and Williams, K.A. (1996) *Particle*

Deposition and Aggregation, Butterworth–Heinemann, Oxford.

71 Adamczyk, Z., Jaszczółt, K., Michna, A. et al. (2007) *Colloid Stability—The Role of Surface Forces*, (ed. Th.F. Tadros), vol. 1 Wiley-VCH Verlag Gmbh, Weinheim.

72 Adamczyk, Z., Siwek, B., Weroński, P. and Jaszczółt, K. (2003) *Colloids Surfaces A*, **222**, 15–25.

73 Song, L., Johnson, P.R. and Elimelech, M. (1994) *Environmental Science & Technology*, **28**, 1164–1171.

74 Adamczyk, Z., Siwek, B. and Musiał, E. (2001) *Langmuir*, **17**, 4529–4533.

75 Adamczyk, Z., Jaszczółt, K. and Siwek, B. (2005) *Applied Surface Science*, **252**, 723–729.

12
Effect of the Interaction Between Heavy Crude Oil Components and Stabilizing Solids with Different Wetting Properties

Simone Less, Andreas Hannisdal, Heléne Magnusson, and Johan Sjöblom

12.1
Introduction

Water-in-oil emulsions are encountered in production, processing and transportation of crude oils. Their stability is an issue of great concern and economic relevance to oil companies. Many are the problems related to the presence of water in such emulsions. Among them are corrosion of pipes, pumps and other processing equipment, increased emulsion viscosity, deactivation of catalysts and the extra cost of transporting water in the pipelines [1]. A vast literature exists on the chemical aspects of emulsion stability and relative characterization techniques [2, 3]. Current opinion is that a complex interplay between indigenous crude oil components (such as asphaltenes, resins, naphthenic acids and waxes) in combination with fine organic and inorganic solids forms a film around the droplets that can hinder destabilization phenomena such as coalescence [2, 4, 5]. The film surrounding the droplets can, due to its immobility and low solubility in both water and oil, create very stable emulsions [6]. The process in which interfacial components act as particles, creating a mechanically stable film on the surface of the droplets, is called mechanical stabilization [6]. There is also evidence that partial surface coverage by solids can result in effective stabilization [7, 8]. The destabilization rate can be reduced because of the higher overall emulsion viscosity due to the presence of these solids. Mechanisms for particle-stabilization of emulsions and the extent to which solids increase emulsion stability depend on several factors such as particle size, shape and morphology, density, concentration, surface coverage and wettability [9]. Wettability is particularly important when the capacity of solids to stabilize emulsions is considered. Kralchevsky *et al.* [10] have described the "particle bridging effect," in which one particle is adsorbed simultaneously at the two film interfaces. Depending on the wetting characteristic of the particles either stable flocculates may form or coalescence occurs.

According to Bancroft's rule [11] hydrophilic particles tend to stabilize oil-in-water emulsions, whereas hydrophobic ones stabilize water-in-oil (W/O) emulsions. For

Highlights in Colloid Science. Edited by Dimo Platikanov and Dotchi Exerowa
Copyright © 2009 WILEY-VCH Verlag GmbH & Co. KGaA, Weinheim
ISBN: 978-3-527-32037-0

the native solids encountered in oilfield emulsions, particles possess hydrophilic characteristics in the form of exposed aluminosilicate surfaces and hydrophobic characteristics in the form of adsorbed humic and petroleum materials [9]. The presence of an adsorbed layer of asphaltenes from crude oil on finely divided solids was reported to alter the wettability and other characteristics of these solids dramatically, thus enabling them to stabilize water-in-crude oil emulsions [12]. Gonzalez et al. [13] have shown that asphaltenes play an important role in the stabilization of w/o emulsions. Asphaltenes alone show a moderate capacity to reduce the oil/water interfacial tension and stabilize emulsions. However, in the presence of other fine particles, asphaltenes enhance emulsion formation. In general, Langmuir-type adsorption isotherms are reported for asphaltene adsorption from toluene solution at low bulk concentrations, with multiple steps or other deviations from the simple model at higher concentrations. This has been explained by surface phase reorientation, multilayer formation or aggregation. Acevedo et al. [14, 15] have reported the formation of asphaltene multilayers or aggregates at silica surfaces from concentrated asphaltene/toluene solutions. The tendency towards multilayer formation of asphaltenes has been related to the history of deposition problems in their original oils. Desorption of asphaltenes from a silica surface under atmospheric conditions is slow enough to be neglected [15].

Hannisdal et al. [16] have studied the effects on the emulsion stability of the addition of silica particles, because their properties resemble those of natural inorganic colloids like clays. They coated some commercially available dry silica nanoparticles with heavy crude oil components (asphaltenes and resins) to investigate the performance of these solids as stabilizers in water-in-model oil emulsions.

Adsorption studies performed with the QCM-D technique showed multilayer or aggregate formation of asphaltenes at the silica surface. Asphaltenes were irreversibly adsorbed as a rigid film. Visual inspection of the resin-coated silica particles clearly showed that the initially hydrophilic silica (Aerosil 200 and Aerosil 7200) adsorbed considerably more resins than the hydrophobic silica (Aerosil 972). By analyzing the toluene solutions for remaining asphaltenes with near-infrared (NIR) spectroscopy, they could conclude that the initially hydrophilic particles had adsorbed considerably more asphaltenes than the hydrophobic particles. All the products exhibited less adsorption of resins than of asphaltenes and the influence of their adsorption was dramatic, especially on the wettability of the hydrophilic silica. Very hydrophilic particles like the unmodified 200 or very hydrophobic particles like the 972 preferred the water and oil phase, respectively. In contrast, the Aerosil 7200 seemed to be equally partitioned between the two phases, indicating intermediate wetting properties. Generally, the stabilization efficiency was enhanced by adsorption of crude oil components onto very hydrophilic or very hydrophobic silica.

The present study used the same silica particles, but the reference system was moved from model oil to crude oil emulsions and the effect of the addition of different amounts and typology of asphaltenes investigated in detail. The aim is to contribute to a better understanding of the interplay between indigenous crude oil components and solids with regard to emulsion stability.

12.2
Experimental

12.2.1
Extraction of Asphaltenes from the Crude Oils

Asphaltenes were obtained from two different crude oils, one coming from the Norwegian Continental Shelf (crude A) and one coming from South West Africa (crude B). Their precipitation was performed in a 1 : 20 excess of *n*-pentane with a double filtration through a 0.45 µm Millipore filter. Asphaltenes were washed carefully with *n*-pentane to remove impurities. The original crude oils were analyzed with respect to SARA components, density, viscosity, acidity and water content. Table 12.1 lists their characteristics. Details about the SARA fractionation are reported elsewhere [17].

12.2.2
Silica Particles: Characterization and Properties

Aerosil particles from Degussa are fumed silica, a synthetic amorphous silicon dioxide manufactured by continuous flame hydrolysis of silicon tetrachloride. Siloxane and silanol groups are present on the surface of unmodified Aerosil particles, where silanol groups account for the hydrophilic behavior. Reactions of the silanol groups with various organosilanes and siloxanes can chemically modify the surface of the particles. Aerosil 200, Aerosil 7200 and Aerosil 972 were used in this study. Aerosil 200 is an unmodified silica, while Aerosil 7200 is a structure modified (3-methacryl-oxypropyl-trimethoxysilane) product based on Aerosil 200. The hydrophobic Aerosil 972 is a fumed silica after-treated with dimethyldichlorosilane. Table 12.2 summarizes the properties of these three products. Importantly, the

Table 12.1 Physicochemical properties of the two crude oils.

Property	Crude A	Crude B
SARA fractionation		
Saturates (wt%)	47.5	46.6
Aromatics (wt%)	40.1	37.6
Resins (wt%)	11.5	14.0
Asphaltenes (wt%)	0.76	1.65
Emulsified water[a]	0.05	0.05
Density (40 °C)[b] (g cm^{-3})	0.9125 (23–24 API)	0.8985 (25–26 API)
Viscosity (40 °C)	49.11 cP	35.55 cP
Acidity (TAN) (mg-KOH mg^{-1})	2.93	1.36
MW[c]	319.005	308.685

[a] Karl-Fischer titration.
[b] AP PAAR density meter DMA 48.
[c] Determined from the freezing point depression of benzene.

Table 12.2 Characterization of silica particles.

Product		After-treated with	Specific surface area (BET) ($m^2 g^{-1}$)	Tapped density[a] ($g L^{-1}$)	Primary particle size (nm)
Aerosil 200	Hydrophilic	—	200 ± 25	50	12
Aerosil 7200	Hydrophilic	3-methacryl-oxypropyl-trimethoxysilane	150 ± 25	230	—
Aerosil 972	Hydrophobic	Dimethyldichlorosilane	110 ± 20	50	16

[a]Volume of product is measured after a standard vibration condition is applied (ISO 787/11).

fumed silica products do not appear as individual particles in aqueous solution, but flocculate to aggregates or agglomerates that can be several hundred nanometers in size. This common feature of fumed silica is due to siloxane (Si–O–Si) linkages between primary particles. The difference between fused silica (Aerosil) and silica produced by alkaline hydrolysis of sodium silicate solutions via nucleation and growth is discussed by Binks and Lumsdon [18].

12.2.3
Preparation of the Emulsions

The aim of this study was to investigate the differences in stability between water-in-crude oil emulsions with and without the addition of solid particles. Both crude A and crude B emulsions were used for the testing procedures. The effect of increasing the amount of crude A asphaltenes to the same as in crude B (from 0.76 to 1.65%) was also examined.

Simple water-in-crude oil emulsions were prepared by mixing the same volume of crude oil and synthetic sea water (3.5 wt% NaCl) at 40 °C. Oil and water were poured into a glass tube (total amount 25 mL). Each emulsification was performed with an Ultra Turrax (IKA, T18 with 10 and 6.75 mm of stator and rotor diameters, respectively) at 21 500 rpm and 40 °C for 3 min, just before proceeding with droplet size determination and stability measurements.

To prepare modified emulsions, first asphaltenes and then solid particles were added to the oil phase, before emulsifying with water. When no addition of asphaltenes was required, particles were dispersed in the crude oil. Subsequently, the glass tube was placed on a shaking table for 5 h at 300 min^{-1} and then in a oven at 40 °C for 36 h to allow the adsorption of crude oil components to proceed. In the case where the total amount of asphaltenes of crude A was increased to 1.65%, asphaltenes extracted from crude A were added to the oil phase before dispersing the particles. The sample was then placed on a shaking table at 300 min^{-1} for 5 h, then for 36 h in an oven at 60 °C and, finally, cooled to 40 °C before particle addition and subsequent mixing with water.

Addition of crude A asphaltenes and particles was also performed on model oil emulsions of 50% water cut. The model oil phase consisted of a 70 : 30 v/v exxolD80 : toluene mixture. All procedures were carried out as previously specified. All samples were analyzed by microscopy and the asphaltenes found solubilized.

12.2.4
Emulsion Stability Measurements and Drop Size Determination

The emulsion stability was measured by two different techniques, at rest by Turbiscan analysis and under intense shear conditions by rheometer, with the simultaneous application of an electric field to enhance coalescence.

The Turbiscan Lab reading head acquires transmission and backscattering data every 40 μm while moving along the 55 mm cell height. The acquisition is repeated with a programmable frequency to monitor changes in the emulsion with time, such as variations in particle size due to coalescence or flocculation. The migration of particles leading to sedimentation or creaming is analyzed by direct calculation of the migration velocity and the thickness of the sediment or cream phases. All data were processed with the software Turbiscan EasySoft Formulaction (2005).

Stability to high shear condition and electrical destabilization was measured by looking at the variation of the rheological properties of the emulsions under evaluation. For this purpose a Physica MCR-301 rotational rheometer was used. Data were acquired in controlled shear mode and processed with the RheoPlus/32 software (V 2.65). The rheometer is equipped with an electro-rheological temperature device (ERD, PTD200/E) with a high DC voltage supply HCL 14-12 500MOD. This apparatus allows a variable voltage, 0–12 500 V, to be applied to the sample. The radii of the measuring bob and cup are 13.32 and 14.46 mm, respectively. The cup is earthed. The rotating bob is also the high potential electrode, since a spring directly connected to the generator slides on it. At low shear rates, some interference therefore affects the quality of the measurements. In the present study this problem can be neglected, since all experiments were run at a high shear rate (300 rpm). The original measuring bob had a couple of sharp edges that were proven to lead to electric field enhancements with consequent agglomeration of water drops due to dielectrophoresis in their proximity. The effects in these areas interfered with the effects in the long parallel gap where the electric field was more uniform. To avoid this inconvenience the lower and upper edges were rounded with a curvature radius of 5 mm. By doing so, the absolute viscosity values are slightly lower than the actual ones. Nevertheless, the viscosity variations are correct and the electric field lines more evenly distributed, such that the results can be considered reliable.

To detect any instability, the viscosities of all emulsions were first monitored at 40 °C, under the influence of shear only for 130 s. After 130 s a voltage was applied and increased from 0 to 1000 V at a rate of 4 V s^{-1}, corresponding to an electric field of about 8.8 kV cm^{-1} reached in 250 s. Short circuit limited the maximum value to 6 kV cm^{-1}.

To describe the stability of the emulsions in terms of the change in viscosity we introduced an adimensional parameter called the viscosity ratio. The viscosity ratio was defined according to Equation 12.1, where v_1 is the viscosity before and v_2 that after the application of the electric field:

$$\bar{v} = \frac{v_1 - v_2}{v_1} \qquad (12.1)$$

The procedure was validated by comparison with the dewatering efficiency measured by Karl-Fischer titration. The dewatering efficiency was calculated as shown in Equation 12.2, where 1 and 2 denote the water content before and after the application of the electric field:

$$\Delta w_\% = 1 - \frac{w_{\%,2}}{w_{\%,1}} \quad (12.2)$$

Droplet size distributions of the emulsions were determined by visual process analysis, ViPA (Visual Process Analyser, JORIN, UK). A small sub-sample of the stable emulsion was diluted in toluene and analyzed immediately to avoid changes of the droplet distribution. Droplets were selected according to a specific range of the shape factor, to avoid counting droplet flocculates as single droplets. The shape factor was defined as $(4\pi \times \text{area})/\text{perimeter}^2$ with a possible range from 0 to 1, where the shape factor of a perfect circle was one. About 50 000 droplet counts produced the reported distribution functions.

12.3
Results and Discussion

12.3.1
Droplet Size Distributions

To verify the reproducibility of the emulsification process, the volumetric water droplet size distribution was measured immediately after emulsification. All emulsions showed similar characteristics, with a log–normal volumetric droplet size distribution centered on a mean size of about 8 µm. However, samples characterized by a particle amount equal to or greater than 1 wt% contained a very small number of bigger droplets with diameter up to 60 µm. The differences between the samples are too small to play an important role in the differences in stability between the samples. Because very small droplets form very stable emulsions, we use systems with small droplet sizes compared to what is commonly encountered in real processes in our study. The effect of electrical destabilization then becomes dominant since other phenomena can be neglected. In addition, the use of a large energy in the mixing process enables samples that are characterized by different physicochemical properties to show homogeneity with regard to droplet size distribution.

12.3.2
Viscosity Observations

Figures 12.1–12.3 show the rheological investigations performed on emulsions stabilized by very hydrophobic, mildly hydrophilic and very hydrophilic silica particles, respectively. From Figure 12.1 it is clear that emulsions stabilized by means of very hydrophobic particles are very stable when no voltage is applied. However, the stability of these emulsions towards electrostatic destabilization drops

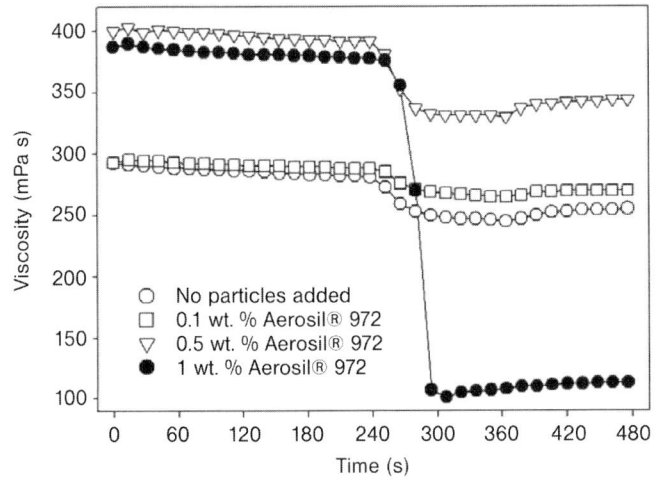

Figure 12.1 Viscosity as a function of time during rheological analysis. An electric field is applied from $t = 120$ to 370 s. The time scale is extended to 480 s to reach equilibrium. The curves refer to 50% water cut emulsions with crude A as oil phase with the addition of very hydrophobic particles.

when the particle concentration exceeds 1 wt%. With increasing hydrophilicity of the particles, their stabilizing effect decreases (Figures 12.2 and 12.3). Emulsions stabilized by mildly hydrophilic particles are very stable up to a particle concentration of 1 wt%. For higher amounts an evident instability takes place and the effect of shear

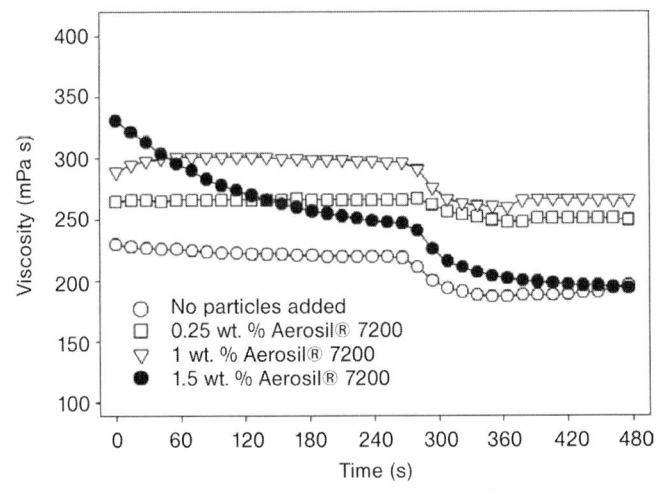

Figure 12.2 Viscosity as a function of time. An electric field is applied from $t = 120$ to 370 s. The time scale is extended to 480 s to reach equilibrium. The curves refer to 50% water cut emulsions with crude B as oil phase and with the addition of mildly hydrophilic particles.

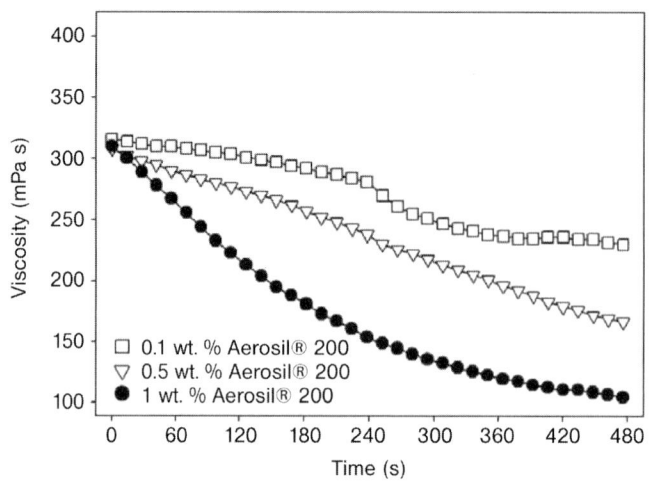

Figure 12.3 Viscosity as a function of time. An electric field is applied from $t = 120$ to 370 s. The time scale is extended to 480 s to reach equilibrium. The curves refer to 50% water cut emulsions with crude A as oil phase and with the addition of very hydrophilic particles.

and electric field only increases the separation rate (Figure 12.2). Emulsions prepared with very hydrophilic particles are generally very unstable. The application of an electric field increases the destabilization up to 0.1 wt% of particles only. For higher particle concentrations its effect is negligible – the separation rate already very fast (Figure 12.3).

12.3.3
Stability Measurements

Figure 12.4 shows the comparison between viscosity ratio and dewatering efficiency for the case of water-in-crude oil A emulsions with the addition of very hydrophobic particles (Aerosil 972). The two variables follow the same trend. Therefore, it is possible to relate the viscosity changes due to the application of a shear stress and an electric field to the emulsions stability and the emulsion stability can be described in terms of the viscosity ratio.

According to Tadros [19], at any given volume fraction of oil, an increase in the droplet size results in a viscosity reduction. This is particularly the case with concentrated emulsions. Thus, by monitoring the decrease in emulsion viscosity with time one may obtain information on its coalescence. However, other simultaneous phenomena, for example, flocculation, sedimentation or creaming, also affect the viscosity of an emulsion. Therefore, we also used Turbiscan analysis to monitor the emulsion stability.

The results from this study show many similarities with what was previously reported by Hannisdal *et al.* [16]. For water-in-crude oil emulsions, the stability is very

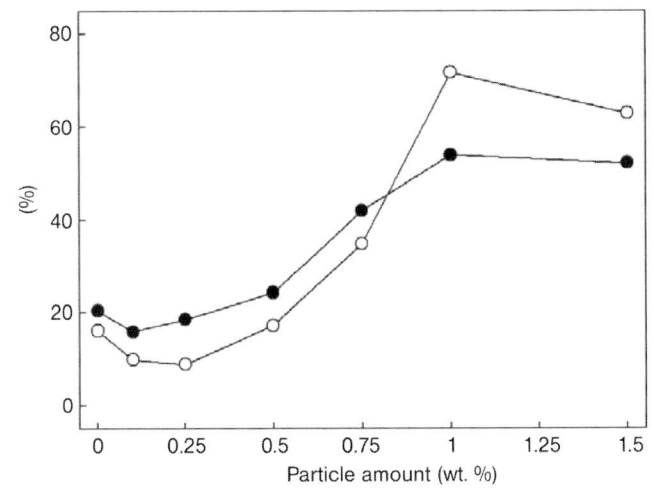

Figure 12.4 Dewatering efficiency (●) and viscosity ratio (○) in 50% water-in-crude oil A emulsions for different amounts of hydrophobic Aerosil 972 particles dispersed in the samples.

dependent on the wetting properties of the added silica particles. Moreover, the fundamental role played by the asphaltene fraction in the adsorption process (and consequently on emulsions stability), compared to that played by other indigenous crude oil components as resins, is further stressed. Crude oil emulsions modified by addition of silica particles generally show increased stability with increasing asphaltene content in the crude oil. Figures 12.5 and 12.6 show the viscosity ratio defined by

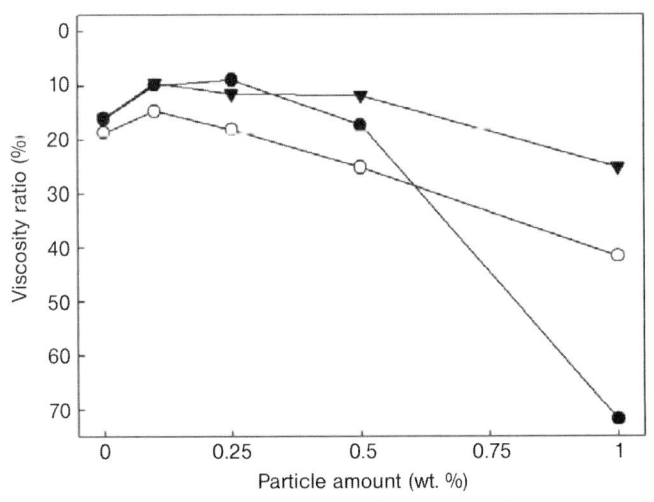

Figure 12.5 Viscosity ratio as a function of the amount of very hydrophobic silica particles (Aerosil 972) for 50% water cut emulsions from crude A (●), crude B (○) and crude A with increased asphaltenes content (▼).

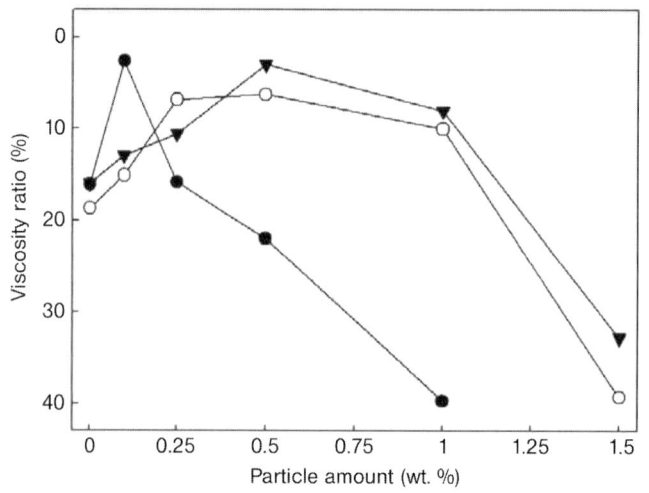

Figure 12.6 Viscosity ratio as a function of the amount of mildly hydrophilic silica particles (Aerosil 7200) for 50% water cut emulsions from crude A (●), crude B (○) and crude A with increased asphaltenes content (▼).

Equation 12.1 as function of the amount of particles added. To show the viscosity ratio as a stability index the y-axis was reversed. Therefore, a positive slope of the curve indicates an increase in stability.

Figure 12.5 illustrates the effect of very hydrophobic particles (Aerosil 972) on emulsion stability. The stability is constant or increases for a wide range of particle concentrations. The hydrophobicity of the particles promotes the stability of the water-in-oil emulsions. In all cases where the particle concentration exceeds a critical value the stability drops, especially for water-in-crude A emulsions (●). The emulsions prepared with crude B (○) and crude A with increased asphaltenes content (▼) show similar behavior and are stable up to a higher particle concentration than the emulsions containing less asphaltenes. In this case the drop in viscosity is impressive, and the viscosity ratio increases more than 70%. This behavior is attributed to the interaction between hydrophobic particles and indigenous crude oil components, especially asphaltenes. As stated in several studies [9, 20–22], the particle size plays an important role in emulsion stability. As the particle size decreases, the surface area per unit mass of the solid increases as a cube of the particle radius. Therefore, more interfacially active material is adsorbed on the particles, increasing their stabilizing capacity. In the present study the silica particles are several orders of magnitude smaller than the water droplets and very stable emulsions can, therefore, be formed. This effect remains up to a weight fraction of particles of about 0.5 wt%. Here, enough particles are present to cover the whole W/O interface. At the same time there are enough asphaltenes present to both cover the particle interface and adsorb at the W/O interface. At higher particle concentrations the interplay between particles and asphaltenes becomes counterproductive for emulsion stability. Under these conditions the particles present at the W/O interface are only partly covered with asphaltenes, therefore destabilizing the film. Also, the asphaltenes are mainly adsorbed on the particles and not at the W/O interface, thus facilitating film rupture. The consequent dramatic decrease in stability is shown in Figure 12.5.

Figure 12.6 shows the effect of the addition of the Aerosil 7200 silica particles, which have intermediate wetting characteristics. Here the role of the asphaltenes is most prominent. In fact at any particle concentration their higher content strongly increases the stability, as demonstrated by a decrease of the viscosity ratio from 15 to 3%. When the concentration of particles is increased above a critical value the stability is reduced strongly and the emulsion collapses. For all emulsions the viscosity ratio increases up to 40%. When more asphaltenes are present more particles can be present before destabilization occurs. Crude B (○) and crude A with increased asphaltene content (▼) are almost superimposed, indicating the strong interplay between particles and asphaltenes.

As previously stated, the addition of very hydrophilic particles (Aerosil 200) to the emulsions does not give stable emulsions. The viscosity of most of these emulsions experiences a monotonic decrease; however, emulsions prepared with crude B and crude A with increased asphaltenes content and with 0.1 wt% added particles require electrical destabilization to phase separate. Here enough asphaltenes are present to adsorb onto the hydrophilic particles, rendering them hydrophobic enough to stabilize water-in-oil emulsions. As the particle amount is increased the coverage of the particles surface decreases and the stability of the emulsions decreases. On visual inspection these emulsions showed a phase separation immediately after the Rheology test. In summary, an increase in stability is generally achieved at the lowest particle concentrations. As more particles are present the asphaltenes are less and less available for adsorption and also removed from the W/O interface, such that stability decreases. In all systems this collapse is shifted to higher concentrations of silica particles when more asphaltenes are available in the systems.

To confirm these hypotheses, additional studies were performed on model oil emulsions. The model oil phases consisted of a 70 : 30 v/v exxolD80 : toluene mixture. Since the emulsions stabilities for this oil-phase only were not sufficiently high to achieve reliable results, asphaltenes precipitated from crude A were added to the model emulsions in the same amount as in the actual crude oil. Figures 12.7 and 12.8 show the viscosity ratio as a function of particle concentration for the model emulsions stabilized with very hydrophobic and mildly hydrophilic silica particles, respectively. Model oil and crude oil emulsions show many similarities. For all model emulsions the stability in the absence of particles is lower, because the absolute viscosity is much lower than the one encountered in crude oil emulsions. Consequently, the effect of the shear rate on the destabilization is more pronounced. However, as particles are added in larger amounts, crude oil and model oil emulsions behave in the same way. Nevertheless, in Figure 12.8 the viscosity ratios are shifted towards lower values since a strong flocculation phenomenon enhanced by the low viscosity of the samples brings coalescence.

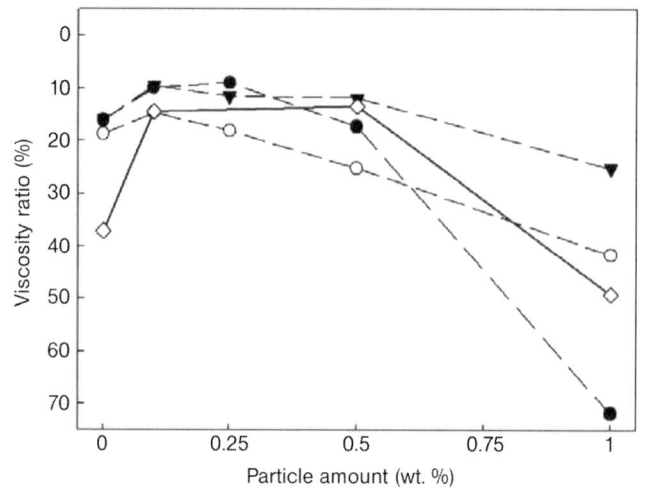

12 Effect of the Interaction Between Heavy Crude Oil Components and Stabilizing Solids

Figure 12.7 Viscosity ratio as a function of the amount of very hydrophobic silica particles (Aerosil 972) for 50% water cut emulsions prepared with an oil phase consisting of a 70:30 v/v exxol:toluene mixture (◇, solid line). For an easier comparison the dashed lines describe the behavior of water-in-crude oil emulsions: (●) crude A emulsions, (○) crude B emulsions and (▼) crude A emulsions with increased asphaltenes content.

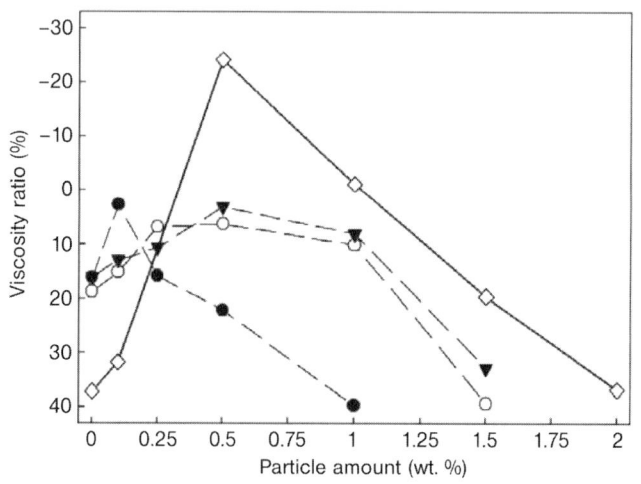

Figure 12.8 Viscosity ratio as a function of the amount of mildly hydrophilic silica particles (Aerosil 7200) for 50% water cut emulsions prepared with an oil phase consisting of a 70:30 v/v exxol:toluene mixture (◇, solid line). For an easier comparison the dashed lines describe the behavior of water-in-crude oil emulsions: (●) crude A emulsions, (○) crude B emulsions and (▼) crude A emulsions with increased asphaltenes content.

To monitor the stability of the systems in the absence of any external destabilizing technique Turbiscan analyses were performed on water-in-crude B emulsions with no particles added, on emulsions containing 1 wt% particles of each type and on the emulsions that showed the lowest viscosity ratio from rheological inspection (the ones containing 0.1 wt% of mildly hydrophilic particles). The results indicated complete stability during the whole experiment (90 min). Only two samples showed a variation of the delta backscattering over the time, indicating some instability. The emulsion containing 1 wt% of hydrophilic silica particles showed some variation in the delta backscattering (1.7%), which we attribute to creaming, with 70% occurring during the first 5 min after emulsification. In contrast, for the emulsion containing 1 wt% of hydrophobic silica particles the stability kinetic curve increased for the whole length of the test and, although with decreasing slope, reached 3.3% of DeltaBST. The backscattering spectrum showed that neither sedimentation nor creaming took place, but coalescence occurred to a small extent. In general the results achieved by Turbiscan analysis confirm the responses obtained by rheological investigation.

Figure 12.9 shows the stability of emulsions prepared with deasphalted crude oils (filled and open circles denote crude A and B, respectively) and crude oil emulsions with 1 wt% of hydrophobic particles added (filled and open triangles denote crude A and B, respectively). In both cases the final value of the viscosity of the systems without asphaltenes and with addition of large amounts of hydrophobic particles is the same. This result indicates that a large addition of very hydrophobic particles in a water-in-crude oil emulsion efficiently remove all the asphaltenes from the W/O

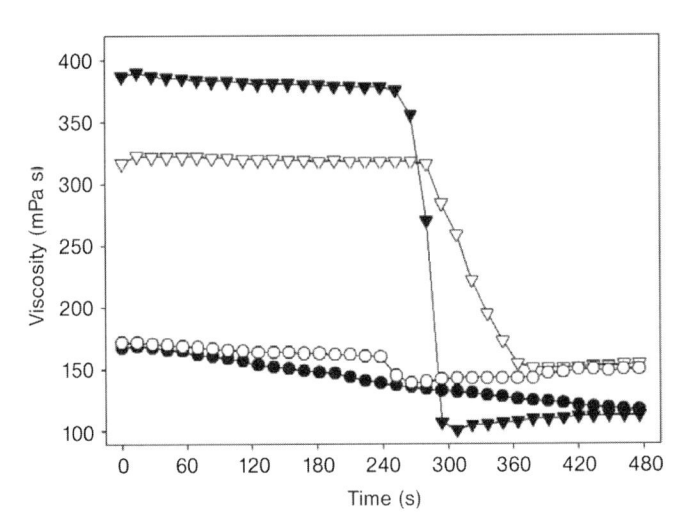

Figure 12.9 Viscosity as a function of time. An electric field is applied from $t = 120$ to 370 s. The time scale is extended to 480 s to reach equilibrium. Empty and filled symbols refer to 50% water cut emulsions prepared with crude A and crude B as oil phases, respectively. Triangles refer to emulsions with 1 wt% of very hydrophobic silica particles (Aerosil 972) added while circles refer to the situation in which the crudes are deasphalted prior to emulsification.

interface. Therefore, the effect of electrostatic destabilization converges to that achieved when deasphalted crude oils are used to make the emulsions. For other particle types the same behavior could not be encountered.

12.4
Conclusions

In the oil industry the operational problems related to the presence of inorganic solids in oilfield fluids are extremely burdensome. On the one hand, the smallest particles contribute to the formation of extremely stable emulsions by enhancing the mechanical properties of the interfacial films. On the other hand, bigger particles can accumulate in the process equipment or erode pipes.

This study describes how the addition of silica nanoparticles to water-in-crude oil emulsions affects their stability. A new parameter, the viscosity ratio of the emulsions, was introduced. The viscosity ratio was correlated to the dewatering efficiency and then used as a measure of the emulsion stability. Two different crude oils (A and B) were used to prepare the emulsions. To elucidate the role of the asphaltene fraction, a third system was prepared by increasing the amount of asphaltenes of the crude oil A to the same as that of crude oil B. Turbiscan analyses and rheological experiments highlighted the importance of asphaltenes in the stability behavior of particle stabilized emulsions. It was revealed that the addition of very hydrophilic particles (Aerosil 200) did not give stable emulsions. However, when crude B and crude A with increased asphaltenes content were mixed with water and 0.1 wt% of particles, the resulting emulsions require electrical destabilization to phase separate. For the emulsions prepared with very hydrophilic particles, Turbiscan data indicated that creaming was responsible for the viscosity decrease. Here enough asphaltenes can adsorb on the particle, modifying their hydrophilic features, and rather stable water-in-oil emulsions may therefore form. As the particle amount was increased, the amount of asphaltenes was not sufficient to cover the particles. Therefore, there was an increased amount of particles at the W/O interface whose character was less and less hydrophobic as their weight fraction increased. Consequently, the water-in-oil emulsions became very unstable. In contrast, when very hydrophobic particles (Aerosil 972) were used, the stability increased or was constant for a much wider range of particle concentrations. This occurred because the hydrophobicity of the particles is favorable for the stability of the water-in-oil emulsions. However, in all cases when the particle amount exceeded a critical value the stability decreased dramatically, especially for crude oil A emulsions. Emulsions prepared with crude B, and crude A with increased asphaltene content, preserved their stability much longer. Up to a volume fraction of 0.5 wt% enough coated particles were present to cover the whole W/O interface. At the same time, the asphaltenes adsorbed at the particle surface and directly at the W/O interface enhanced stability. However, at higher particle concentrations the fraction of uncoated particles increased and asphaltenes were desorbed from the W/O interface. Removal of these surface-active components caused destabilization of the film that surrounds the water droplets and destabilized the emulsions. When mildly hydrophilic particles (Aerosil 7200) are added a strong increase in stability occurred for all emulsions at intermediate concentrations. A maximum is reached at higher particle concentrations when the amount of asphaltenes in the emulsions is higher. When the concentration of the particles was brought above a critical value the stability dropped and the behaviors of the emulsions prepared with crude B and crude A with increased asphaltenes content were almost superimposed. As the particle concentration increased the asphaltenes are less and less available for adsorbing onto the particles. By adding even more particles the probability that modified particles lie at the interface decreases. Therefore, when the volume fraction of the particle is high it is very likely that unmodified hydrophilic particles surround the water droplets. Their lower activity and stabilizing capability together with the absence of asphaltenes at the W/O interface compromise the stability, which decreased dramatically. Moreover, a large fraction of the asphaltenes that was originally at the interface irreversibly adsorbs on the particles, compromising the rigidity of the interfacial film. Notably, emulsions containing more than 1 wt% of this type of particle showed some instability even in the absence of an electric field.

Additional studies performed on model oil emulsions confirmed the above hypotheses. By taking into account differences due to the low viscosity of the model systems the general trends converged nicely with the one observed with crude oil emulsions.

Since all the emulsions presented very similar DSD, with a mean diameter centered at about 8 µm the water droplet size was not considered as an important factor affecting the stability behaviors of the emulsions stabilized by different silica nanoparticles.

Abbreviations

$\bar{\upsilon}$ viscosity ratio, defined by Equation 12.1
υ_1 fluid viscosity before electric field application (mPa s)
υ_2 fluid viscosity after electric field application (mPa s)
$\Delta w_\%$ dewatering efficiency, defined by Equation 12.2
$w_{\%,1}$ water content in the emulsion before electric field application (%)
$w_{\%,2}$ water content in the emulsion after electric field application (%)
DSD   droplet size distribution
TAN   total acidic number [mg-KOH g^{-1}] measured by spectroscopic techniques or by titration

Acknowledgments

S.L., H.M. and J.S. gratefully acknowledge support from the project "Electrocoalescence – Criteria for an efficient process in real crude oil systems" co-ordinated by SINTEF Energy Research. The project is supported by the Research Council of Norway, under the contract number 169466/S30, and by industrial partners.

References

1 Eow, J.S. and Ghadiri, M. (2001) Electrostatic enhancement of coalescence of water droplets in oil: a review of the technology. *Chemical Engineering Journal*, **85**, 357–368.

2 Siöblom, J., Aske, N., Auflem, I.H. et al. (2003) Our current understanding of water-in-crude oil emulsions. Recent characterization techniques and high pressure performance. *Advances in Colloid and Interface Science*, **100–102**, 399–473.

3 Sjöblom, J., Øye, G., Glomm, W.R. et al. (2006) *Modern Characterization Techniques for Crude Oils, their Emulsions, and Functionalized Surfaces*, Taylor & Francis.

4 Fordedal, H., Nodland, E., Sjoblom, J. and Kvalheim, O.M. (1995) A multivariate-analysis of W/O emulsions in high external electric-fields as studied by means of dielectric time-domain spectroscopy. *Journal of Colloid and Interface Science*, **173**, 396–405.

5 Fordedal, H., Schildberg, Y., Sjoblom, J. and Volle, J.L. (1996) Crude oil emulsions in high electric fields as studied by dielectric spectroscopy. Influence of interaction between commercial and indigenous surfactants. *Colloids and Surfaces A-Physicochemical and Engineering Aspects*, **106**, 33–47.

6 Arntzen, R. and Andersen, P.A.K. (2001) Three-phase wellstream gravity separation, in *Encyclopedic Handbook of Emulsion Technology*, (ed. J. Siöblom), Marcel Dekker, Inc., Chapter 27.

7 Binks, B.P. and Kirkland, M. (2002) Interfacial structure of solid-stabilised emulsions studied by scanning electron microscopy. *Physical Chemistry Chemical Physics*, **4**, 3727–3733.

8 Vignati, E., Piazza, R. and Lockhart, T.P. (2003) Pickering emulsions: interfacial tension, colloidal layer morphology, and trapped-particle motion. *Langmuir*, **19**, 6650–6656.

9 Sztukowski, D.M. and Yarranton, H.W. (2005) Oilfield solids and water-in-oil emulsion stability. *Journal of Colloid and Interface Science*, **285**, 821–833.

10 Kralchevsky, P.A., Ivanov, I.B., Ananthapadmanabhan, K.P. and Lips, A. (2005) On the thermodynamics of particle-stabilized emulsions: curvature effects and catastrophic phase inversion. *Langmuir*, **21**, 50–63.

11 Holmberg, K., Jönsson, B., Kronberg, B. and Lindman, B. (2003) *Surfactants and Polymers in Aqueous Solution* 2nd edn, John Wiley & Sons.

12 Menon, V.B. and Wasan, D.T. (1986) Particle fluid interactions with application to solid-stabilized emulsions 1. The effect of asphaltene adsorption. *Colloids and Surfaces*, **19**, 89–105.

13 Gonzalez, G. and Travolloni-Louvisee, A. (1993) Adsorption of asphaltenes and its effect on oil production. *SPE Production and Facilities*, **8**, 91–96.

14 Acevedo, S., Ranaudo, M.A., Escobar, G. et al. (1995) Adsorption of asphaltenes and resins on organic and inorganic substrates and their correlation with precipitation problems in production well tubing. *Fuel*, **74**, 595–598.

15 Acevedo, S., Ranaudo, M.A., Garcia, C. et al. (2000) Importance of asphaltene aggregation in solution in determining the adsorption of this sample on mineral surfaces. *Colloids and Surfaces A-Physicochemical and Engineering Aspects*, **166**, 145–152.

16 Hannisdal, A., Ese, M.H., Hemmingsen, P.V. and Sjoblom, J. (2006) Particle-stabilized emulsions: effect of heavy crude oil components pre-adsorbed onto stabilizing solids. *Colloids and Surfaces A-Physicochemical and Engineering Aspects*, **276**, 45–58.

17 Hannisdal, A., Hemmingsen, P.V. and Sjoblom, J. (2005) Group-type analysis of heavy crude oils using vibrational spectroscopy in combination with

multivariate analysis. *Industrial & Engineering Chemistry Research*, **44**, 1349–1357.
18 Binks, B.P. and Lumsdon, S.O. (1999) Stability of oil-in-water emulsions stabilised by silica particles. *Physical Chemistry Chemical Physics*, **1**, 3007–3016.
19 Tadros, T. (2004) Application of rheology for assessment and prediction of the long-term physical stability of emulsions. *Advances in Colloid and Interface Science*, **108–09**, 227–258.
20 Ali, M.F. and Alqam, M.H. (2000) The role of asphaltenes, resins and other solids in the stabilization of water in oil emulsions and its effects on oil production in Saudi oil fields. *Fuel*, **79**, 1309–1316.
21 Aveyard, R., Binks, B.P. and Clint, J.H. (2003) Emulsions stabilised solely by colloidal particles. *Advances in Colloid and Interface Science*, **100**, 503–546.
22 Lee, R.F. (1999) Agents which promote and stabilize water-in-oil emulsions. *Spill Science & Technology Bulletin*, **5**, 117–126.

13
Impact of Micellar Kinetics on Dynamic Interfacial Properties of Surfactant Solutions

Reinhard Miller, Boris A. Noskov, Valentin B. Fainerman, and Jordan T. Petkov

13.1
Introduction

Many modern technologies depend on the optimum use of surfactants. The applied concentrations are often above the critical micelle concentration (CMC) and special effects are directly related to the presence of micelles. This is true for example in cleaning and detergency [1], encapsulation of drugs in micelles [2, 3] or microemulsions [4], and many others [5]. The important parameters of micellar solutions are the CMC and the aggregation number n. The formation and dissolution of aggregates or the release or incorporation of single molecules are controlled by the relaxation times of slow and fast processes. Their values, however, depend on the models applied.

Examples where micelles play a significant role are the various types of very complex processes in detergency. Wetting of the fabrics, interaction of the detergent with the stains, their removal and the soil stabilization in the bulk are controlled by surfactants and their micelles [6, 7]. The creation of a new area, at the liquid–liquid or liquid–solid interface, requires transport of surfactants from the bulk. Depletion of single surfactant molecules due to adsorption at the respective interfaces leads to micelle breakdown. Hence, surfactants aggregated in micelles serve as an additional reservoir.

The efficiency of surfactant delivery via disintegration of micelles depends on their dynamics. Wetting of the fabrics as one of the initial stages of a washing process represents an excellent example of the importance of micelle lifetime. When fabrics come into contact with the washing liquor the solution penetrates into the fabric structure and wets the fiber surface. The adsorbing surfactant monomers interact with the hydrophobic sites of the fabric, rendering the surface hydrophilic and consequently supporting the wetting process by reducing the interfacial tension. The kinetics of this process depend strongly on how fast monomers are released from the micelles, leading to an acceleration of the wetting [8]. Slowly disintegrating micelles cannot supply enough monomers. A better understanding of micelle kinetics is therefore a paramount prerequisite for efficient detergency.

Highlights in Colloid Science. Edited by Dimo Platikanov and Dotchi Exerowa
Copyright © 2009 WILEY-VCH Verlag GmbH & Co. KGaA, Weinheim
ISBN: 978-3-527-32037-0

The first models for describing micelle break-up from experimental studies were published by Aniansson et al. [9, 10]. Their model assumed essentially the release of single molecules from a micelle, given by the relaxation time τ_1. The complete aggregate dissolution was assumed as a second relaxation process, characterized by τ_2 [9, 10]. The relaxation mechanisms were discussed in many reviews, for example in a book chapter [11], and are not covered here.

Recently, Colegate and Bain [12] and also Song et al. [13] discussed the possibility of direct micelle adsorption and disintegration at the interface. There is actually no driving force for a micelle to adsorb at a water/air interface. However, their work shows that micelles, once transported into the subsurface adjacent to the interface, could behave dynamically different as compared to the bulk, that is, their kinetic constants could differ by orders of magnitude. Hence, it appears to be interesting in the future to examine micelle kinetics under conditions far from equilibrium and in a strongly inhomogeneous system.

There are various direct measurements of micellar solutions giving access to the dynamics rate constants – mainly based on disturbance of the equilibrium state by imposing various types of perturbations, such as stop flow, ultrasound, temperature and pressure jump [14, 15]. This aspect is also not further elaborated here; we focus instead on the impact of micellar kinetics on interfacial properties, to demonstrate that tensiometry and dilational rheology are suitable methods to probe the impact of micellar dynamics. The first work on this subject was published by Lucassen already in 1975 [16] and he showed that the presence of micelles in the bulk have a measurable impact on the adsorption kinetics, and hence on the dilational elasticity, when measured by a longitudinal wave damping technique. Subsequent work demonstrated the effect of micellar dynamics on non-equilibrium interfacial properties [17–29]. The physical idea of the impact of micellar dynamics on the dynamic properties of interfacial layers can be easily understood from the scheme given in Figure 13.1.

When surfactant molecules adsorb at the interface, the concentration in the subphase decreases and a concentration gradient of monomers results. Such a molecular distribution is not in local equilibrium with micelles in the bulk, which tend to release single molecules or may even disintegrate completely, to re-establish the local equilibrium. This change in the local micelle concentration also generates a concentration gradient of the micelles and a diffusion process of micelles set in. Hence, the adsorption of surfactants at an interface generates a diffusion process of monomers, micelles and a disintegration of micelles, all processes directed to re-establish equilibrium conditions over the whole system: locally in the bulk between monomers and micelles, and close to the interface between monomers in the sublayer and at the interface.

In rheological experiments, that is, when the surface layer is periodically compressed and expanded around the equilibrium state, we also meet the situation that molecules desorb from the interface, increasing the local concentration of monomers such that micelles have to either take up molecules or form new micelles. In this case, a diffusion flux of monomers and micelles from and to the interface exists, depending on the respective situation at the interface [16, 25]. The peculiarities of the micellar kinetics in various systems are discussed in several papers; however, the general principles hold [30–34].

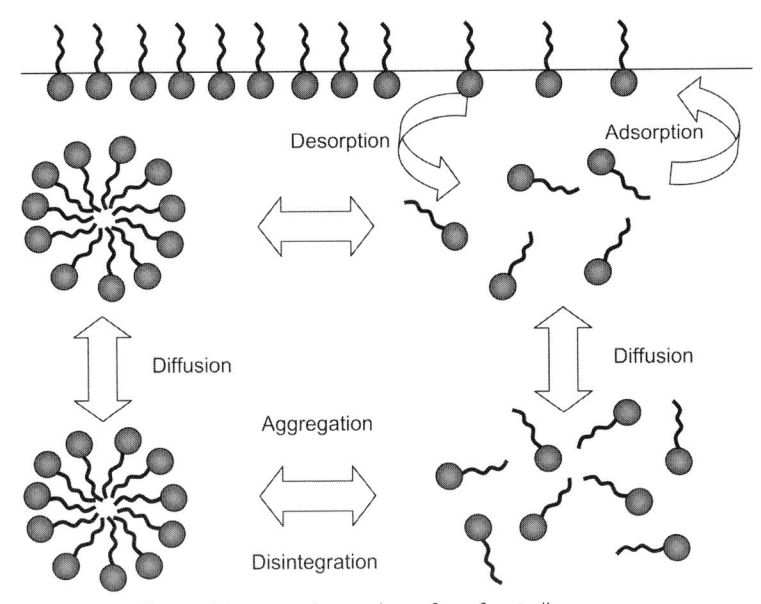

Figure 13.1 Scheme of the region close to the surface of a micellar solution, including the main processes going on in the subphase close to the adsorption layer.

During the last decade, some new and very efficient experimental tools were developed, which led to a renaissance of interfacial investigations on micellar solutions. One such tool is maximum bubble pressure tensiometry, which now works routinely at adsorption times as short as 0.1 ms [35–45]. Also, new instrumentation for dilational rheology, available in addition to wave damping methods [16, 21, 24, 25], based on oscillating drops or bubbles work over a broad frequency range and represent an excellent methodology to study the effect of micellar dynamics on interfacial properties [46–51]. While there will be no detailed description of the experimental tools, we will discuss a few examples that allow insight into the approach.

13.2
Micellization Kinetics Mechanisms

In this chapter we mainly concentrate on the mechanism proposed by Anniansson and Wall [9, 10] and will not discuss other mechanisms [12, 13, 52]. Relaxation after a small perturbation of the aggregation equilibrium in micellar solutions consists of two well-separated steps. The fast process can be reduced to the release/incorporation of a single molecule X_1 from/by the micelle, symbolized by:

$$X_{j-1} + X_1 \underset{k_j^-}{\overset{k_{j-1}^+}{\Leftrightarrow}} X_j$$

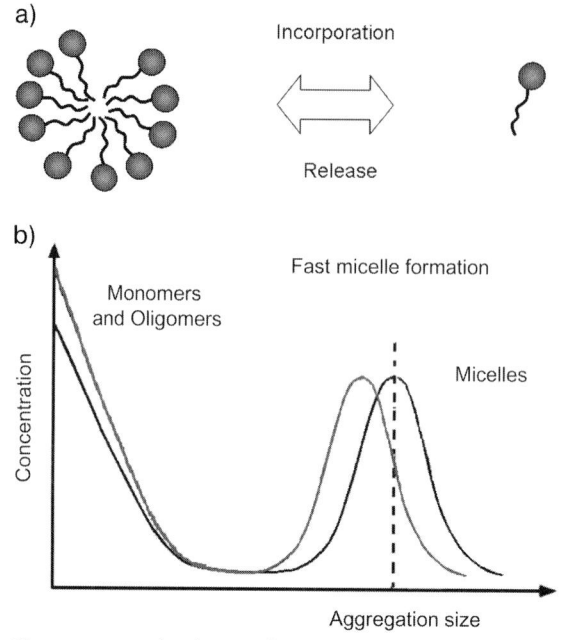

Figure 13.2 Size distribution of aggregates after perturbation of a micellar solution governed by fast micellar kinetics.

while j is a number close to the average aggregation number n. The coefficients k_j^- and k_{j-1}^+ are the respective rate constants (Figure 13.2a). Owing to this release slightly smaller micelles result (Figure 13.2b). Note that the number of micelles remains constant during the first step.

The second, slow step of the relaxation in micellar solutions consists in the dissolution/formation of several micelles and can be represented formally as a slow diffusion of an aggregate along the aggregation size axis through the local minimum (Figure 13.2b) [9]. Although the slow step can be depicted formally as complete micellar dissolution/formation $nX_1 \Leftrightarrow X_n$, it has nothing to do with the step shown in Figure 13.3a and is a complex process based on the release/incorporation of single molecules. This step leads only to the change of the number of micelles and monomers; however, the average aggregation number remains the same (Figure 13.3b).

13.3
Impact of Micelles on Adsorption Kinetics

The maximum bubble pressure method (MBPM) has been demonstrated as suitable for studies of micellar solutions. Various aspects of this methodology have been reviewed in detail [35, 40], including the physical processes taking place during the formation and growth of a bubble at the tip of a capillary and its separation, the

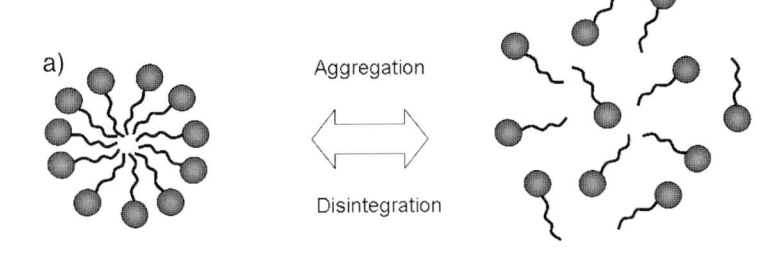

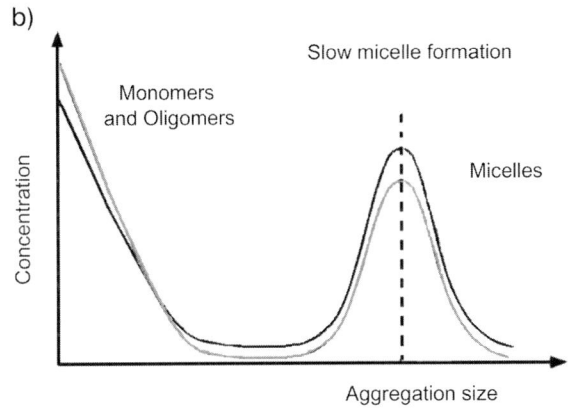

Figure 13.3 Size distribution of aggregates after perturbation of a micellar solution governed by slow micellar kinetics.

problems of measuring bubble pressure, lifetime and so-called dead time. The recently developed BPA device provides a surface lifetime range from 0.1 ms to 50 100 s, that is, an interval spanning over six orders of magnitude [11]. This broad range is sufficient to measure the dynamic surface tensions of micellar solutions and analyze the micellar kinetics. In Ref. [42] some non-ionic ethoxylated surfactants (Triton X-100, Triton X-45 and $C_{14}EO_8$) were studied at concentrations 20–200 times higher than the respective CMC, and it was shown that rather simple models for the adsorption kinetics, including the kinetics of micelle formation and dissolution, which give access to the rate constants of micelle formation and dissolution. The adsorption behavior of these surfactants is well known and has been summarized [37].

We report here on dynamic surface tensions recently obtained from BPA (from SINTERFACE Technologies, Berlin, Germany) measurements on micellar solutions of $C_{14}EO_8$.

A linear dependence of the fast micelle dissolution rate constant k_f on CMC for over 30 different surfactants (taken from Ref. [19]) and for non-ionic ethoxylated surfactants (taken from Ref. [42]) is obtained when plotting the data on a logarithmic scale as presented in Figure 13.4. These results were obtained essentially from

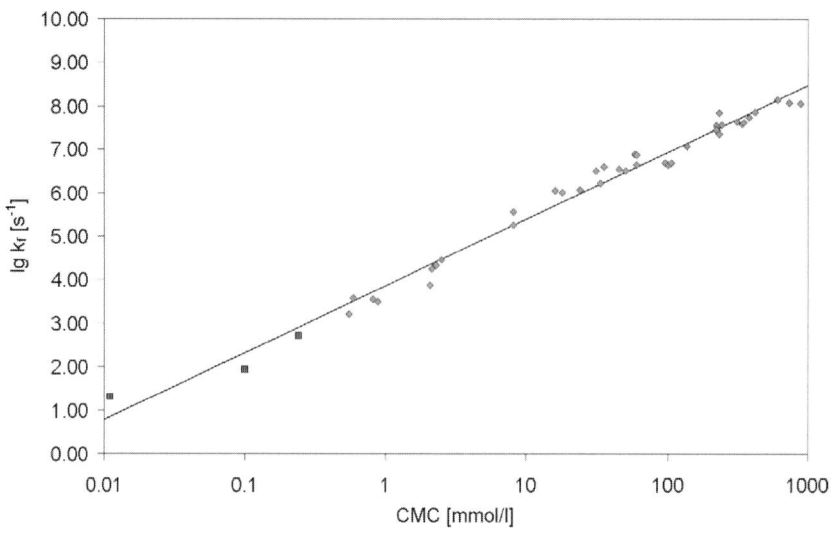

Figure 13.4 Variation of the fast micelle dissolution rate constant k_f with the critical micellization concentration (CMC) for various surfactants; (■) data from [42]; (♦) data from [19].

dynamic adsorption data, assuming the same fast micellar mechanism. There can be strong deviations from such a behavior, as discussed recently in [12, 13]. In this respect it seems important to distinguish between the possible situations, namely, close or very far from the equilibrium interfacial state.

In Ref. [42] it was shown that at short times and under the assumption $k_f t \ll 1$ ($k_f = k_j^-$) the following equation (obtained in Ref. [17]):

$$\Pi = 2RTc_k \left(\frac{Dt}{\pi}\right)^{1/2} \left[1 + \frac{2}{3}\left(\frac{c_0 - c_k}{c_k}\right) k_f t\right] \tag{13.1}$$

agrees well with experimental data for $C_{14}EO_8$, using $k_f = 20\,s^{-1}$, only for relatively low concentrations, below $20 \times$ CMC (Figure 13.5). In this equation $c_k =$ CMC, c_0 is the total concentration of the surfactant, D is the diffusion coefficient of surfactant monomers (here taken as $4 \times 10^{-10}\,m^2\,s^{-1}$). The limiting equation for the relatively fast dissolution of micelles ($k_f t > 1$) for the adsorption dynamics and dynamic surface pressure (diffusion mechanism), as derived in Ref. [17] reads:

$$\Pi = RT\Gamma = 2TRc_0 \left(\frac{Dt}{\pi}\right)^{1/2} \tag{13.2}$$

is in good agreement with experimental dependences shown in Figure 13.5 for $t > 1$ ms and concentrations of $C_{14}EO_8$ between $20 \times$ CMC and $200 \times$ CMC for $D = 2-4 \times 10^{-10}\,m^2\,s^{-1}$.

Thus, data for $C_{14}EO_8$ from Ref. [42], as well as from Ref. [12] for this surfactant, show faster dissociation kinetics of micelles at concentration above $20 \times$ CMC. To reach agreement between theory and experiment at higher concentrations, k_f has to be increased by approximately 100 times.

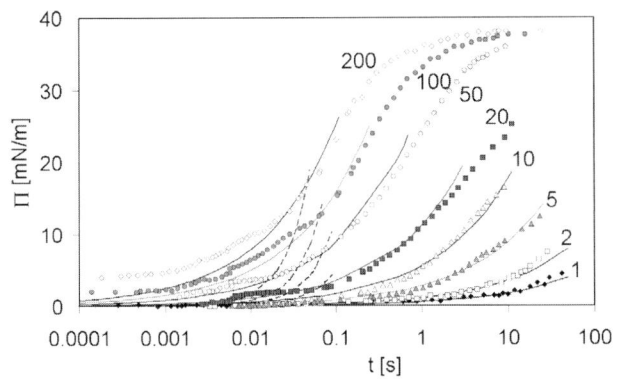

Figure 13.5 Dynamic surface tension for $C_{14}EO_8$ solutions as a function of the effective surface lifetime, solid lines calculated from Equations 13.2 or 13.3 using $D = 4 \times 10^{-10}\,m^2\,s^{-1}$; dashed lines calculated from Equation 13.1 using $k_f = 20\,s^{-1}$; data partially from Ref. [42]; the numbers correspond to the concentrations ($\mu mol\,L^{-1}$).

Notably, an expression quite similar to Equation 13.3 for low surface pressures Π follows from the model that accounts for the influence of micelles (assuming fast micelle disintegration kinetics) on the effective monomer diffusion coefficient [25, 42]:

$$\Pi = 2TRc_k \left(\frac{D^*t}{\pi}\right)^{1/2} \tag{13.3}$$

An equation for this effective diffusion coefficient D^* for the fast micellization process was proposed by Joos [20]:

$$D^* = D(1+\beta)(1+\alpha\beta), \tag{13.4}$$

with $\beta = (c_0 - c_k)/c_k$ and $\alpha = D_m/D \approx 0.25$; D_m is diffusion coefficient of micelles.

At very high concentration the micelles probably dissociate in the subsurface layer [13]. The close distance to the surface could influence the free energy of micelle formation and kinetics of dissociation of micelles. To calculate the standard free energy of micelle formation for non-ionic surfactants, the expression:

$$\Delta G_m^0 = RT \ln x_{cmc} = RT \ln(c_{cmc}/\rho) \tag{13.5}$$

can be used [37, 53–55]. Here R is the gas law constant, T is the temperature, x_{cmc} is the molar fraction of surfactant in solution bulk at CMC, ρ is the ratio of the solvent (water) mass of 1 L to its molecular weight, $\rho \approx 1000/18 = 55.6\,mol\text{-}H_2O\,L^{-1}$. The standard free energy of adsorption can be expressed via the standard energy of micelle formation as [37, 53, 55, 56]:

$$\Delta G^0 = \Delta G_m^0 - (\gamma_0 - \gamma_{cmc})\omega \tag{13.6}$$

where γ_0 and γ_{cmc} are the surface tension of the solvent and solution at CMC, respectively, and ω is the partial molar area of the surfactant at the surface. Equation 13.6 directly indicates that the free energy of adsorption is much higher (as an absolute value) than the free energy of micelle formation. This difference is defined by $-(\gamma_0 - \gamma_{cmc})\omega$, and for $C_{14}EO_8$ it is equal to 12 kJ mol^{-1}. If micelles are found near the surface, at $c_0 \gg$ CMC [13], the release of single molecules needs about 12 kJ mol^{-1} less free energy. This decreased value of free energy of micelle formation corresponds, according to Figure 13.4, to a higher k_f (approximately 10^4 s^{-1}). At this value of k_f the experimental results in Figure 13.5 and the data in Ref. [12] would be in agreement with the theoretical diffusion adsorption models (2) or (3) for adsorption times >0.1 ms. Note, to reach agreement with experiments given in Ref. [12] and Figure 13.5 the model of Equation 13.2 requires adsorption times of $t >$ 1 ms, that is, it needs a smaller k_f (10^3 s^{-1}) and differences between the free energies in Equation 13.6.

13.4
Impact of Micelle Kinetics on Interfacial Dilational Visco-Elasticity

Using the drop and bubble profile analysis tensiometer PAT-1 (SINTERFACE Technologies, Berlin, Germany) it is possible to perform dilational rheology studies of surfactant solutions at low frequencies.

Expressions for the dilational visco-elasticity for non-micellar solutions $\varepsilon(i\omega)$ were obtained by Lucassen [57, 58]:

$$\varepsilon(i\omega) = \varepsilon_0[1+(1-i)\zeta]^{-1} \tag{13.7}$$

with:

$$\zeta = \sqrt{\omega_D/2\omega} \tag{13.8}$$

the limiting (high frequency) elasticity:

$$\varepsilon_0(c) = -d\gamma/d\ln\Gamma \tag{13.9}$$

and the diffusion relaxation frequency:

$$\omega_D(c) = D \cdot (d\Gamma/dc)^{-2} \tag{13.10}$$

$\omega = 2\pi f$ is the angular frequency of the generated oscillation at frequency f. Equation 13.7 can be transformed into expressions for the visco-elasticity modulus ε and phase angle ϕ between stress (dγ) and strain (dA) [57, 58]:

$$\varepsilon = \varepsilon_0(1+2\zeta+2\zeta^2)^{-1/2}, \quad \phi = \arctan[\zeta/(1+\zeta)] \tag{13.11}$$

where A is the interfacial area. Theoretical models for the complex elasticity $\varepsilon(i\omega)$ for different micellar kinetics mechanisms were discussed in Refs. [11, 16, 21, 25, 59]. We want to restrict ourselves here to a simple case, proposed by P. Joos, that is, the fast dissolution of micelles with an equilibrium between monomers and micelles [59]:

$$\varepsilon(i\omega) = \varepsilon_0[1+(1-i)\zeta\sqrt{(1+\beta)(1+\alpha\beta)}]^{-1} \tag{13.12}$$

Equation 13.12 is identical to Equations 13.7 and 13.11, supposing the diffusion coefficient D is replaced by an effective diffusion coefficient of monomers D^*, according to Equation 13.4 proposed by Joos [20, 59].

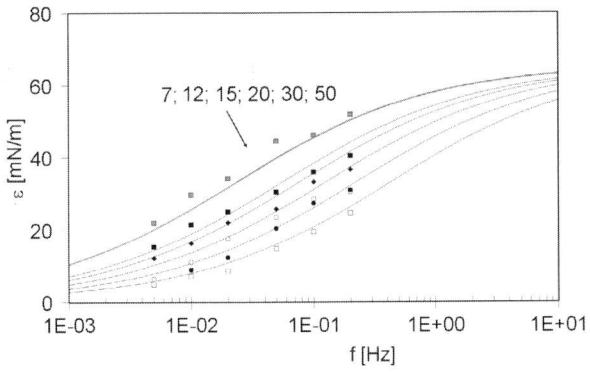

Figure 13.6 Visco-elasticity module ε as a function of frequency f for various $C_{14}EO_8$ concentrations at the CMC = 7 μmol L^{-1} and slightly above the CMC; thick (top) line: calculated according to the theory of Lucassen [57, 58]. Equation 13.11 for CMC; thin lines: calculated for c_0 > CMC according to Equation 13.12 with D^* calculated from Equation 13.4; the numbers correspond to the concentrations [7 (■) to 50 (□) μmol L^{-1}].

Thus, to describe the dilatational visco-elasticity modulus of micellar solutions the model Equations 13.7–13.11 can be used. In this case we have to assume c_0 = CMC for c_0 > CMC, and have used instead of D an effective diffusion coefficient of monomers D^*.

Figures 13.6 and 13.7 present the dependences of the visco-elasticity module ε on frequency f.

For calculations of the rheological dependences for $C_{14}EO_8$ according to the theory of Lucassen [57, 58] for non-micellar solution we applied a combined reorientation model with two-dimensional compressibility in state with minimal molar area ω_2 as

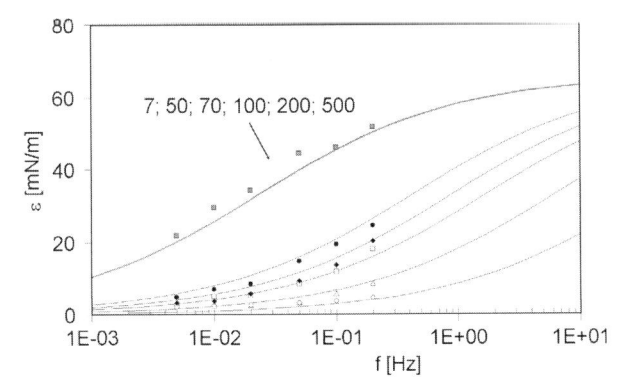

Figure 13.7 Visco-elasticity module ε as a function of frequency f for various $C_{14}EO_8$ concentrations far above the CMC (50–500 μmol L^{-1}, corresponding to 7–70 × CMC); thick (top) line: calculated from the theory of Lucassen [57, 58] (Equation 13.11) for c = CMC; thin lines: calculated for c_0 > CMC according to Equation 13.12 with D^* calculated from Equation 13.4; the numbers correspond to the concentrations [7 (■) to 500 (○) μmol L^{-1}].

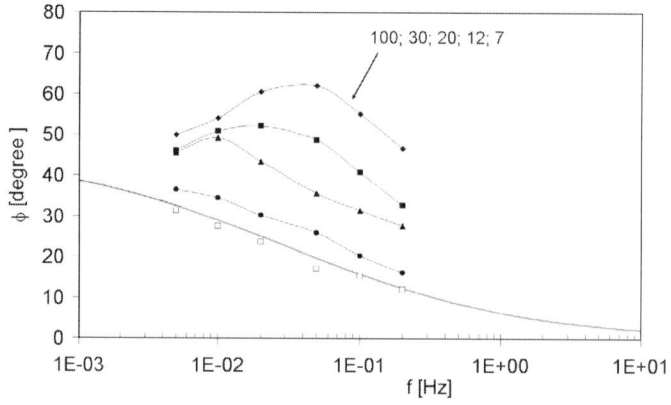

Figure 13.8 Phase angle ϕ as a function of frequency f for $C_{14}EO_8$ concentrations between 7 and 100 μmol L^{-1}; theoretical calculations for CMC = 7 μmol L^{-1} (thick, bottom, line) are made with the model of Lucassen (Equation 13.11) using a diffusion coefficient of 4×10^{-10} $m^2 s^{-1}$; thin lines are a guide for eyes; the numbers correspond to the concentrations [7 (□) and 100 (♦) μmol L^{-1}].

proposed in Ref. [51] and D^* values according to Equation 13.4. The programme IsoFit [60] yields the following model parameters [51]: $\omega_2 = 4.4 \times 10^5$ $m^2 mol^{-1}$, the maximal molar areas $\omega_1 = 1.0 \times 10^6$ $m^2 mol^{-1}$, the intermolecular interaction constant $\alpha = 0.9$, the adsorption equilibrium constant $b_2 = 1.0 \times 10^4$ $m^3 mol^{-1}$ and the relative two-dimensional compressibility coefficient $e = 0.009$ m mN^{-1}. The diffusion coefficient of monomers used here was taken to be 4×10^{-10} $m^2 s^{-1}$, in agreement with the data discussed in Refs [42, 51].

As one can see, this theoretical model describes the visco-elasticity modulus very well, even at very high concentrations, as shown in Figure 13.7 for 50 to 500 μmol L^{-1}, corresponding to 7–70 × CMC. For comparison, the data for 7 and 50 μmol L^{-1} from Figure 13.6 are also shown.

Figure 13.8 presents the dependences of the phase angle ϕ on the frequency f for various $C_{14}EO_8$ concentrations. With increasing concentration, of course the phase angle ϕ increases significantly, up to 70° (Figure 13.8). The external behavior of the phase angle, with maximal values reaching up to 90°, is in full agreement with Lucassen's theory for micellar solutions given in Ref. [16].

13.5
Summary

Using the surfactant $C_{14}EO_8$ we have demonstrated how the effect of micellar kinetics on the dynamic properties of adsorption layers can be experimentally studied. Although the simplest theoretical models are applied, a rather good agreement with the experimental data is observed. In adsorption dynamics from solutions far above the CMC we observe adsorption rates that cannot be explained by the usual micelle formation constants k_f. For the relaxation behavior (at low frequencies) no peculiarities are observed, that is the frequency behavior of the viscoelasticity modulus can be understood by using the effective diffusion coefficient of monomers proposed by the model of Joos.

Acknowledgments

The work was financially supported by a project of the European Space Agency (FASES MAP AO-99-052), the DFG SPP 1273 (Mi418/14-1) and the COST actions P21 and D43.

References

1. von Rybinski, W. (2001) Surface chemistry in detergency, in *Handbook of Applied and Colloid Chemistry* (ed. K. Holmberg), John Wiley & Sons, Ltd.
2. Rangel-Yagui, C.O., Pessoa, A. and Tavares, L.C. (2005) *The Journal of Pharmacy and Pharmaceutical Sciences*, **8**, 147–163.
3. Cevc, G. (2004) *Advanced Drug Delivery Reviews*, **56**, 675–711.
4. Kogan, A. and Garti, N. (2006) *Advances in Colloid and Interface Science*, **123–126**, 369–385.
5. Patist, A., Kanicky, J.R., Shukla, P.K. and Shah, D.O. (2002) *Journal of Colloid and Interface Science*, **245**, 1–15.
6. Cutler, W.G. and Kissa, E. (1987) *Detergency – Theory and Applications*, Marcel Dekker, New York.
7. Lange, K.R. (1994) *Detergents and Cleaners*, Carl – Hanser – Verlag, Munich.
8. Carter, D.L., Draper, M.C., Peterson, R.N. and Shah, D.O. (2005) *Langmuir*, **21**, 10106–10111.
9. Aniansson, E.A.G. and Wall, S.N. (1974) *Journal of Physical Chemistry*, **78**, 1024–1030.
10. Aniansson, E.A.G., Wall, S.N.W., Almgren, M., et al. (1976) *Journal of Physical Chemistry*, **80**, 905–922.
11. Noskov, B.A. and Grigoriev, D.O. (2001) Adsorption from micellar solutions, in *Surfactants – Chemistry, Interfacial Properties and Application, Studies in Interface Science* (eds V.B. Fainerman, D. Möbius and R. Miller), Vol. 13, Elsevier, pp. 402–510.
12. Colegate, DM. and Bain, CD. (2005) *Physical Review Letters*, **95**, 198302-1–198302-4.
13. Song, Q., Couzis, A., Somasundaran, P. and Maldarelli, C. (2006) *Colloids and Surfaces A: Physicochemical and Engineering Aspects*, **282–283**, 162–182.
14. Kahlweit, M. and Teubner, M. (1980) *Advances in Colloid and Interface Science*, **13**, 1–64.
15. Gradzielski, M. (2004) *Current Opinion in Colloid and Interface Science*, **9**, 256–263.
16. Lucassen, J. (1975) *Faraday Discussions of the Chemical Society*, **59**, 76–87.
17. Fainerman, V.B. (1981) *Colloid Journal*, **43**, 94–100.
18. Miller, R. (1981) *Colloid and Polymer Science*, **259**, 1124–1128.
19. Fainerman, V.B. (1981) *Colloid Journal (Russian)*, **43**, 717–725; Fainerman, V.B. (1981) *Kolloidn Zh (Russian)*, **43**, 926–932.
20. Joos, P. and van Hunsel, J. (1988) *Colloids Surfaces*, **33**, 99–108.
21. Noskov, B.A. (1989) *Fluid Dynamics*, **24**, 251–260.
22. Fainerman, V.B. (1992) *Colloids Surfaces*, **62**, 333–347.

23 Fainerman, V.B. and Makievski, A.V. (1993) *Colloids Surfaces*, **69**, 249–263.
24 Noskov, B.A. (1996) *Advances in Colloid and Interface Science*, **69**, 63–129.
25 Noskov, B.A. (2002) *Advances in Colloid and Interface Science*, **95**, 237–293.
26 Liao, Y.C., Basaran, O.A. and Franses, E.I. (2003) *AIChE Journal*, **49**, 3229–3240.
27 Nyrkova, I.A. and Semenov, A.N. (2005) *Macromolecular Theory and Simulations*, **14**, 569–585.
28 Danov, K.D., Kralchevsky, P.A., Denkov, N.D. et al. (2006) *Advances in Colloid and Interface Science*, **119**, 1–16.
29 Danov, K.D., Kralchevsky, P.A., Denkov, N.D. et al. (2006) *Advances in Colloid and Interface Science*, **119**, 17–33.
30 Patist, A., Axelberd, T. and Shah, D.O. (1998) *Journal of Colloid and Interface Science*, **208**, 259–265.
31 Patist, A., Oh, S.G., Leung, R. and Shah, D.O. (2001) *Colloids and Surfaces A: Physicochemical and Engineering Aspects*, **176**, 3–16.
32 Danov, K.D., Valkovska, D.S. and Kralchevsky, P.A. (2003) *Journal of Colloid and Interface Science*, **251**, 18–25.
33 Kjellin, U.R.M., Reimer, J. and Hansson, P. (2003) *Journal of Colloid and Interface Science*, **262**, 506–515.
34 Danov, K.D., Kralchevsky, P.A., Ananthapadmanabhan, K.P. and Lips, A. (2006) *Colloids and Surfaces A: Physicochemical and Engineering Aspects*, **282–283**, 143–161.
35 Fainerman, V.B. and Miller, R. (1998) The maximum bubble pressure technique, monograph in "drops and bubbles in interfacial science", in *Studies of Interface Science* (eds D. Möbius and R. Miller), Vol. 6, Elsevier, Amsterdam, pp. 279–326.
36 Eastoe, J. and Dalton, J.S. (2000) *Advances in Colloid and Interface Science*, **85**, 103–144.
37 Fainerman, V.B., Möbius, D. and Miller, R.(eds) (2001) *Surfactants – Chemistry, Interfacial Properties and Application, Studies in Interface Science*, Vol. 13, Elsevier.

38 Prosser, A.J. and Franses, E.I. (2001) *Colloids and Surfaces A: Physicochemical and Engineering Aspects*, **178**, 1–40.
39 Eastoe, J., Rankin, A., Wat, R. and Bain, C.D. (2001) *International Reviews in Physical Chemistry*, **20**, 357–386.
40 Fainerman, V.B. and Miller, R. (2004) *Advances in Colloid and Interface Science*, **108–109**, 287–301.
41 Fainerman, V.B., Makievski, A.V. and Miller, R. (2004) *Review of Scientific Instruments*, **75**, 213–221.
42 Fainerman, V.B., Mys, V.D., Makievski, A.V. et al. (2006) *Journal of Colloid and Interface Science*, **302**, 40–46.
43 Frese, C., Ruppert, S., Sugar, M. et al. (2004) *Journal of Colloid and Interface Science*, **267**, 475–482.
44 Delgado, C., Lopez-Diaz, D., Merchan, M.D. and Velazquez, M.M. (2006) *Tenside Surfactants Detergents*, **43**, 192–196.
45 Zhmud, B.V., Tiberg, F. and Kizling, J. (2000) *Langmuir*, **16**, 2557–2565.
46 Loglio, G., Pandolfini, P., Miller, R. et al. (2001) Drop and bubble shape analysis as tool for dilational rheology studies of interfacial layers, in *Novel Methods to Study Interfacial Layers*, (ed. D. Möbius and R. Miller), Elsevier, Amsterdam, pp. 439–484.
47 Wantke, K.D. and Fruhner, H. (2001) *Journal of Colloid and Interface Science*, **237**, 185–199.
48 Örtegren, J., Wantke, K.D. and Motschmann, H. (2003) *Review of Scientific Instruments*, **74**, 5167–5172.
49 Ravera, F., Ferrari, M. and Liggieri, L. (2006) *Colloids and Surfaces A: Physicochemical and Engineering Aspects*, **282–283**, 210–216.
50 Benjamins, J., Lyklema, H. and Lucassen-Reynders, EH. (2006) *Langmuir*, **22**, 6181–6188.
51 Fainerman, V.B., Zholob, S.A., Petkov, J.T. and Miller, R. (2007) *Colloids and Surfaces A: Physicochemical and Engineering Aspects*, **323**, 56.
52 Kahlweit, M. (1982) *Journal of Colloid and Interface Science*, **90**, 92–101.

53 Miller, R., Fainerman, V.B. and Möhwald, H. (2002) *Surfactants Detergents*, **5**, 281–286.

54 Rosen, M.J. (1978) *Surfactants and Interfacial Phenomena*, John Wiley & Sons, New York.

55 Rosen, M.J., Cohen, A.W., Dahanayake, M. and Hua, X.-Y. (1982) *Journal of Physical Chemistry*, **86**, 541–545.

56 Aratono, M., Uryu, S., Hayami, Y. *et al.* (1984) *Journal of Colloid and Interface Science*, **98**, 33–38.

57 Lucassen, J. and van den Tempel, M. (1972) *Chemical Engineering Science*, **27**, 1283.

58 Lucassen, J. and Hansen, R.S. (1967) *Journal of Colloid and Interface Science*, **23**, 319.

59 Joos, P. (1999) *Dynamic Surface Phenomena*, VSP, Dordrecht, The Netherlands.

60 Access to software packages is free of charge via www.sinterface.com or www.mpikg.mpg.de/gf/miller/, last accessed 22/07/2008.

14
Aggregation of Colloids: Recent Developments in Population Balance Modeling

Ponisseril Somasundaran and Venkataramana Runkana

14.1
Introduction

Aggregation and dispersion of solid, liquid or gas particulates in colloidal suspensions are critical phenomena that control many natural and industrial processes. Examples include removal of suspended solids in water treatment [1], selective beneficiation of colloidal mineral suspensions to recover valuable minerals [2–5], manufacture of pulp and paper [6], fabrication of ceramic components [7], food processing, personal and home care product formulation and transport and the fate of suspended contaminant particles in aquatic systems [8]. Aggregation or dispersion is usually induced by natural or synthetic inorganic electrolytes, polymers and surfactants [4]. A colloidal suspension is a complex multiphase multicomponent system of solid/liquid/gas particulates, solvent, electrolyte species and polymers and as such the stability of a suspension depends strongly on the interactions amongst the particulates, the solvent and the dissolved species. Since the particles are small in size, surface forces play a dominant role in determining the behavior of the colloidal suspensions and the stability and the rate of aggregation. These forces include van der Waals attraction, electrical double layer repulsion/attraction, hydration repulsion, hydrophobic attraction and, in the presence of polymers, steric repulsion or bridging attraction [9, 10]. The nature and magnitude of these forces depend on variables such as particulate size and shape, pH, temperature, type and concentration of added or dissolved electrolyte species, the polarity of the solvent and concentration and properties of polymer or surfactant used. Aggregation is essentially a two-step process: diffusion of particles or clusters and collision leading to attachment or detachment, depending on the net energy of interaction and the strength of attachment. To control or enhance the frequency of collisions and the detachment, shear or vibration is usually applied in many industrial applications to improve the rate of aggregation or dispersion. Shear and vibration also lead to fragmentation of clusters and a state of dynamic equilibrium is usually attained in electrolyte-induced coagulation wherein the rate of fragmentation is close to that of aggregation [11]. In

Highlights in Colloid Science. Edited by Dimo Platikanov and Dotchi Exerowa
Copyright © 2009 WILEY-VCH Verlag GmbH & Co. KGaA, Weinheim
ISBN: 978-3-527-32037-0

the case of polymer-induced flocculation, the rate of fragmentation appears to be relatively higher than that of aggregation during later stages of flocculation due to polymer scission and degradation [12]. While polymers are usually employed to stabilize or flocculate emulsions or suspensions, surfactants are used to stabilize them in non-aqueous environments [13].

The key indicator of the state of aggregation of the suspension is the aggregate or floc size and shape distribution, which can be measured using techniques that utilize light scattering, sedimentation or diffusion characteristics or osmotic pressure of the suspension [14, 15]. Population balances are usually employed to model kinetics of aggregation of dilute suspensions and predict the temporal evolution of aggregate size distribution as a function of process variables such as primary particle size or initial particle size distribution, temperature, shear rate, pH, ionic strength of the solution, polymer concentration, and so on. Population balance (PB) framework is commonly applied to mathematically represent processes dealing with discrete entities such as particles, bubbles and droplets [16]. Important assumptions made in formulating population balances and their implications on simulation results, numerical solution of population balance equations, mechanisms of aggregation, and so on have been discussed in detail elsewhere [1, 17–19]. Aggregation and dispersion of colloidal suspensions has been studied extensively because of its widespread industrial applications and its role in natural processes. Recent research efforts have focused on three important aspects of modeling aggregation of colloidal suspensions: the role of surface forces on aggregation kinetics, evolution of aggregate structure and the influence of shear rate distribution in turbulent environments on aggregate size distribution. These developments are reviewed in this chapter. Since aggregation takes place either in quiescent (Brownian aggregation) or in shear environments, this chapter is broadly divided into two sections and developments in Brownian and shear aggregation are discussed in detail separately.

14.2
Aggregation in Quiescent Environments

In the absence of applied shear, aggregation of colloids occurs mainly by Brownian motion and to give clusters or aggregates or flocs. As aggregation proceeds, differential sedimentation also contributes to the rate of aggregation as flocs grow in size and settle at different velocities, depending on floc size, structure and density. The population balance is a statement of continuity for particulate systems and the size-continuous form of population balance equation (PBE) for aggregation of dilute colloidal suspensions in a batch process is given by [20]:

$$\frac{\partial n(v, t)}{\partial t} = -\int_0^\infty \alpha(v, u)\, \beta(v, u)\, n(v, t)\, n(u, t)\, du \\ + \frac{1}{2}\int_0^v \alpha(v-u, u)\, \beta(v-u, u)\, n(v-u, t)\, n(u, t)\, du \quad (14.1)$$

where n is number concentration of aggregates (or particles), v and u denote aggregate volume, t is aggregation time, β is collision frequency factor and α is collision efficiency factor. The first term on the right-hand side accounts for the disappearance of aggregates of size v due to their interaction with aggregates of all sizes. The second term represents the rate of formation of aggregates of size v due to interaction between aggregates of smaller sizes. Since it is relatively difficult and perhaps not possible to obtain a close form analytical solution, numerical methods are usually applied to solve Equation 14.1. Generally, Equation 14.1 is first discretized into a set of nonlinear ordinary differential equations and grouped, either uniformly or geometrically, into different sections and solved by a suitable numerical technique [21]. The rate of change of particle or aggregate number concentration due to aggregation alone is given by the following discretized PBE [22]:

$$\frac{dN_i}{dt} = N_{i-1} \sum_{j=1}^{i-2} 2^{j-i+1} \alpha_{i-1,j} \beta_{i-1,j} N_j + \frac{1}{2} \alpha_{i-1,i-1} \beta_{i-1,i-1} N_{i-1}^2 \\ - N_i \sum_{j=1}^{i-1} 2^{j-i} \alpha_{i,j} \beta_{i,j} N_j - N_i \sum_{j=i}^{max_l} \alpha_{i,j} \beta_{i,j} N_j \qquad (14.2)$$

where N_i is number concentration of particles or aggregates in a section i and max_l is maximum number of sections used to represent the complete size spectrum. The two important parameters in the PBE are the collision frequency factor and the collision efficiency factor. All the collisions between particles do not necessarily lead to aggregation if repulsive forces are stronger than attractive forces. While the influence of surface forces was not incorporated previously, efforts have been made recently [18, 23–25] to integrate fundamental theories of surface forces with the population balance paradigm and these developments are discussed in this section.

14.2.1
Models Incorporating Surface Forces

Depending on the surface and colloid chemistry of the system, the nature and magnitude of various interaction forces between particles will change and influence colloid stability and the rate of aggregation. The influence of these surface forces is taken into account by the collision efficiency factor in the population balance. The collision efficiency factor for aggregates is computed as reciprocal of the modified Fuchs' stability ratio W for two primary particles k and l [26–29]:

$$W_{k,l} = \frac{\int_{r_{0_k}+r_{0_l}}^{\infty} D_{k,l} \frac{\exp(V_T/k_B T)}{s^2} ds}{\int_{r_{0_k}+r_{0_l}}^{\infty} D_{k,l} \frac{\exp(V_{vdW}/k_B T)}{s^2} ds} \qquad (14.3)$$

where k_B is Boltzmann constant, T is temperature, $D_{k,l}$ is a hydrodynamic correction factor [30], V_T is total energy of interaction, V_{vdW} is van der Waals energy of attraction between two primary particles of radii r_{0_k} and r_{0_l} (assumed spherical), s is distance between particle centers ($s = r_{0_k} + r_{0_l} + h_0$); h_0 is the distance of closest approach between particle surfaces. It is assumed that the efficiency of aggregate collisions depends mainly on the interaction between particles lying on the surface of the aggregates. The total interaction energy is generally assumed to be a sum of all the surface forces expected to be present between particles under the given conditions of pH, temperature, type and concentration of electrolyte species and type and concentration of coagulant or flocculant employed and its properties. Runkana *et al.* [23–25] incorporated the above equation into the population balance for aggregation in the presence of inorganic electrolytes and polymers. The results for electrolyte-induced aggregation are discussed first followed by the results for polymer-induced flocculation.

14.2.1.1 Aggregation in the Presence of Inorganic Electrolytes

Colloidal aggregation is usually characterized in terms of two mechanisms, diffusion-limited aggregation (DLA) and reaction-limited aggregation (RLA). DLA takes place when repulsive forces between particles are negligible. It is generally fast and follows power-law growth kinetics. In contrast, RLA follows slow, exponential growth kinetics [31]. The aggregates are commonly referred to as mass fractals because aggregate mass M_i is related to its collision radius r_c through the mass fractal dimension d_F [32]:

$$M_i \propto r_{c_i}^{d_F} \tag{14.4}$$

The aggregate fractal dimension d_F is a function of size distribution of constituting particles, solids concentration, temperature, pH, ionic strength and polymer concentration [14, 33–35]. Computer simulation has been applied extensively [36] to study both DLA and RLA and predict aggregate structure by manipulating simulation parameters such as the random walk size [37, 38].

Smoluchowski [39] derived the collision kernel or frequency factor for aggregation of spherical primary particles under the DLA regime. Since collisions mainly involve clusters or aggregates, except during the initial stages of aggregation, the equation for spherical particles derived by Smoluchowski needs to be modified to incorporate the collision radii of interacting aggregates. This is relatively straightforward and requires substitution of the aggregate collision radius, which is a function of its mass fractal dimension, into Smoluchowski's original expression. The collision frequency factor for fractal aggregates under the DLA regime $\beta_{i,j}^{DLA}$ is:

$$\beta_{i,j}^{DLA} = \frac{2k_B T}{3\mu}\left(\frac{1}{r_{c_i}} + \frac{1}{r_{c_j}}\right)(r_{c_i} + r_{c_j}) \tag{14.5}$$

where μ is solvent viscosity. The classical DLVO theory [40, 41] is applicable for most electrolyte-induced coagulation processes and Somasundaran and Runkana [18] incorporated it into the PBE and simulated aggregation of colloidal hematite suspensions. The total collision frequency factor was calculated as a sum of collisions due to Brownian motion and due to differential sedimentation [42]. The predicted time evolution of mean aggregate diameter at different temperatures is compared with experimental data [14] in Figure 14.1.

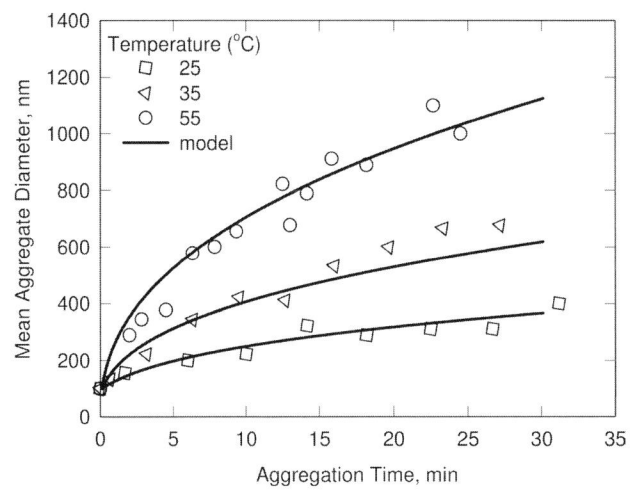

Figure 14.1 Comparison of predicted (solid lines) and experimental (symbols) time evolution of mean diameter of hematite aggregates at different temperatures; mean primary particle diameter: 100 nm; solids concentration: 2.25×10^{16} m^{-3}; [KCl] = 50 mM [18]. (Experimental data from Ref. [14].)

Smoluchowski's equation, however, is not applicable for RLA because it occurs in the presence of repulsive forces and follows slow exponential growth kinetics. With RLA, aggregation takes place only after several collisions occur between particles or aggregates due to the repulsive forces. Though several empirical expressions have been proposed in the literature for the collision frequency factor for the RLA regime, none of them predict the aggregation kinetics accurately. Recently, Runkana et al. [24] derived the following kernel based on theoretical arguments proposed by Ball et al. [43] for cluster–cluster collisions in the RLA regime:

$$\beta_{i,j}^{RLA} = K_{RLA} r_0^2 \left(\frac{1}{C_L}\right)^{2/d_F} \left(n_{0_i}^{1/d_F} + n_{0_j}^{1/d_F}\right)^2 \tag{14.6}$$

where $\beta_{i,j}^{RLA}$ is the collision frequency factor for RLA kinetics, C_L is the aggregate structure prefactor, K_{RLA} is a proportionality constant or lumped parameter, which takes into account the effect of temperature and viscosity of the suspension, and n_{0_i} and n_{0_j} are the number of primary particles in aggregates belonging to sections i and j, respectively. This kernel was found to be accurate for RLA of γ-alumina suspensions in the presence of various electrolytes [24].

RLA occurs when repulsive forces are stronger than attractive forces. In addition to the electrical double layer force, under certain conditions of pH and electrolyte concentration, a thin hydration layer develops around the particle surface and results in short-range hydration repulsion, which could cause a reduction in collision efficiency between particles and leads to slow aggregation kinetics [44, 45]. Based on empirical models derived from force–distance measurements using atomic force microscopy, Runkana et al. [24] derived the following expression for hydration interaction energy (V_{hyd}) between two unequal spherical particles and incorporated it, along with the DLVO theory, into the population balance model (PBM) to simulate reaction-limited aggregation of γ-alumina suspensions in the presence of KCl, NaCl and KNO$_3$ electrolytes:

$$V_{hyd} = \left(\frac{2\pi r_{0_k} r_{0_l}}{r_{0_k}+r_{0_l}}\right) P_{hyd} \lambda_{hyd}^2 \exp(-h_0/\lambda_{hyd}) \tag{14.7}$$

where P_{hyd} is a structural or hydration force constant and λ_{hyd} is the hydration decay length. The collision frequency factor was calculated as a sum of contributions due to differential sedimentation and due to Brownian motion, Equation 14.6, for RLA. Figure 14.2 compares their simulation results with the experimental data of Beattie et al. [44].

14.2.1.2 Aggregation in the Presence of Polymers

Aggregation in the presence of polymers can occur by any one or more of the four well-known mechanisms: simple charge neutralization, charge patch neutralization, polymer bridging and polymer depletion [46, 47]. Aggregation by the first three mechanisms takes place due to polymer adsorption while depletion flocculation occurs in the presence of non-adsorbing polymers. Owing to polymer adsorption, the electrochemical nature of particle surfaces are modified, which leads to a change in the nature and magnitude of van der Waals attraction and electrical double layer force. In addition, polymer-induced forces, namely, steric repulsion or bridging attraction come into effect and influence suspension stability and aggregation kinetics. Polymer adsorption density and its conformation at the particle–solution interface depend on pH, temperature, nature of solvent, surface charge distribution on

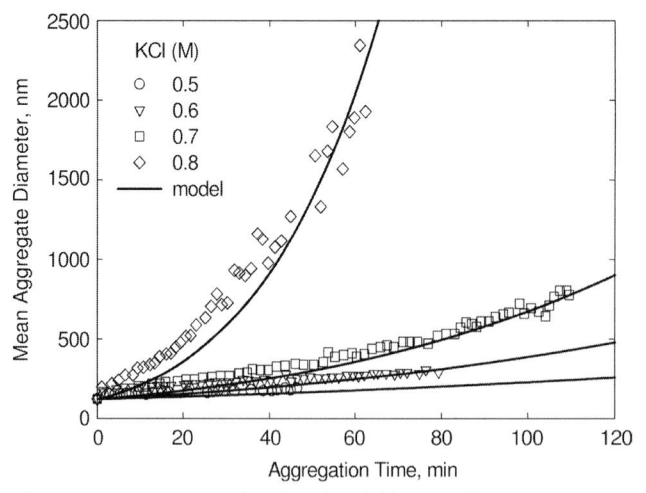

Figure 14.2 Comparison of predicted (solid lines) and experimental (symbols) time evolution of mean hydrodynamic diameter of γ-alumina aggregates in KCl solutions at pH 4.5 [24]. (Experimental data from Ref. [44].)

particles, charge density distribution of polymer chains, ionic strength, hydrodynamic conditions, and so on. Polymer chains may adsorb in the form of thin layers or trains, coils or loops and tails at the solid–liquid interface, and undergo relaxation and reconformation with time [48, 49]. The collision frequency of polymer-covered particles is also enhanced due to the adsorbed polymer layers.

Runkana et al. [23, 25] developed population balance models (PBMs) for polymer-induced flocculation by two well-known mechanisms, simple charge neutralization [23] and bridging [25]. They assumed that polymer adsorption on oppositely charged particle surfaces is very fast and equilibrium conformation is achieved before collisions between particles take place. It was also assumed that polymer adsorbs uniformly and polymer surface coverage and adsorbed layer thickness are the same for all particles. The composite polymer-coated particle radius was estimated by adding adsorbed layer thickness to the solid particle radius.

For aggregation by simple charge neutralization, it is sufficient to modify the expression for van der Waals attraction and estimate the total interaction energy as a sum of van der Waals attraction and electrical double layer repulsion. Runkana et al. [23] incorporated Vincent's expression [50] for van der Waals attraction between particles with adsorbed layers into the PBM and simulated aggregation of hematite suspensions in the presence of polyacrylic acid (PAA). Their simulation results for evolution of mean aggregate diameter with flocculation time at some typical PAA concentrations are compared with experimental data of Zhang and Buffle [34] in Figure 14.3.

In the cases of aggregation by charge patch neutralization or bridging, it is necessary to include steric or bridging forces. Bridging flocculation takes place when polymer chains adsorb on more than one particle and act as bridges. Invoking

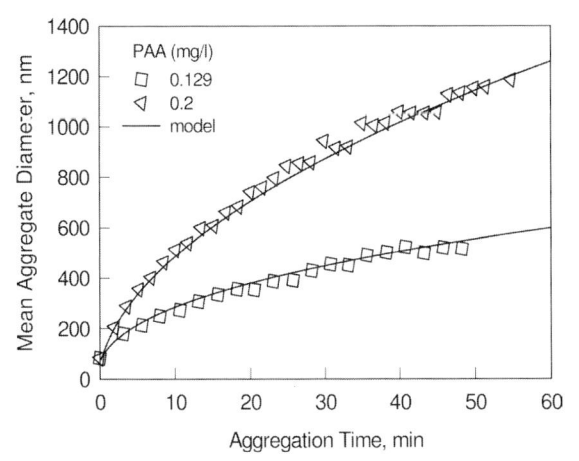

Figure 14.3 Comparison of predicted (solid lines) and experimental (symbols) time evolution of mean diameter of hematite aggregates at some typical concentrations of 1.36×10^6 g mol^{-1} PAA (pH 3.0; ionic strength 1 mM; temperature 25 °C) [23]. (Experimental data from Ref. [34].)

de Gennes' scaling theory [51] for polymer adsorption, Runkana et al. [25] derived the following expression for interaction energy between two unequal polymer-coated spherical particles V_S:

$$V_S = \left(\frac{2\pi r_{0_k} r_{0_l}}{r_{0_k}+r_{0_l}}\right)\left(\frac{\alpha_{Sc} k_B T}{a_m^3}\right)\Phi_{S_0}^{9/4}$$

$$\times D_{Sc}\left\{-\frac{16\Gamma D_{Sc}}{\Gamma_0}\ln\left(\frac{2\delta}{h_0}\right)+\frac{4D_{Sc}^{5/4}}{2^{5/4}}\left(\frac{8\Gamma}{\Gamma_0}\right)^{9/4}\left[\frac{1}{h_0^{1/4}}-\frac{1}{(2\delta)^{1/4}}\right]\right\}$$

(14.8)

where Γ is the total amount of polymer adsorbed on a single surface, Γ_0 is the adsorbed amount at saturation, δ is the adsorbed polymer layer thickness, α_{Sc} is a numerical constant, a_m is the effective monomer size and Φ_{s_0} is the polymer concentration at a single saturated surface. D_{sc} is the scaling length, a measure of segment–surface attraction and is related to γ_1^S, the local solute–interface interaction energy per unit area [51]. Runkana et al. [25] simulated bridging flocculation by solving the discretized population balance equation and assuming the total interaction energy to be a sum of van der Waals attraction, electrical double layer repulsion and bridging attraction or steric repulsion due to adsorbed polymer. Their model was tested with experimental data published by Biggs et al. [52] for flocculation of anionic polystyrene latex particles by a high molecular weight (16×10^6 g mol^{-1}) cationic quaternary ammonium based derivative of polyacrylamide. Figure 14.4 compares the predicted floc size distribution with experimental data.

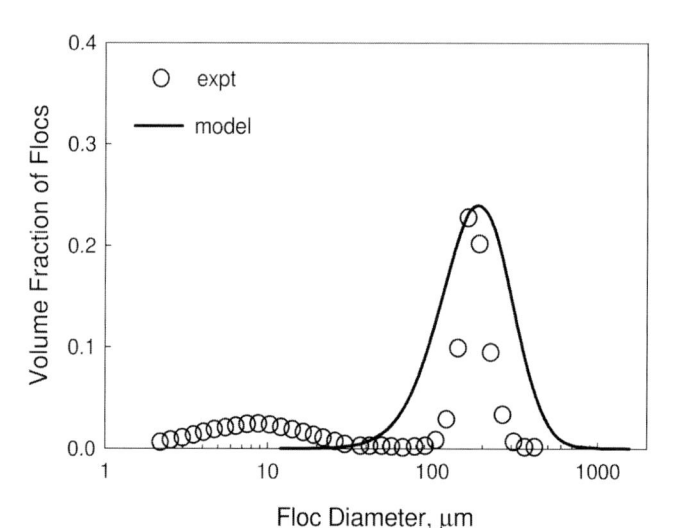

Figure 14.4 Comparison of predicted (solid line) and experimental (symbols) size distributions of polystyrene latex aggregates obtained after 10 min of aggregation using 20 ppm cationic quaternary ammonium based derivative of polyacrylamide [25]. (Experimental data from [52].)

14.3
Aggregation in Shear Environments

Industrial-scale coagulation and flocculation processes involve application of shear as it enhances the rate of aggregation, which in turn helps in improving the rate of sedimentation of aggregates. However, application of shear results in fragmentation of aggregates concurrently. The PBE for simultaneous aggregation and fragmentation is given by [16]:

$$\frac{\partial n(v,t)}{\partial t} = -\int_0^\infty \alpha(v,u)\,\beta(v,u)\,n(v,t)\,n(u,t)\,du$$
$$+ \frac{1}{2}\int_0^v \alpha(v-u,u)\,\beta(v-u,u)\,n(v-u,t)\,n(u,t)\,du - S(v)\,n(v,t) \quad (14.9)$$
$$+ \int_v^\infty S(u)\,\gamma(v,u)\,n(u,t)\,du$$

where γ is the breakage distribution function and S is the specific rate constant of floc fragmentation. The first two terms on the right-hand side account for aggregation. The third term accounts for the loss of aggregates due to fragmentation while the last term represents the generation of primary particles or smaller aggregates due to breakage or erosion of larger aggregates. The rate of fragmentation has a first order dependence on the solids concentration, unlike the rate of aggregation, which has a second order dependence. Analytical solutions do not exist for the above equation and several techniques for its discretization have been proposed in the literature [17]. The discretized population balance equation for simultaneous aggregation and fragmentation is obtained by applying discretization procedures for the aggregation and fragmentation terms separately. The rate of change of particle number concentration during simultaneous aggregation and fragmentation is given by the following discretized PBE [53]:

$$\frac{dN_i}{dt} = N_{i-1}\sum_{j=1}^{i-2} 2^{j-i+1}\alpha_{i-1,j}\,\beta_{i-1,j}\,N_j + \frac{1}{2}\alpha_{i-1,i-1}\,\beta_{i-1,i-1}\,N_{i-1}^2$$
$$- N_i\sum_{j=1}^{i-1} 2^{j-i}\alpha_{i,j}\,\beta_{i,j}\,N_j - N_i\sum_{j=i}^{\max_1}\alpha_{i,j}\,\beta_{i,j}\,N_j - S_i N_i + \sum_{j=i}^{\max_2}\gamma_{i,j}\,S_j\,N_j \quad (14.10)$$

where $\max_2$ is the largest section from which aggregates in the current section are produced by fragmentation. The first and second terms on the right-hand side account for growth while the third and fourth terms represent loss by aggregation, respectively. The fifth and sixth terms represent loss and growth of flocs by fragmentation, respectively.

Several developments have taken place recently for enhancing the capabilities of population balance models for processes involving simultaneous aggregation and fragmentation. Some of the key developments such as incorporating surface force theories into the PB paradigm, modeling the evolution of aggregate structure and coupling the PBM with computational fluid dynamics (CFD) models for fluid flow are discussed in this section.

14.3.1
Models Incorporating Surface Forces

Experimental studies on flocculation in shear environments [54, 55] have shown that the rate of aggregation depends strongly on the surface and colloid chemistry of the system. Shear enhances the rate of aggregation when the repulsive forces are absent or weak. It does not appear to influence the rate of aggregation when electrical double layer forces are strong. Taking these into consideration, Somasundaran et al. [56, 57] have developed a population balance model (PBM) that rigorously captures the dynamics of aggregation and fragmentation and incorporates the influence of microscopic interaction forces between the particles in the presence of inorganic electrolytes and polymers. Somasundaran et al. [56] simulated aggregation in the presence of inorganic electrolytes in a Couette apparatus and a stirred tank. The collision efficiency factor was calculated using the classical DLVO theory and the collision frequency factor was estimated as a sum of the contributions due to Brownian motion, applied shear and differential sedimentation. They have incorporated the drag force correction factor and the fluid collection efficiency of an aggregate into the expression for collision frequency factor [58], to account for open and

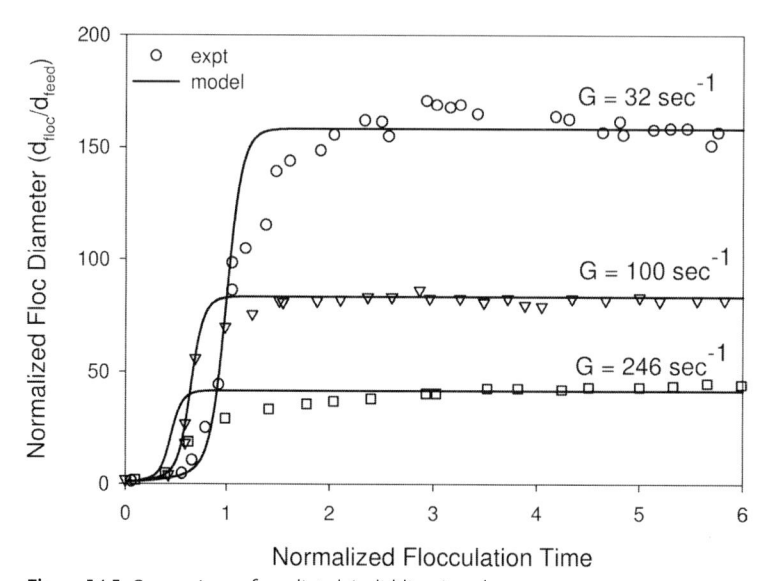

Figure 14.5 Comparison of predicted (solid lines) and experimental (symbols) time evolution of the mass mean aggregate diameter in a Couette flow apparatus at different global average shear rates; Mean primary particle diameter: 0.81 μm; solids volume fraction: 3.76×10^{-5}; pH 9.15; [$MgCl_2$] = 0.05 M [56]. (Experimental data from [59].)

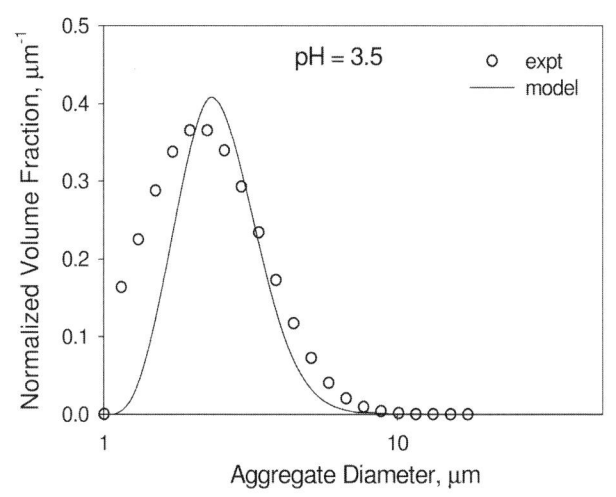

Figure 14.6 Comparison of predicted (solid line) and experimental (symbols) size distributions of alumina aggregates in a stirred tank after 6 min of aggregation. (Mean primary particle diameter: 700 nm; pH 3.5; 0.03 M NaCl) [56]. (Experimental data from [60].)

irregular structure of aggregates. Both the drag force correction factor and the fluid collection efficiency of an aggregate depend on aggregate permeability which in turn is a function of solid volume fraction of the aggregate. The predicted evolution of mean diameter of polystyrene latex aggregates with time in a Couette flow apparatus in the presence of 0.05 M $MgCl_2$ at pH 9.15 are compared with experimental data of Selomulya et al. [59] in Figure 14.5. Somasundaran et al. [56] have also applied their PBM to simulate aggregation of alumina in a stirred tank at different pH. Figure 14.6 compares the predicted aggregate size distribution at pH 3.5 with the experimental data of Das [60].

In a later development, Somasundaran et al. [57] developed a PBM for aggregation by polymers in shear environments. The DLVO theory was extended for this case, as discussed in the previous section, by using the modified expression for van der Waals attraction for particles covered with polymers and the expression for bridging attraction or steric repulsion derived from the scaling theory [25]. Their model was tested qualitatively with experimental data for the flocculation of colloidal alumina suspensions in the presence of PAA and was found to reproduce the observed experimental trends [60] reasonably well.

14.3.2
Models Incorporating Evolution of Aggregate Structure

One of the key assumptions made in population balance models is that the fractal dimension and permeability of an aggregate remain constant during aggregation. It is also generally assumed that aggregates of all sizes have the same fractal dimension. The structure of an aggregate is not only a function of variables such as pH, ionic strength and temperature but also the size distribution of the constitutive particles. Starting from spherical or near-spherical primary particles, aggregates develop as open and irregular objects. Floc structure evolves during aggregation and floc restructuring takes place due to aggregation, fragmentation and elongation or compaction induced by fluid stresses. Selomulya *et al.* [59] found that the scattering exponent of aggregates was initially low and reached a steady value after some time.

To take floc restructuring into account, Thill *et al.* [61] assigned different values for the aggregate fractal dimension for different stages of aggregation. In contrast, Selomulya *et al.* [62] developed an empirical expression for the evolution of floc fractal dimension and incorporated it into the PBM. Lattuada *et al.* [63] obtained the fractal dimension of an aggregate through off-lattice simulations and derived an empirical particle–particle correlation function that was then employed, along with aggregate number concentrations predicted by a PBM for aggregation, to calculate the aggregate structure factor.

The rate of aggregation is a function of the aggregate structure because the frequency of collisions between aggregates depends not only on the shape of the aggregate but also on the viscous drag it experiences. Taking note of this Somasundaran *et al.* [56] incorporated two factors, the drag force correction factor and the fluid collection efficiency, in the expression for the collision frequency factor. Both these parameters depend on the floc permeability, which was estimated using Happel's cell model [64]. They, however, assumed the aggregate mass fractal dimension to be constant during aggregation. An important future development in this direction could be a theoretical model for predicting the evolution of aggregate fractal dimension and its permeability as a function of pH, temperature, electrolyte or polymer concentration, number of particles and their size distribution as aggregate size distribution evolves during the process.

14.3.3
Coupled Population Balance – Fluid Flow Models

One of the important issues in modeling aggregation in shear environments is how to incorporate the inhomogeneous nature of fluid flow. In general, the global average shear rate is employed to compute the rate constants of aggregation and fragmentation. The local rates of shear are significantly different from the global average [65, 66] and the local turbulent energy dissipation rate depends not only on the impeller type and surface but also on the tank dimensions [65]. In addition, the number of impellers and baffles also affects the shear rate distribution.

Some approximate methods have been applied previously to deal with the inhomogeneous nature of flow patterns in stirred tanks [67, 68]. Koh [67] divided the stirred tank into three compartments, the impeller zone, the bulk zone and a dead space, and assigned different shear rates for each compartment. Furthermore, Koh *et al.* [69] ignored the dead space, but split the impeller zone into impeller tip zone and impeller zone. Ducoste [68] essentially followed the same approach, dividing the suspension volume into two zones, the impeller discharge zone and the bulk zone.

Instead of assigning different shear rates, he employed different breakage rate expressions for the two zones. The problem of coupling population balance models with fluid flow models has received some attention recently and coupled PB-CFD models have been developed for a wide variety of processes such as fluidization [70], gas–liquid reactions in bubble columns [71] and nanoparticle synthesis in flame aerosol reactors [72]. Complete description of aggregation in turbulent environments requires simultaneous solution of basic balance equations for mass, momentum, energy and concentration of species present along with population balances for particles/aggregates of different size classes.

In one of the earliest attempts, Schuetz and Piesche [73] calculated the flow field in a stirred tank first to determine the local energy dissipation rate and then solved the PBEs using the finite volume method [74] to predict the local aggregate size distribution. Heath and Koh [75] have solved the population balances as scalar equations in the commercial CFD software CFX for simulating flocculation of suspensions by polymers. They employed 35 discrete sectional equations to represent the aggregate size distribution.

A coupled PB-CFD model with a relatively large number of size classes requires excessive computational time and memory because scalar equations for all the size classes have to be solved in every cell of the computational domain. To overcome this problem, efforts have been made recently to reduce the number of equations required to represent the aggregate size distribution. One of the standard techniques is to formulate the PBM based on the method of moments (MOM) in which the time evolution of only a few lower order moments of the size distribution is tracked. Although the MOM does not capture the complete aggregate size distribution accurately, it can be used to predict important properties such as aggregate size with reasonable accuracy. One of the issues related to the standard MOM is that the transport equations of the moments have to be formulated in a closed form, that is, in terms of the moments themselves. McGraw [76] has proposed a quadrature approximation to overcome this problem and developed what is now popularly known as the quadrature method of moments (QMOM). The QMOM has been applied to solve several problems related to aggregation [77], precipitation [78] and nanoparticle synthesis [79]. Population balance models based on QMOM were coupled with CFD software such as FLUENT [78] and PHOENICS [80] to simulate the influence of local shear rate distribution on aggregate size distribution. For example, Marchisio et al. [78] implemented the QMOM-based PBM in FLUENT to study aggregation in a Taylor–Couette apparatus, while Prat and Ducoste [80] studied flocculation in a square stirred-tank reactor coupling QMOM-based PBM with the CFD software PHOENICS.

Coupling the PBM with CFD tools is an important advancement in modeling aggregation. However, these coupled models are still computationally intensive. As such, the QMOM itself may require excessive computational time and face numerical difficulties if the variations in moments are large [81]. One of the drawbacks of QMOM-based PBMs is that the complete aggregate size distribution cannot be reconstructed accurately using the moments. Generally, fluid flow models are computationally intensive and discretized PBEs require less computational time. It is worth identifying ways and means of reducing the computational requirements of fluid flow models and employ discretized PBEs so that the complete aggregate size distribution can be predicted accurately at reduced computational costs. Moreover, none of the coupled PB-CFD models incorporate the influence of surface forces.

Clearly, current aggregation models can reliably predict simple effects of salt and some effects of polymers, but additional work is warranted to develop the capability to predict aggregation and dispersion of particulates of all types (solid, gas and liquid) under real-life conditions of varying concentrations of flocculants or dispersants, hydrodynamic perturbation and local pressure and temperature variation in addition to the effects of any magnetic or sonic fields.

14.4
Summary and Suggestions for Future Research

The development of population balance models (PBMs) for aggregation of colloidal suspensions is an active area of research because of its diverse industrial applications and role in environmental pollution. The capabilities of PBMs have improved considerably because of several recent advances, which include the incorporation of fundamental theories of surface forces into the population balance paradigm, coupling PBMs with commercial CFD tools and development of models to predict evolution of aggregate structure during aggregation. The influence of critical aggregation process variables such as pH, electrolyte concentration and shear rate distribution on the rate of aggregation and aggregate size distribution can now be predicted due to these new developments.

Despite the advances made so far, several issues have to be addressed to make the aggregation models applicable for real-life coagulation and flocculation processes. It is suggested here that focused efforts have to be made in the following research areas to enhance the capabilities of aggregation models and to improve our understanding of aggregation in the presence of coagulants and flocculants in shear environments.

14.4.1
Multidimensional Population Balances

PBMs developed thus far are one-dimensional in nature, that is, they can handle only one internal coordinate (e.g. aggregate volume). Besides volume, aggregate porosity and surface area also change during aggregation. To predict the evolution of all these properties, it is necessary to formulate multidimensional population balances. Moreover, current PBMs treat single component systems only. Colloidal suspensions encountered in natural systems and in industrial processes are usually multicomponent in nature and formulation of PBMs for such systems should be an important area of research.

14.4.2
Polymer Adsorption Dynamics

Although surface force theories have been incorporated into the PBE, the current models require experimentally determined parameters such as solid–liquid interface potential, adsorbed polymer layer thickness and particle surface coverage. Future efforts should focus on integrating polymer adsorption dynamics models with PBMs. These models should be extended subsequently for systems involving a mixture of polymers or polymer–surfactant systems.

14.4.3
Computationally Efficient Population Balance–Fluid Flow Models

Since the current QMOM-based PBMs have limitations in accurate prediction of aggregate size distribution and face numerical difficulties, it is necessary to develop simpler fluid flow models and couple them with discretized PBEs so that the complete aggregate size distribution can be predicted with less computational demands. Additionally, coupled PB-CFD models should also incorporate surface force theories and models for polymer adsorption dynamics. Development of novel algorithms to solve coupled PB-CFD models is another important requirement. The grand challenge here would be to develop a coupled multidimensional PB–CFD model, incorporating coagulant/flocculant adsorption dynamics, influence of surface forces and the evolution of aggregate structure.

14.4.4
Depletion Flocculation

Although depletion is one of the main mechanisms of aggregation, very little effort has been made to develop PBMs for depletion flocculation. This could be an important area of research for application of aggregation models in the processing of agrochemicals and cosmetics.

Acknowledgments

This work was supported by the National Science Foundation (NSF Grants # INT-96-05197 and INT-01-17622) and the NSF Industry/University Cooperative Research Center (IUCRC) for Advanced Studies in Novel Surfactants at Columbia University (NSF Grant # EEC-98-04618). The authors thank the management of Tata Research Development and Design Centre for the permission to publish this chapter. V.R. thanks Professor E. C. Subbarao, Professor Mathai Joseph, Professor P. C. Kapur and Dr Pradip for their advice and encouragement.

References

1 Thomas, D.N., Judd, S.J. and Fawcett, N. (1999) *Water Research*, **33**, 1579–1592.
2 Somasundaran, P., Das, K.K. and Yu, X. (1996) *Current Opinion in Colloid and Interface Science*, **1**, 530–534.
3 Somasundaran, P. and Arbiter, N.(eds) (1979) *Beneficiation of Mineral Fines*, AIME, USA.
4 Moudgil, B.M. and Somasundaran, P. (1994) *Dispersion and Aggregation: Fundamentals and Applications*, Engineering Foundation, USA.
5 Somasundaran, P., Markovic, B., Krishnakumar, S. and Yu, X. (1997) Colloidal systems and interfaces: stability of dispersions through polymer and surfactant adsorption, in *Handbook of Surface and Colloid Chemistry* (ed. K.S. Birdi), CRC Press, USA.
6 Pelton, R.H. (1999) in *Colloid-Polymer Interactions: From Fundamentals to Practice* (eds R.S. Farinato and P.L. Dubin), John Wiley & Sons, Inc., USA, pp. 51–82.
7 Pugh, R.J. and Bergstrom, L. (1994) *Surface and Colloid Chemistry in Advanced Ceramics Processing*, Marcel Dekker, Inc, USA.
8 Jackson, G.A. and Burd, A.B. (1998) *Environmental Science & Technology*, **32**, 2805–2814.
9 Israelachvili, J.N. (1991) *Intermolecular and Surface Forces*, 2nd edn, Academic Press, USA.
10 Somasundaran, P. (1979) Principles of selective aggregation, in *Beneficiation of Mineral Fines* (eds P. Somasundaran and N. Arbiter), AIME, USA.
11 Flesch, J.C., Spicer, P.T. and Pratsinis, S.E. (1999) *AIChE Journal*, **45**, 1114–1124.
12 Heath, A.R., Bahri, P.A., Fawell, P.D. and Farrow, J.B. (2006) *AIChE Journal*, **52**, 1641–1653.
13 Malbrel, C.A. and Somasundaran, P. (1989) *Journal of Colloid and Interface Science*, **133**, 404–408.
14 Amal, R., Raper, J.A. and Waite, T.D. (1990) *Journal of Colloid and Interface Science*, **140**, 158–168.
15 Soos, M., Wang, L., Fox, R.O. et al. (2007) *Journal of Colloid and Interface Science*, **307**, 433–446.
16 Ramkrishna, D. (2000) *Population Balances: Theory and Applications to Particulate Systems in Engineering*, Academic Press, USA.
17 Vanni, M. (2000) *Journal of Colloid and Interface Science*, **221**, 143–160.
18 Somasundaran, P. and Runkana, V. (2003) *International Journal of Mineral Processing*, **72**, 33–55.
19 Taboada-Serrano, P., Chin, C.-J., Yiacoumi, S. and Tsouris, C. (2005) *Current Opinion in Colloid and Interface Science*, **10**, 123–132.
20 Smoluchowski, M.v. (1917) *Zietschrift fur Physikalische Chemie*, **92**, 129–168.
21 Press, W.H., Teukolsky, S.A., Vetterling, W.T. and Flannery, B.P. (1992) *Numerical Recipes in FORTRAN: The Art of Scientific Computing*, 2nd edn, Cambridge University Press, USA.
22 Hounslow, M.J., Ryall, R.L. and Marshall, V.R. (1988) *AIChE Journal*, **34**, 1821–1832.
23 Runkana, V., Somasundaran, P. and Kapur, P.C. (2004) *Journal of Colloid and Interface Science*, **270**, 347–358.
24 Runkana, V., Somasundaran, P. and Kapur, P.C. (2005) *AIChE Journal*, **51**, 1233–1245.
25 Runkana, V., Somasundaran, P. and Kapur, P.C. (2006) *Chemical Engineering Science*, **61**, 182–191.
26 Fuchs, N. (1934) *Zeitschrift fur Physik*, **89**, 736–743.
27 Derjaguin, B.V. and Muller, V.M. (1967) *Doklady Akademii Nauk SSSR*, **176**, 738–741.
28 McGown, D.N. and Parfitt, G.D. (1967) *The Journal of Physical Chemistry*, **71**, 449–450.
29 Spielman, L.A. (1970) *Journal of Colloid and Interface Science*, **33**, 562–571.
30 Honig, E.P., Roebersen, G.J. and Wiersema, P.H. (1971) *Journal of Colloid and Interface Science*, **36**, 97–109.

31 Lin, M.Y., Lindsay, H.M., Weitz, D.A. et al. (1989) *Nature*, **339**, 360–362.
32 Feder, J. (1988) *Fractals*, Plenum Press, USA.
33 Klimpel, R.C. and Hogg, R. (1986) *Journal of Colloid and Interface Science*, **113**, 121–131.
34 Zhang, J. and Buffle, J. (1995) *Journal of Colloid and Interface Science*, **174**, 500–509.
35 Somasundaran, P., Huang, Y.-B. and Gryte, C.C. (1987) *Powder Technology*, **53**, 73–77.
36 Meakin, P. (1988) *Advances in Colloid and Interface Science*, **28**, 249–331.
37 Huang, Y.-B. and Somasundaran, P. (1987) *Physical Review A*, **36**, 4518–4521.
38 Wong, K., Cabane, B. and Somasundaran, P. (1988) *Colloids Surfaces A*, **30**, 355–360.
39 Smoluchowski, M.v. (1916) *Physikalische Zeitschrift*, **17**, 557–571.
40 Derjaguin, B.V. and Landau, L.D. (1941) *Acta Physiochim USSR*, **14**, 633.
41 Verwey, E.J.W. and Overbeek, J.Th.G. (1948) *Theory of the Stability of Lyophilic Colloids*, Elsevier, The Netherlands.
42 Camp, T.R. and Stein, P.C. (1943) *Journal of the Boston. Society of Civil Engineers*, **30**, 219–237.
43 Ball, R.C., Weitz, D.A., Witten, T.A. and Leyvraz, F. (1987) *Physical Review Letters*, **58**, 274.
44 Beattie, J.K., Cleaver, J.K. and Waite, T.D. (1996) *Colloids Surfaces A*, **111**, 131–138.
45 Waite, T.D., Cleaver, J.K. and Beattie, J.K. (2001) *Journal of Colloid and Interface Science*, **241**, 333–339.
46 Napper, D.H. (1983) *Polymeric Stabilization of Colloidal Dispersions*, Academic Press, Inc., USA.
47 Levine, S. and Friesen, W.I. (1987) in *Flocculation in Biotechnology and Separation Systems* (ed. Y.A. Attia), Elsevier Science Publishers B.V., The Netherlands, pp. 3–20.
48 Chander, P., Somasundaran, P., Turro, N.J. and Waterman, K.C. (1987) *Langmuir*, **3**, 298–300.
49 Fleer, G.J., Cohen Stuart, M.A., Scheutjens, J.M.H.M. et al. (1993) *Polymers at Interfaces*, Chapman and Hall, Inc., USA.
50 Vincent, B. (1973) *Journal of Colloid and Interface Science*, **42**, 270–285.
51 de Gennes, P.G. (1982) *Macromolecules*, **15**, 492–500.
52 Biggs, S., Habgood, M., Jameson, G.J. and Yan, Y-d. (2000) *Chemical Engineering Journal*, **80**, 13–22.
53 Spicer, P.T. and Pratsinis, S.E. (1996) *AIChE Journal*, **42**, 1612–1620.
54 Warren, L.J. (1975) *Journal of Colloid and Interface Science*, **50**, 307–318.
55 Tjipangandjara, K., Huang, Y.-B., Somasundaran, P. and Turro, N.J. (1990) *Colloids Surfaces A*, **44**, 229–236.
56 Somasundaran, P., Runkana, V. and Kapur, P.C. (2006) Paper No. 35244, CD-ROM Proceedings, 5th World Congress on Particle Technology (WCPT5-2006), Orlando, USA.
57 Somasundaran, P., Kapur, P.C. and Runkana, V. (2006) Paper No. F159, International Mineral Processing Conference (IMPC-2006), Istanbul, Turkey.
58 Veerapaneni, S. and Weisner, M.R. (1996) *Journal of Colloid and Interface Science*, **177**, 45–57.
59 Selomulya, C., Bushell, G., Amal, R. and Waite, T.D. (2002) *Langmuir*, **18**, 1974–1984.
60 Das, K.K. (1998) Investigations on the Polyelectrolyte Induced Flocculation/Dispersion of Colloidal Alumina Suspensions, D.E.Sc. Thesis, Columbia University, USA.
61 Thill, A., Moustier, S., Aziz, J. et al. (2001) *Journal of Colloid and Interface Science*, **243**, 171–182.
62 Selomulya, C., Bushell, G., Amal, R. and Waite, T.D. (2003) *Chemical Engineering Science*, **58**, 327–338.
63 Lattuada, M., Wu, H. and Morbidelli, M. (2003) *Journal of Colloid and Interface Science*, **268**, 106–120.
64 Happel, J. (1958) *AIChE Journal*, **4**, 197–201.

65 Ducoste, J.J., Clark, M.M. and Weetman, R.J. (1997) *AIChE Journal*, **43**, 328–338.
66 Kramer, T.A. and Clark, M.M. (1997) *Journal of Environmental Engineering*, **123**, 444–452.
67 Koh, P.T.L. (1984) *Chemical Engineering Science*, **39**, 1759–1764.
68 Ducoste, J.J. (2002) *Chemical Engineering Science*, **57**, 2157–2168.
69 Koh, P.T.L., Andrews, J.R.G. and Uhlherr, P.H.T. (1984) *Chemical Engineering Science*, **39**, 975–985.
70 Fan, R., Marchisio, D.L. and Fox, R.O. (2004) *Powder Technology*, **139**, 7–20.
71 Wang, T. and Wang, J. (2007) *Chemical Engineering Science*, **62**, 7107–7118.
72 Johannessen, T., Pratsinis, S.E. and Livbjerg, H. (2000) *Chemical Engineering Science*, **55**, 177–191.
73 Schuetz, S. and Piesche, M. (2002) *Chemical Engineering Science*, **57**, 4357–4368.
74 Patankar, S.V. (1980) *Numerical Heat Transfer and Fluid Flow*, 1st edn, Hemisphere Publishing Corp., USA.
75 Heath, A.R. and Koh, P.T.L. (2003) 3rd International Conference on CFD in Minerals and Process Industries, Melbourne, Australia, pp. 339–344.
76 McGraw, R. (1997) *Aerosol Science and Technology*, **27**, 255–267.
77 Marchisio, D.L., Vigil, R.D. and Fox, R.O. (2003) *Journal of Colloid and Interface Science*, **258**, 322–334.
78 Marchisio, D.L., Pikturna, J.T., Fox, R.O. et al. (2003) *AIChE Journal*, **49**, 1266–1276.
79 Rosner, D.E. and Pyykonen, J.J. (2002) *AIChE Journal*, **48**, 476–491.
80 Prat, O. and Ducoste, J.J. (2006) *Chemical Engineering Science*, **61**, 75–86.
81 Su, J., Gu, Z., Li, Y. et al. (2007) *Chemical Engineering Science*, **62**, 5897–5911.

15
Cubosomes as Delivery Vehicles
Nissim Garti, Idit Amar-Yuli, Dima Libster, and Abraham Aserin

15.1
Introduction

Lyotropic liquid crystalline phases are formed in concentrated mixtures of amphiphilic molecules and water. The common and well-studied lyotropic liquid crystalline phases are lamellar (L_α), hexagonal (normal, H_1 or inverted H_2) and normal or inverted cubic (bicontinuous or micellar) structures.

Cubic phases exhibit the most complex organization of all known lyotropic liquid crystals. These structures do not display an optical texture. To date, seven different cubic structures have been studied (Q^{212}, Q^{223}, Q^{224}, Q^{225}, Q^{227}, Q^{229} and Q^{230}). They can be either micellar or bicontinuous and possess either normal (type I, oil-in-water) or inverse (type II, water-in oil) topology. The primitive (P) type [body-centered lattice (Im3m, denoted Q^{224})], diamond (D) type (primitive lattice Pn3m, denoted Q^{229}) and, most frequently, gyroid (G) type (Ia3d, denoted Q^{230}) have an inverse and bicontinuous structures (Figure 15.1) [1–3].

Structure P contains two aqueous channels that are separated by a bilayer. The unit cell has three mutually perpendicular aqueous channels that are connected to contiguous unit cells, forming a cubic array. Structure D also contains a bilayer that separates two interpenetrating aqueous channels that form a diamond lattice. In this structure four aqueous channels of the D-surface meet at a tetrahedral angle of 109.5°. The aqueous compartments of a G-surface consist of two separate, left-handed and right-handed helical channels. The aqueous channels can extend through the matrix, as in the P-surface, but the centers of the water channels never intersect. Rather they are connected to give rise to a helical arrangement. The Ia3d structure is the least hydrated of the three and has helical channels possessing opposing chiralities [4, 5].

Numerous efforts were recently dedicated to study the structural properties of liquid crystalline cubic phases. Various aspects related to their structures and phase transformations in the presence of different guest molecules within the cubic phases have been studied.

Highlights in Colloid Science. Edited by Dimo Platikanov and Dotchi Exerowa
Copyright © 2009 WILEY-VCH Verlag GmbH & Co. KGaA, Weinheim
ISBN: 978-3-527-32037-0

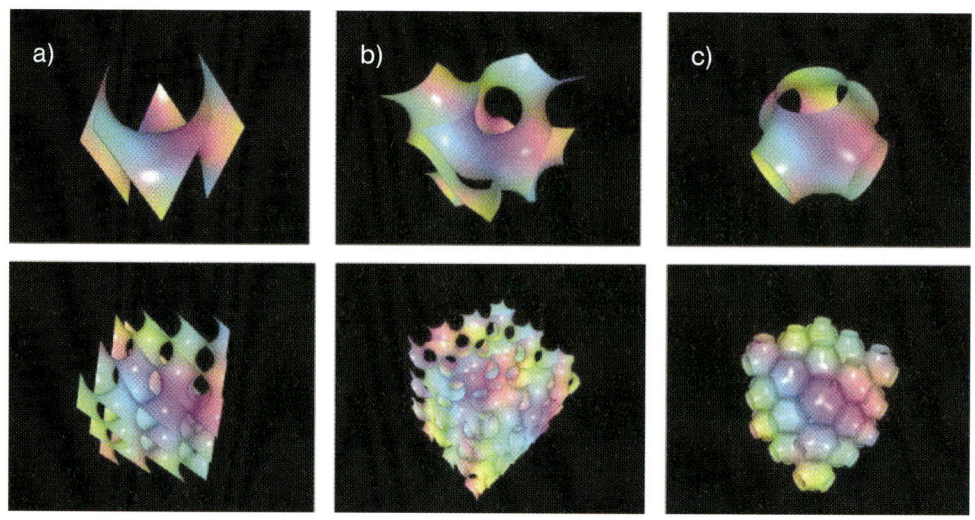

Figure 15.1 Structure of bicontinuous cubic phases based on periodic minimal surfaces: (a) diamond; (b) gyroid; (c) primitive. The top panels show one unit cell representation whilst the bottom panels show these surfaces extending in space to form infinite periodic minimal surfaces [2, 3]. (Images courtesy of R. Enlow.)

When the cubic phases are further dispersed in aqueous phase containing a stabilizer (usually amphiphilic polymer) nano-soft particles termed cubosomes are formed (Figure 15.2) [6].

In the dispersed soft matter particles the nanostructure of the original cubic inner structure remains intact despite the dispersion process. The three-dimensional

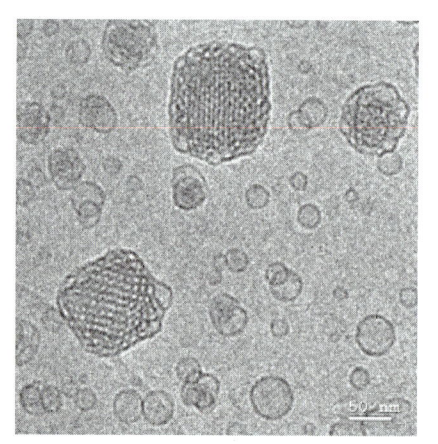

Figure 15.2 Cryo-TEM image of cubosomes and vesicles formed by sonicating monoolein in aqueous Poloxamer 407 solution [6].

symmetry of the cubic phases, combined with their large interfacial area and balanced content of hydrophobic and hydrophilic domains, make them very promising universal drug carriers, with numerous advantages over most other systems used at present. These exceptional physical and chemical properties stimulated the study of the liquid crystal dispersions. Thus, one can take advantage of the complex phases for controlled delivery, while simultaneously having the benefit of the low viscosity of a system of soft nanoparticles dispersed in a continuous aqueous phase. The preparation of stable colloidal dispersions of the cubic liquid crystalline phases has opened up new exciting opportunities for applications of lyotropic liquid crystals [7–9]. This short chapter focuses on the potential of cubosomes as delivery vehicles.

15.2
Preparation Techniques

Several techniques have been proposed for the preparation of cubosomes.

The original dispersion agents for soft nanoparticles were bile salts and caseins, which were assumed to form an envelope on the surface of GMO-based cubic phases forming the cubosomes [9, 10]. A few years later amphiphilic block copolymers, especially F127 [with 99 units of poly(ethylene oxide) in each of the two tails and 67 units of poly(propylene oxide), $PEO_{99}PPO_{67}PEO_{99}$], were found to provide very powerful steric stabilization of the dispersed cubic phase particles [11–17]. The formation of relatively stable cubosomes was assumed to be related to the specific and unique adsorption of F127 on the surface of the particles. Furthermore, the observation of vesicular structures at the periphery of the particles in the dispersions, as observed by cryo-TEM (cryogenic-transmission electrons microscopy) images, suggested that further stabilization was provided by coexisting lamellar structures Figure 15.2). Simple magnetic stirrer agitation, followed by heat treatment during the addition of the polymer, yielded a wide range of particle sizes from 1 to 100 microns [17]. High shear homogenization techniques that were adopted later formed soft particles of 100–500 nm [7, 12–16, 18, 19].

Earlier studies of the GMO/F127/water phase diagram led to the selection of preferred lipid to polymer weight ratios to produce cubosome particles [11]. It was recognized that the polymer caused phase transformations among cubic phases. Upon addition of polymer, up to an 80 : 20 weight ratio of GMO : F127, diamond (Q^{224}) and gyroid (Q^{230}) cubic phases, which were formed in its absence, transformed into the primitive cubic phase Q^{229}. With 70–90 wt% water, lamellar and sponge phases were formed. This phase diagram indicates that to produce stable cubosome particles the GMO : F127 weight ratio should be kept in the range of about 94 : 6–80 : 20 where the primitive cubic phase or a mixture of the primitive cubic, sponge or lamellar phases are formed in the presence of excess aqueous solution [11]. Cryo-TEM images of cubosomes confirm the coexistence of soft cubic particles with a lamellar envelope and multilamellar vesicles [9, 10]. Spicer and Hyden proposed a method based on a dilution process of an ethanolic solution of monoolein with an aqueous solution of Poloxamer. Ethanol was used as a hydrotrope to create a liquid precursor, spontaneously forming cubosomes after dilution [19–21]. Owing to their exceptional physical properties, cubosomes can be tailored for specific drug delivery applications, including but not limited to oral, topical and intravenous techniques.

15.3
Drug Delivery Applications

Bioactive molecules require a range of delivery approaches and vehicles due to their variable solubility and availability. Cubosomes offer the opportunity to accommodate hydrophilic, hydrophobic and amphiphilic compounds in a submicron form for oral/periodontal delivery, as implants and as bioadhesives [22]. Several reports have suggested that protein incorporation into lipid/water mixtures caused fragmentation of a bulk cubic phase, which assisted in cubosomes formation [6, 14, 23–27].

Angelov et al. [27] have investigated the diamond-type (D-type) cubic structure of functionalized cubosomes. The latter were created by full hydration of monoolein as a main component mixed with a synthetic lipid with a tri(ethylene glycol) polar head group (MTEG = 1-{8-[4-(p-maleimidophenyl)butaroylamino]-3,6-dioxaloctyl}-2,3-distearyl glyceryl DL-ether), followed by encapsulation of protein (Fab fragments of immunoglobulin). By means of freeze–fracture electron microscopy the structural modifications induced by the guest biomolecules (MTEG lipid and Fab protein), which embedded into the cubic phase, were analyzed. The authors investigated both experimentally and theoretically the smallest cubosome structural unit built-up from D-type cubosome nanochannels.

The cubosome structures modeled and depicted in Figure 15.3a–g describe the major stages in the growth of the nanochannel architecture, starting from a single nanovesicle and building up to a cubosome assembly [27]. It was revealed that some of the growing scaffolds (e.g. those at $N=71$ and at $N=191$) involved extension resulting in unstable lipid membrane configurations. To maintain the cubosomes stability during their growth, the generated surfaces should not be terminated by closed water channels.

The investigators utilized high-resolution FF-EM (freeze–fracture electron microscopy) [28–31] for direct visualization of the internal structure of the 3D cubosomes structure of a double diamond type formed upon spontaneous lipid/protein assembly in excess of water. They indicated that the self-assembly cubic organization could possibly locally deviate from a "perfect" 3D structure due to the distortions caused by the entrapped protein macromolecules. Angelov et al. [27] compared the high-resolution FF-EM images of the single-crystal systems based on pure monoolein and in the presence of guest biomolecules. In the presence of guest biomolecules, the cubic membrane architecture is no longer perfect (Figure 15.4).

It was concluded that the guest MTEG molecules caused changes in the local curvature of the lipid bilayer. Notably, these distortions were generated on continuous interfaces rather than on single or punctual defects in the cubic lattice. These results implied that the aqueous nanochannels dynamically could open or close, hence affecting the guest molecules exchange between the cubosome interior and the

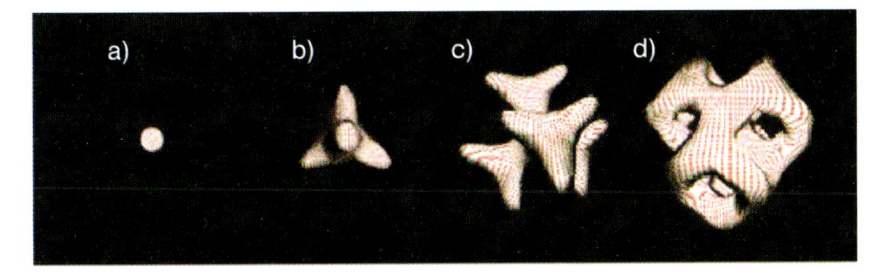

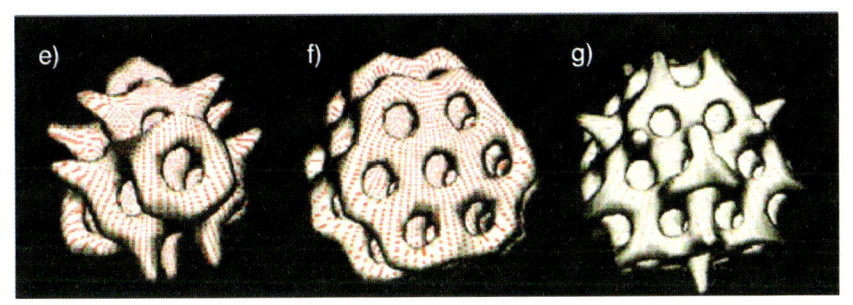

Figure 15.3 "Bottom-up" mechanism of the 3D growth of a cubosomic nano-object represented by nodal surfaces that form diamond-type skeletons of nanochannels. Note that the cubosome, generated from a curved lipid bilayer, adopts only discrete sizes upon growth. The latter is determined by geometrical constraints for preservation of the lipid bilayer integrity in the cubic lattice skeleton. The presented stages of cubosomal nanostructure growth correspond to $N =$ (a) 1, (b) 5, (c) 17, (d) 29, (e) 71, (f) 147 and (g) 191 [where N is the number of nodes (repeat volumes) in the nano-object]. (Reprinted with permission from Ref. [27].)

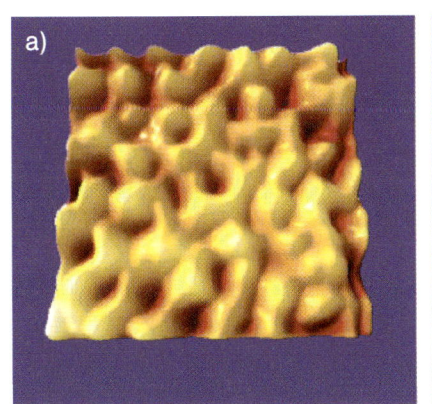

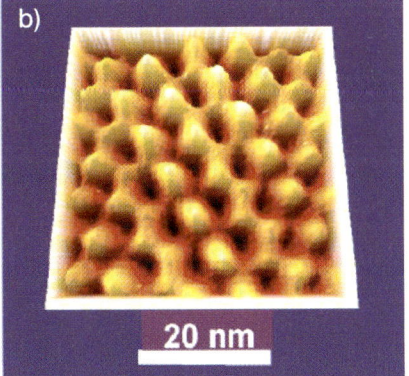

Figure 15.4 3D reconstructions of freeze–fracture electron microscopy images of a chemically functionalized cubic lipid membrane MO/MTEG (MO is monoolein and MTEG is a double-chain synthetic lipid, 1-{8-[4-(*p*-maleimidophenyl)butaroylamino]-3,6-dioxalocty}-2,3-distearyl glyceryl DL-ether). The cubic structure was assembled at full hydration in excess aqueous buffer phase. (a) Raw image showing the presence of undulations in the complex fluid 3D cubic membrane; (b) Fourier-filtered image corresponding to the underlying matrix structure of the cubic liquid-crystalline freeze–fracture section.

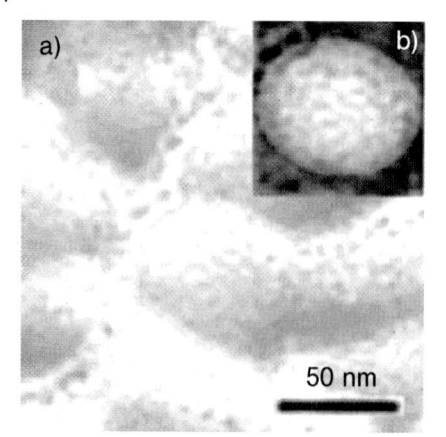

Figure 15.5 (a) FF-EM image of a nanocubosome in a proteocubosome sample (MO/MTEG/Fab immunoglobulin fragments/buffer) at full hydration. The inset (b) shows a selected area of the raw image (a), in which the image contrast was processed to remove the experimental shadows and to reveal the contour of the nano-object. (Reprinted with permission from Ref. [27].)

continuous aqueous phase. Furthermore, the internal cubosome nanostructure was explored in the presence of water-soluble fragments (Fab) of the protein immunoglobulin (extended length ∼7 nm). Notably, its size was larger than the cubosomic aqueous nanochannel diameter ($D_w = 3.6$ nm) in MO/MTEG cubic assemblies. The authors experimentally proved that the steric incompatibility induced local perturbations of the overall cubic supramolecular organization. The local cubic topology modifications (Figures 15.4 and 15.5) resulted in protein-induced nanocubosomic patterns. The structural distortions caused by the guest lipid MTEG alone were unable to initiate the formation of a nanocubosome defect network in the generated patterns. In contrast, the incorporated large proteins can induce network defects that result in fragmentation of the bicontinuous cubic phase into nanosized entities. Regarding the locus of the solubilization of immunoglobulin in the cubosomes, it was suggested that it is incorporated along the contact interfaces of the adjacent cubosomic nanodroplets (Figure 15.5).

The authors of this work think that the location of a protein in the cubic phase is a function of its hydrophobic–hydrophilic balance and affinity for interaction with the lipid bilayer. It was previously demonstrated in the literature that small proteins of amphiphilic nature can adsorb or penetrate into the lipid bilayer, affecting the mean interfacial curvature and inducing phase transformations. The effect could be stabilizing, for instance with the protein cytochrome c embedded in monoolein cubic mesophases [32, 33], or could induce structural variations as in the case of protein transfer [34]. In addition, it was suggested that the steric parameters, the interfacial activity of soluble proteins and their effect on the lipid hydration are important factors in determining the partitioning of the proteins in the cubic phase.

These nano-soft particles dispersions can provide enhanced drug solubility, protection of the solubilized drugs, controlled release of drugs avoiding substantial side effects, and relatively long shelf-life. It was demonstrated that better biodistribution of peptide pharmaceutical molecules, and enhanced ability to target these molecules toward specific tissues *in vivo*, can be achieved with such liquid crystalline vehicles. Moreover, when a drug molecule is administered within a protective carrier, the drug clearance decreases (half-life increases) and the volume of its distribution decreases [35]. This protective capability can be further enhanced if the liquid crystalline materials are to be used for transdermal delivery, where peptide degradation and low absorption rates are much less relevant.

Kim *et al.* [36] claimed that slow release of a lipophilic compound can be achieved from cubosomes. The authors measured the *in vitro* release of rifampicin from cubosomes prepared by the hydrotrope precursor method [19]. The experiments revealed zero-order release of rifampicin from the cubosome dispersion over a period of 10 to 12 days compared to the rapid release rate of the drug (over only a few days) using the control solution [21].

The peptide somatostatin has been studied *in vivo* in a rabbit model intravenously, either as a bolus dose of free peptide or as an injection of peptide-loaded cubosome dispersion [19]. The cubosome dispersion provided sustained somatostatin plasma levels for up to 6 h compared to free somatostatin given as a bolus dose in solution, which rapidly eliminated from the circulation (half-life of less than a minute).

It was suggested that after an early "burst release" process, which was attributed to free somatostatin in the dispersion, the remaining drug was apparently retained inside the cubosomes and cleared more slowly from the circulation. Engström *et al.* [37] surmised that the sustained plasma levels of somatostatin may be attributed to long circulation times of the surface stabilizer providing a coating on the cubosome surface, and hence avoiding reticuloendothelial system uptake.

A similar tendency of sustained release from cubosomes *in vitro* studies was detected in insulin-loaded systems given orally to diabetic rats [38]. The investigators demonstrated that after oral administration of insulin in cubosomes a sustained hypoglycemic effect occurred for over 6 h, while the same dose of insulin as an oral solution had a minor effect on blood glucose levels.

It was surmised that the mucoadhesive nature of GMO and protection of insulin from proteolytic enzymes were responsible for the sustained release phenomenon[21].

Boyd has examined the delivery profile from cubosomes of various hydrophobic drugs, including griseofulvin, rifampicin, diazepam and propofol [39]. In each case only burst release was found and the release reached a plateau within 20 min (Figure 15.6).

The authors assumed that the partition coefficient of the lipophilic drug dictates the extent of release compared to other drugs at the same dilution. The rank order for the extent of release for the three unionized drugs, griseofulvin, diazepam and propofol was consistent with the octanol–water partition coefficient (log $K_{o/w}$) listed in Table 15.1. It was reasonable to assume that the lower affinity of griseofulvin for the hydrophobic phase allowed the greatest extent of release, and conversely for propofol.

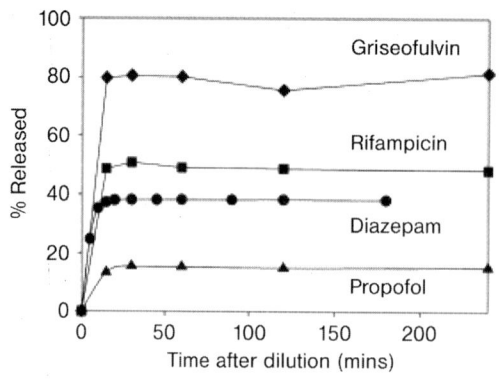

Figure 15.6 Release of lipophilic drugs from cubosome dispersions by pressure ultrafiltration. Dispersions were loaded with griseofulvin (♦), rifampicin (■), diazepam (●) and propofol (▲). (Reprinted with permission from Ref. [39].)

Rifampicin was expected to be ionized to some extent in the neutral dispersions ($pK_a \sim 7.9$).

By examining the dispersions mentioned above, Boyd [39] aimed to distinguish between delivery systems that exhibited burst release and those that provided therapeutically useful controlled- or sustained-release over a period of time. When the drug release was completed within 1 h, the delivery system was considered as a burst release profile type. The main outcome was that upon dilution of the cubosomes containing lipophilic drugs only burst release phenomenon was observed, leading to the conclusion that any therapeutic benefit, by means of controlled release, could not be obtained using cubosomes as a delivery system for lipophilic drugs.

Nevertheless, modifying the physical properties of the cubosome such as charge, viscosity and structure can improve the release kinetics [44–47].

Lynch et al. [46] demonstrated that negatively charged water-soluble active ketoprofen was incorporated with positively charged surfactants into a cubic liquid crystal. This study revealed that cationic surfactants such as dioctadecyl(dimethyl) ammonium chloride (DODMAC) and (dioctadecyl)ammonium chloride (DOAC) and di(canola ethyl ester)(dimethyl)ammonium chloride (DEEDAC) can be incorpo-

Table 15.1 Extent of release at equilibrium for 50-fold dilution of cubosome dispersions and literature octanol–water partition coefficients.

Drug	Extent of release (%)	Log $K_{o/w}$	Reference
Griseofulvin	81	2.2	[40]
Diazepam	38	2.9	[41]
Propofol	15	3.8	[42]
Rifampicin	48	4.2	[43]

Reprinted with permission from Ref. [39].

Table 15.2 Effect of alkyl chain length on the characteristics of release from the cubic phase of tryptophan and its derivatives.

Number of carbons in alkyl chain	Release in first 24 h (%)	Time for 20% release, $t_{20\%}$ (h)
0	94	0.5
1	75	1.3
2	61	3.3
4	14	35
8	2	300
18	0.01	≫300
(N-18) 2[a]	0	—

Reprinted with permission from [47].
[a] Refers to N-oleoyl-DL-tryptophan ethyl ester.

rated into bicontinuous cubic phase liquid crystals, creating a positively charged matrix that increased the attraction and loading of materials and effectively lowered the release rate. As a result, the drug uptake was customized to a far greater extent.

Clogston et al. [47] studied the dual hydrophobic and hydrophilic properties of the cubic phase by "hydrophobically" anchoring the tryptophan as a model water-soluble drug. Thus, their location within the cubic phase was altered. Control was achieved by alkylating the drug with chains of different lengths. Thus, the release of water-soluble drugs from the cubic phase was manipulated by selective alkylation. The longer the alkyl chain, the slower the release from the cubic phase. The data presented in Table 15.2 demonstrate that lengthening the alkyl chain attached to the water-soluble tryptophan served to anchor it more securely and increased its residence time in the apolar compartment of the cubic phase. Hence, the concentration of an additive in the aqueous channel was reduced and the driving force for release was lowered.

In our most recent work we learned how to prepare discontinuous cubic micellar mesophases (termed Q_L) that are very fluid and almost Newtonian. The Q_L structures

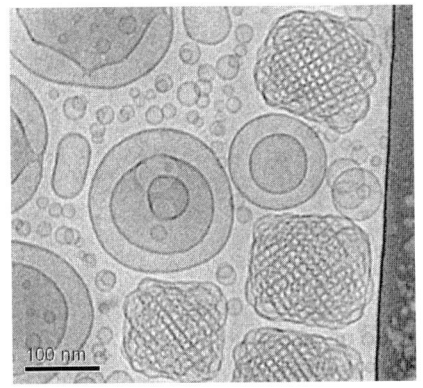

Figure 15.7 Cryo-TEM image of micellosomes and vesicles formed by homogenization of monoolein in aqueous Poloxamer 407 solution [53]. (Image courtesy of R. Efrat.)

are prepared in the presence of third component (ethanol) by much simpler methods than those of the discontinuous structures. The fluidic Q_L mesophase has a somewhat less ordered structure and offers some very significant new applications for oral and transdermal applications [48–52].

The Q_L phase can be further dispersed into aqueous phases to form small (300 nm) soft particles that were termed micellosomes (Figure 15.7). The solubilization capacity of these new structures is unique since they can embed synergistically both lipophilic as well as hydrophilic drugs and bioactives [53].

15.4
Summary

Considering the cubic phase liquid crystals for drug delivery and other applications, nanostructures with a cubic inner phase dispersed in water can be formulated, offering the advantage of low viscosity. Studies published in the last decade demonstrate that cubosomes can be loaded with both hydrophilic and hydrophobic guest molecules with variable sizes, including macromolecules such as proteins. It was reported that the release profiles of the solubilized drugs are highly dependent on the hydrophilic–hydrophobic balance and interfacial properties of the molecules. Moreover, it was demonstrated that the physical properties of the host cubic phase could be altered significantly by solubilization of proteins. In this respect the protein can assist in the preparation of cubosomes, opening up new perspectives in this field.

References

1 Gruner, S.M., Tate, M.W., Kirk, G.L. et al. (1988) *Biochemistry*, **27**, 2853–2866.
2 Rizwan, S.B., Dong, Y.D., Boyd, B.J. et al. (2007) *Micron*, **38**, 478–485.
3 Enlow, J.D., Enlow, R.L., McGrath, K.M. and Tate, M.W. (2004) *Journal of Chemical Physics*, **20**, 1981–1989.
4 Hyde, S.T. (2001) in *Handbook of Applied Surface and Colloid Chemistry* (ed. K. Holmberg), John Wiley & Sons, New York, Ch 16.
5 Rummel, G., Hardmeyer, A., Widmer, C. et al. (1998) *Journal of Structural Biology*, **121**, 82–91.
6 Spicer, P.T. (2005) *Current Opinion in Colloid and Interface Science*, **10**, 274–279.
7 Drummond, C.J. and Fong, C. (2000) *Current Opinion in Colloid and Interface Science*, **4**, 449–456.
8 Gustafsson, J., Ljusberg-Wahren, H., Almgren, M. and Larsson, K. (1997) *Langmuir*, **13**, 6964–6971.
9 Lopes, L.B., Collett, J.H. and Bently, M.V.L.B. (2005) *European Journal of Pharmaceutics and Biopharmaceutics*, **60**, 25–30.
10 Larsson, K. (1989) *The Journal of Physical Chemistry*, **93**, 7304–7314.
11 Buchein, W. and Larsson, K. (1987) *Journal of Colloid and Interface Science*, **117**, 582–583.
12 Landh, T. (1994) *The Journal of Physical Chemistry*, **98**, 8453–8467.
13 Nakano, M., Teshigawara, T., Sugita, A. et al. (2002) *Langmuir*, **18**, 9283–9288.
14 Nakano, M., Sugita, A., Matsuoka, H. and Handa, T. (2001) *Langmuir*, **17**, 3917–3922.

15 Monduzzi, M., Ljusberg-Wahren, H. and Larsson, K. (2000) *Langmuir*, **16**, 7355–7358.

16 Neto, C., Aloisi, G., Baglioni, P. and Larsson, K. (1999) *The Journal of Physical Chemistry. B*, **103**, 3896–3899.

17 Larsson, K. (1999) *Journal of Dispersion Science and Technology*, **20**, 27–34.

18 Kamo, T., Nakano, M., Leesajakul, W. et al. (2003) *Langmuir*, **19**, 9191–9195.

19 Spicer, P.T., Hyden, K.L., Lynch, M.L. et al. (2001) *Langmuir*, **17**, 5748–5756.

20 Garg, G., Saraf, S. and Saraf, S. (2007) *Biological & Pharmaceutical Bulletin*, **30**, 350–353.

21 Boyd, B.J. (2005) *Bicontinuous Liquid Crystals, Surfactant Science Series*, Vol. 127 (eds M.L. Lynch and P.T. Spicer), Taylor & Francis, CRC Press, Boca Raton, FL, USA, Chapter 10.

22 Hansen, J., Nielsen, L.S. and Norling, T. (1999) U.S. Patent 5955502.

23 Spicer, P.T., Small, W.E., Lynch, M.L. and Burns, J.L. (2002) *Journal of Nanoparticle Research*, **4**, 297–311.

24 Gustafsson, J., Ljusberg-Wahren, H., Almgren, M. and Larsson, K. (1996) *Langmuir*, **12**, 4611–4613.

25 Almgren, M., Edwards, K. and Karlsson, G. (2000) *Colloids and Surfaces A: Physicochemical and Engineering Aspects*, **174**, 3–21.

26 Barauskas, J., Johnsson, M., Joabsson, F. and Tiberg, F. (2005) *Langmuir*, **21**, 2569–2577.

27 Angelov, B., Angelova, A., Papahadjopoulos-Sternberg, B. et al. (2006) *Journal of the American Chemical Society*, **128**, 5813–5817.

28 Sternberg, B., Moody, M.F., Yoshioka, T. and Florence, A.T. (1995) *Nature*, **378** (6552), 21.

29 Sternberg, B., Hong, K., Zheng, W. and Papahadjopoulos, D. (1998) *Biochimica et Biophysica Acta*, **1375**, 23–35.

30 Torchilin, V.P., Lukyanov, A.N., Gao, Z. and Papahadjopoulos-Sternberg, B. (2003) *Proceedings of the National Academy of Sciences of the United States of America*, **100**, 6039–6044.

31 Torchilin, V.P., Levchenko, T.S., Rammohan, R. et al. (2003) *Proceedings of the National Academy of Sciences of the United States of America*, **100**, 1972–1977.

32 Kraineva, J., Narayanan, R.A., Kondrashkina, E. et al. (2005) *Langmuir*, **21**, 3559–3571.

33 Razumas, V., Larsson, K., Miezis, Y. and Nylander, T. (1996) *The Journal of Physical Chemistry*, **100**, 11766–11774.

34 Angelova, A., Angelov, B., Papahadjopoulos-Sternberg, B. et al. (2005) *Journal of Drug Delivery Science and Technology*, **15**, 108–112.

35 Gabizon, A., Shmeeda, H. and Barenholz, Y. (2003) *Clinical Pharmacokinetics*, **42**, 419–436.

36 Kim, J.S., Kim, H., Chung, H. et al. (2000) *Proceedings of the International Symposium of Controlled Release Bioactive Materials*, **27**, 1118–1119.

37 Engström, S., Ericsson, B. and Landh, T. (1996) *Proceedings of the International Symposium of Controlled Release Bioactive Materials*, **23**, 89–90.

38 Chung, H., Kim, J., Um, J.Y. et al. (2002) *Diabetologia*, **45**, 448–451.

39 Boyd, B.J. (2003) *International Journal of Pharmaceutics*, **260**, 239–247.

40 Leo, A., Hansch, C. and Elkins, D. (1971) *Chemical Reviews*, **71**, 525–616.

41 Taillardat-Bertschinger, A., Marca Martinet, C.A., Carrupt, P.-A. et al. (2002) *Pharmaceutical Research*, **19**, 729–737.

42 Calculated using the KowWin software program located at www.esc.syrres.com (2003).

43 Hansch, C., Leo, A. and Hoekman, D.H. (1995) *Exploring QSAR*, American Chemical Society, Washington, D.C.

44 Lindell, K., Engblom, J., Jonstromer, M. et al. (1998) *Progress in Colloid and Polymer Science*, **108**, 111–118.

45 Puvvada, S., Qadri, S.B., Naciri, J. and Ratna, B. (1993) *The Journal of Physical Chemistry*, **97**, 11103–11107.

46 Lynch, M.L., Ofori-Boateng, A., Hippe, A. et al. (2003) *Journal of Colloid and Interface Science*, **260**, 404–413.

47 Clogston, J., Craciun, G., Hart, D.J. and Caffrey, M. (2005) *Journal of Controlled Release*, **102**, 441–461.

48 Garti, N., Efrat, R. and Aserin, A. (2005) World Patent WO 2005063370 A1.

49 Efrat, R., Aserin, A., Danino, D. et al. (2005) *Australian Journal of Chemistry*, **58**, 762–766.

50 Efrat, R., Aserin, A., Shalev, D.E. et al. (2007) Abstracts of Papers, 233rd ACS National Meeting, Chicago, IL, United States, March 25–29.

51 Efrat, R., Aserin, A., Kesselman, E., Danino, D., Wachtel, E.J. and Garti, N. (2007) *Colloids and Surfaces A: Physicochemical and Engineering Aspects*, **299**, 133–145.

52 Efrat, R., Aserin, A. and Garti, N. (2007) *Food Colloids: Self-Assembly and Material Science*. Royal Society of Chemistry, Special Publication 302, pp. 87–102.

53 Efrat, R., Aserin, A., Danino, D. and Garti, N. (2008) Langmuir, accepted.

16
Highly Concentrated (Gel) Emulsions as Reaction Media for the Preparation of Advanced Materials

Conxita Solans and Jordi Esquena

16.1
Introduction

Highly concentrated emulsions, also referred to in the literature as high-internal-phase ratio (HIPRE) emulsions [1, 2], gel emulsions [3–5], and so on, have received a great deal of attention in recent decades due to their intrinsic theoretical interest and numerous technological applications (as pharmaceutical, cosmetic, food, and so on, formulations). Highly concentrated emulsions are characterized by an internal phase volume fraction exceeding 0.74, the critical value of the most compact arrangement of uniform, undistorted spherical droplets [1, 2]. Accordingly, their structure consists of compact polyhedrical and/or polydisperse droplets separated by a thin film of continuous phase, a structure resembling gas-liquid foams (Figure 16.1a). Rheologically, highly concentrated emulsions are non-Newtonian fluids characterized by a yield stress below which they show a solid-like behavior [2, 6–9] (as seen in Figure 16.1b).

In this context, special mention is made of the studies of the viscoelastic behavior of highly concentrated emulsions by Th. F. Tadros [8], which allowed the correlation of structural parameters with rheological behavior, thus providing good insight into the structure of these systems.

Highly concentrated emulsions can be prepared by different methods. The conventional method, schematically represented in Figure 16.2a, consists of dissolving a suitable emulsifier in the component that will constitute the continuous phase followed by stepwise addition of the component that will constitute the dispersed phase, with continuous moderate stirring. The multiple emulsification method [3–5] (Figure 16.2b) consists of weighing all the components at the final composition, followed by shaking or stirring the sample and, as the name indicates, at a certain step of the emulsification process the mixture consists of a multiple W/O/W emulsion [5].

The above-mentioned emulsification methods result in rather big (droplets over a micrometer in diameter) and polydisperse emulsions, as illustrated in Figure 16.1a. An interesting method, based on the phase inversion temperature (PIT) emulsification

Highlights in Colloid Science. Edited by Dimo Platikanov and Dotchi Exerowa
Copyright © 2009 WILEY-VCH Verlag GmbH & Co. KGaA, Weinheim
ISBN: 978-3-527-32037-0

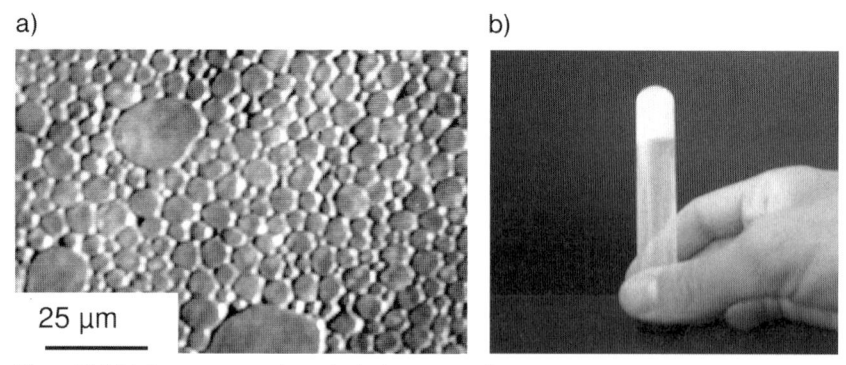

Figure 16.1 (a) Aspect, as seen by optical microscopy, of a water-in-oil highly concentrated emulsion with 90 wt% of dispersed phase obtained by the conventional method, which leads to relatively large and highly polydispersed droplets; (b) visual aspect of the emulsion, which does not flow when turned upside down.

method is the so-called "spontaneous" formation method [10–12], which produces emulsions with smaller droplets and narrower size distributions (Figure 16.3) than those obtained by the other methods.

The PIT, introduced by Shinoda [13], is the temperature at which polyoxyethylene nonionic surfactants change their preferential solubility from water (oil) to oil (water) and inversion from O/W to W/O emulsions or vice versa is produced. The PIT emulsification method consists of preparing the emulsions at the HLB temperature (taking advantage of the low interfacial tension achieved) followed by a rapid cooling or heating of the samples to produce O/W or W/O emulsions, respectively. It takes advantage of the phase transitions produced during the emulsification process. Highly concentrated emulsions have been formed by the "spontaneous" emulsification method, without the need for mechanical stirring, by quickly cooling (or heating) a water-in-oil (or oil-in-water) microemulsion from a temperature higher (or lower) than the HLB temperature of the system to a temperature below (or above)

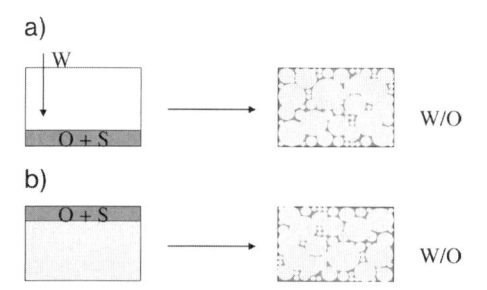

Figure 16.2 Schematic representation of two preparation methods for water-in-oil (W/O) highly concentrated emulsions: (a) Slow addition of dispersed phase to the continuous phase (conventional method); (b) weighting and shaking all components together (multiple emulsification method).

Figure 16.3 Aspect, as seen by optical microscopy, of a water-in-oil highly concentrated emulsion with 90 wt% of dispersed phase obtained by the "spontaneous" emulsification method, which leads to small and low polydispersed droplets.

it [5, 10, 12]. Figure 16.4 shows the phase transitions taking place by heating an O/W microemulsion to obtain a W/O highly concentrated emulsion [11].

One of the most promising applications of highly concentrated emulsions is their use as reaction media [14–23]. In this chapter, the preparation of low-density organic macroporous materials and inorganic materials with dual meso-macroporosity is described.

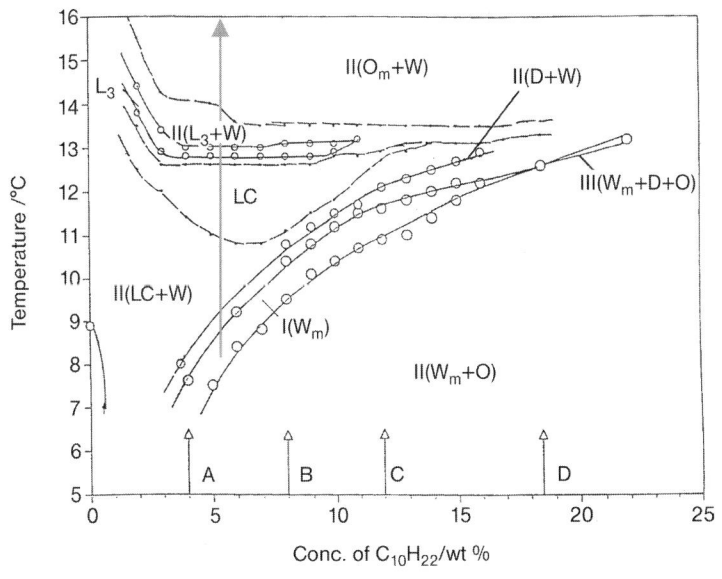

Figure 16.4 Phase diagram of 0.1 M NaCl/$C_{12}(EO)_4$/decane system as a function of temperature. The arrow indicates the changes that an oil-in-water microemulsion, region I(W_m), undergoes when the temperature is increased to reach the two-phase region, II(O_m+W), in which a water-in-oil (W/O) highly concentrated emulsion is formed. During the process of "spontaneous" formation of the emulsion, the curvature of surfactant aggregates changes from positive, region I(W_m), to negative, region II(O_m+W), through zero curvature in the lamellar liquid crystalline (Lα) and bicontinuous (L_3) phases. (From Ref. [11] with permission.)

16.2
Highly Concentrated Emulsions as Templates for Low-Density Macroporous Materials

The first report on the use of highly concentrated emulsions as templates to obtain macroporous organic foams dates back to 1982 [14]. Since then, a wide variety of different low-density macroporous organic materials, referred to as solid foams or aerogels, have been obtained by polymerization in the continuous phase of highly concentrated emulsions, followed by removal of the dispersed phase [14–16, 20, 23–28]. In this context, it is worth mentioning the contributions of Ruckenstein et al. [15, 26, 27], who reported the preparation of composite polymers by simultaneous polymerization of both hydrophilic and lipophilic monomers. The films that separate adjacent droplets are very thin in highly concentrated emulsions and the polymerization takes place mainly in the emulsion plateau borders, resulting in polyhedral macropores that are interconnected through narrower necks. These high pore volume and very low bulk density (as low as $0.02\,g\,mL^{-1}$) macroporous solid foams have been applied as supports for catalysts, selective membranes, immobilization of enzymes, templates for the preparation of other materials, and so on [15, 16, 27–29].

The properties (average cell size and cell size distribution, surface area, macroscopic density, etc.) of the materials obtained by using the technology based on highly concentrated emulsions were improved with respect to those obtained by other technologies. However, their structures were still difficult to control: the cells were large, with diameters generally bigger than $10\,\mu m$, and highly non-homogeneous (Figure 16.5) due to the high polydispersity of the precursor emulsions [14–16, 24–31].

Notably, the emulsions were prepared by the conventional method (described in Section 16.1 and shown schematically in Figure 16.2a), which leads to highly polydisperse emulsions (Figure 16.1a).

Using the "spontaneous" emulsification method (also described in Section 16.1) based on the PIT, our group has reported [20] the formation of highly concentrated

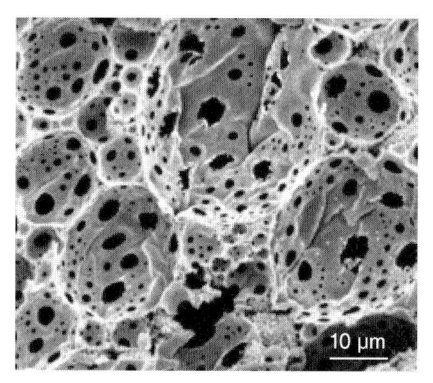

Figure 16.5 Micrograph of a polystyrene foam obtained by polymerization ($60\,°C$, $48\,h$) in the continuous phase of a water-in-styrene emulsion containing $90\,wt\%$ aqueous solution, prepared by the conventional method. The scale bar indicates $10\,\mu m$. (Adapted from Ref. [20], with permission.)

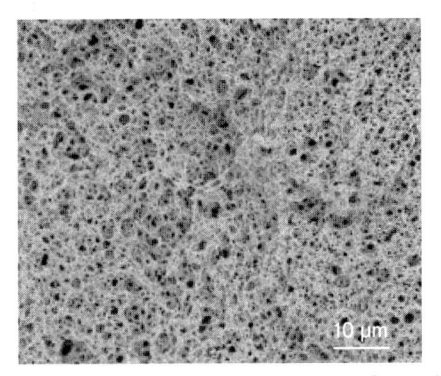

Figure 16.6 Micrograph of a polystyrene foam, obtained by polymerization (60 °C, 48 h) in the continuous phase of a water-in-styrene emulsion with 90% aqueous solution, prepared by the "spontaneous" emulsification method. The scale bar indicates 10 μm. (Adapted from Ref. [20], with permission.)

emulsions, containing an organic monomer in the continuous phase, with smaller and more homogeneous droplet size (as that shown in Figure 16.3). Consequently, the corresponding macroporous foams (Figure 16.6), obtained after polymerization in the continuous phase and removal of the dispersed phase, consist of smaller and less polydispersed cells [20] than foams from emulsions prepared by other methods [14, 24–31].

The template effect of the emulsions is evident by comparing Figures 16.1 and 16.3 with Figures 16.5 and 16.6, respectively.

The macroporous solid foams obtained from highly concentrated emulsions prepared by the "spontaneous" emulsification method, apart from being more homogeneous and with smaller cell size, also show better mechanical properties and lower densities than those prepared by conventional methods. The polystyrene foams were approximately three times stronger and approximately 50% tougher [20].

16.3
Materials with Dual Meso- and Macroporous Structure Templated in Macroporous Foams Obtained From Highly Concentrated Emulsions

Macroporous foams obtained by polymerization in the continuous phase of highly concentrated emulsions have been used as scaffolds for the preparation of meso/macroporous inorganic oxides [21]. Materials with dual meso- and macroporous structure have received widespread attention because they combine the advantages of high specific surface due to the mesopores with the accessible diffusion pathways associated with macroporous structures. These meso/macroporous materials are important in many applications, including catalytic surfaces and supports [32], adsorbents, chromatographic materials, filters [33, 34], light-weight structural materials [35], and so on.

This preparation method for hierarchically textured foams consists of imbibing the solid macroporous foam with alcohol solutions containing inorganic precursors and block copolymer surfactants [21]. The cooperative self-assembly of the surfactant molecules, together with the inorganic oxide precursor species, leads to the formation of ordered mesostructures. The resulting organic–inorganic composite materials are dried and then calcined in air at high temperature to remove all organic components and to obtain the meso/macroporous inorganic materials.

This method allows independent control over both the macropores, which are determined by the emulsion droplets in the first step of the process, and the mesopores, by the supramolecular self-assembly in the second step. Therefore, the two templating systems do not interfere with each other, and consequently the meso- and macro-structures can be optimized separately [21]. Inorganic oxides, such as silica, titania and zirconia, were obtained by this two-step process. The bimodal distribution of pores was characterized by highly ordered mesopores (5–10 nm) and by interconnected macropores (0.1 to 5 μm). These materials possess high pore volumes (17 cm^3 g^{-1}) and specific surface areas, as determined by the BET method applied to the nitrogen adsorption isotherm, of between 250 and 750 m^2 g^{-1}. The higher surface areas were obtained at higher inorganic precursor concentrations and correlate with reduced macropore volumes [21].

16.4
Conclusions

Highly concentrated emulsions are interesting systems for the preparation of low-density macroporous materials by polymerization in the continuous phase of the emulsions followed by the removal of the dispersed phase components. Macroporous solid foams or aerogels produced by this method consist of interconnected sponge-like macropores. The droplet size distribution of the highly concentrated emulsion, which can be controlled by choosing an appropriate emulsification method (e.g. "spontaneous" emulsification method), is a crucial factor in determining the properties of the macroporous monoliths.

Macroporous solid foams produced in highly concentrated emulsions have been used as scaffolds for the preparation of hierarchically textured foams. Such materials possess structural organization at different size ranges: mesopores are templated by supramolecular aggregates and macropores by the emulsion droplets. This strategy allows independent control of the mesopore and the macropore sizes.

Acknowledgments

Financial support by the Spanish Ministry of Science and Education (Grant CTQ2005-08241-C03-01/PPQ) and "Generalitat de Catalunya, DURSI" (Grant 2005 SGR-0018) is acknowledged.

References

1 Lissant, K.J. (1966) *Journal of Colloid and Interface Science*, **22**, 462.
2 Princen, H.M. (1979) *Journal of Colloid and Interface Science*, **71**, 55.
3 Kunieda, H., Solans, C., Shida, N. and Parra, J.L. (1987) *Colloids and Surfaces A: Physicochemical and Engineering Aspects*, **24**, 225.
4 Solans, C., Domínguez, J.G., Parra, J.L. et al. (1988) *Colloid and Polymer Science*, **266**, 570.
5 Solans, C., Pons, R. and Kunieda, H. (1998) in *Modern Aspects of Emulsion Science* (ed. B.P. Binks), Royal Society of Chemistry, Cambridge, pp. 367–394.
6 Princen, H.M. (1983) *Journal of Colloid and Interface Science*, **91**, 160.
7 Princen, H.M. and Kiss, A.D. (1986) *Journal of Colloid and Interface Science*, **112**, 427.
8 Pons, R., Solans, C. and Tadros, Th.F. (1995) *Langmuir*, **11**, 1966–1971.
9 Pons, R., Erra, P., Solans, C. et al. (1993) *The Journal of Physical Chemistry*, **97**, 12320.
10 Pons, R., Carrera, I., Erra, P. et al. (1994) *Colloids and Surfaces A: Physicochemical and Engineering Aspects*, **91**, 259.
11 Kunieda, H., Fukui, Y., Uchiyama, H. and Solans, C. (1996) *Langmuir*, **12**, 2136.
12 Ozawa, K., Solans, C. and Kunieda, H. (1997) *Journal of Colloid and Interface Science*, **188**, 275–281.
13 Shinoda, K. and Saito, H. (1968) *Journal of Colloid and Interface Science*, **26**, 70.
14 Barby, D. and Haq, Z. (1982) European Patent 0060138 (Unilever).
15 Ruckenstein, E. (1997) *Advances in Polymer Science*, **127**, 3–58.
16 Cameron, N.R. and Sherington, D.C. (1996) *Advances in Polymer Science*, **126**, 163–214.
17 Pinazo, A., Infante, M.R., Izquierdo, P. and Solans, C. (2000) *Journal of the Chemical Society, Perkin Transactions*, **2**, 1535–1539.
18 Solans, C., Pinazo, A., Calderó, G. and Infante, M.R. (2001) *Colloids and Surfaces A: Physicochemical and Engineering Aspects*, **176**, 101–108.
19 Clapés, P., Espelt, L., Navarro, M.A. and Solans, C. (2001) *Journal of the Chemical Society, Perkin Transactions*, **2**, 1394.
20 Esquena, J., Sankar, G.S.R.R. and Solans, C. (2003) *Langmuir*, **19**, 2983–2988.
21 Maekawa, H., Esquena, J., Bishop, S. et al. (2003) *Advanced Materials*, **15**, 591–596.
22 Espelt, L., Clapés, P., Esquena, J. et al. (2003) *Langmuir*, **19**, 1337.
23 Solans, C., Esquena, J. and Azemar, N. (2003) *Current Opinion in Colloid and Interface Science*, **8**, 156–163.
24 Williams, J.M. and Wrobleski, D.A. (1988) *Langmuir*, **4**, 656–662.
25 Williams, J.M. (1988) *Langmuir*, **4**, 44–49.
26 Ruckenstein, E. and Park, J.S. (1988) *Journal of Polymer Science Part C-Polymer Letters*, **26**, 529.
27 Ruckenstein, E. and Park, J.S. (1992) *Polymer*, **33**, 405–417.
28 Cameron, N.R., Sherington, D.C., Albiston, L. and Gregory, D.P. (1996) *Colloid and Polymer Science*, **274**, 592–595.
29 Hainey, P., Huxham, I.M., Rowatt, B. et al. (1991) *Macromolecules*, **24**, 117–121.
30 Mercier, A., Deleuze, H. and Mondain-Monval, O. (2000) *Actualité Chimique*, **5**, 10–15.
31 Williams, J.M., Gray, A.J. and Wilkerson, M.H. (1990) *Langmuir*, **6**, 437–444.
32 Harold, M.P., Lee, C., Burggraaf, A.J. et al. (1994) *MRS Bulletin*, **19**, 34–39.
33 Bhave, R.R. (1991) *Inorganic Membranes Synthesis, Characteristics, Applications*, Van Nostrand Reinhold, New York.
34 Ishizuka, N., Minakuchi, H., Nakanishi, K. et al. (1998) *Journal of Chromatography. A*, **797**, 133–137.
35 Wu, M.X., Fujiu, T. and Messing, G.L. (1990) *Journal of Non-Crystalline Solids*, **121**, 407–412.

Index

a

A-B-A block copolymers   100
absolute viscosity values   233
absorption process   25
acid-degradable protein-loaded microgel   33
– schematic representation   33
acoustic impedance   60
AcoustoSizer   62
– measurements   74
acrylic acid, moieties   29
adhesion energy   49
adsorbed polymer   65
– layers   65–70
– particles   65
adsorption-curing process   150
adsorption equilibrium constant   256
adsorption kinetics process   4, 237, 248, 250, 251
– micelles impact   250
Aerosil silica particles   231, 239
affinity chromatography   203
aggregate collision radius   264
aggregate fractal dimension   264
aggregate structure   264, 271
– parameters   264, 272
aggregate volume   263
aggregation process   261, 262, 274
– electrolyte concentration   274
– mechanisms   262
– pH   274
– shear rate   274
– time   263
air-vapor interface   79
air-water interface   79, 88, 90, 92
– stability   79
– structure   79
AKD droplets   14
AKD particles   7, 8, 9
– anionic deposition   7, 9
– cationic deposition   7, 9
alkyltrimethylammonium bromide   190
amphiphilic block copolymers   97, 281
amphiphilic polyelectrolyte, *see* DNA
antigen responsive microgel   24
Antonov's empirical rule   48
aqueous channels   279
aqueous phase   280
asphaltenes extraction   231
asphaltene/toluene solutions   230
atomic force microscopy (AFM)   113, 140, 265
atomic systems   119
– high-temperature   119

b

backscattering spectrum   241
Bancroft's rule   229
batch process   262
block-copolymers   42
Boltzmann constant   21, 263
Boltzmann factors   86
bovine serum albumin (BSA)   32, 33
BPA device   251
Brownian aggregation   262
Brownian motion   119, 262, 264, 270
brush-to-brush interaction   100, 101, 102, 106
bubble pressure tensiometry   249
bubble profile analysis tensiometer   254
burst release phenomenon   286

c

cadmium seleniode (CdSe)   34
cationic gold nanorods   38
cationic polymer   12
– cationic starch   12
– cPAM   12
cationic surfactants   197, 286

Highlights in Colloid Science. Edited by Dimo Platikanov and Dotchi Exerowa
Copyright © 2009 WILEY-VCH Verlag GmbH & Co. KGaA, Weinheim
ISBN: 978-3-527-32037-0

- cetylpyridinium chloride 27
- dioctadecyl/dimethyl ammonium chloride 286

cellulose 135, 143
- β-D-anhydroglucopyranose 135
- carboxymethylation 143

cellulose microfibrils 140, 149
- methods 140

cetyltrimethylammonium bromide (CTAB) 36, 192
- coated QDs 36
- DNA interactions 194
- DNA particles 193

CFD software 273
- FLUENT 273
- PHOENICS 273

charge-balancing counter-ions 22
charged hemi-micelles 79
- stability 79
- structure 79

citrate-stabilized gold NPs 37
Clausius–Clapeyron equation 156, 169
cluster-cluster collisions 265
coagulant/flocculant adsorption dynamics 275
coalescence 229
collision efficiency factor 263
collision frequency factor 263, 265
collision kernel 264

colloid particles 1, 14, 75, 203
- aggregation process 261
- calcium carbonates 1
- clays 1
- deposition 203
- electric forces 75
- ground calcium carbonate (GCC) 14
- introduction 261
- titanium dioxide 1

colloidal aggregation 264
- diffusion-limited aggregation (DLA) mechanism 264
- reaction-limited aggregation (RLA) mechanism 264

colloidal dynamics AcoustoSizer 59
colloidal suspensions system 261, 262, 272
colon-specific drug delivery 23
complementary end-point distributions 87
computational fluid dynamics (CFD) 270
- models 270
- tools 273, 274

core-shell microgel particles 32
Couette flow apparatus 270, 271
cPAM-coated fibers 7
critical aggregation concentration (CAC) 26

critical micelle concentration (CMC) 9, 247, 252, 256

crude oil components 229, 230, 231, 237
- asphaltenes 229, 230
- interaction effects 229
- naphthenic acids 229
- physicochemical properties 231
- resins 229, 230, 237
- waxes 229
- wetting properties 229

crude oil emulsions 237
cryogenic-transmission electrons microscopy (cryo-TEM) images 281
cubic phase 279, 280, 281
- structure 280

cubosomes 279, 280, 281, 282, 284
- 3D structure 282
- cryo-TEM image 280
- diamond-type cubic structure 282
- preparation techniques 281–282

d

damping technique 248
DC mobilities 68
DDA 182
- binding isotherms 182
de Gennes' scaling theory 268
degrees of surface substitution (DSS) 144
- values 144
delivery vehicles 279
- cubosomes 279
deposition rate constant 15
Derjaguin's equation 47, 48
destabilization phenomenon, see coalescence
detachment rate coefficients 12
- polymer rearrangement 12
- polymer transfer 12
dewatering efficiency 236
dielectric constants 73
- barium titanate 73
- titania 73
diethylaminoethyl methacrylate 17
diffuse double layer (DDL) 105
diffusion coefficient 253
diffusion process 248
disk-shaped adsorption sites 206
dispersion stabilizers 149
- biochemical applications 149
- biomedical applications 149
DLVO forces 97, 105, 109, 116
DLVO interactions 108
DLVO theory 102, 104, 105, 109, 111, 219, 264, 266, 270, 271
DNA 179, 180–183, 187, 189, 190, 192

– adsorption 190
– A-forms 180
– B-forms 180
– ds-DNA 179, 181
– hybridization 35
– hydrophobic moieties 192
– ss-DNA 179
– Z-forms 180
DNA-based materials 199
– gels 193, 199
– membranes 199
– particles 199
DNA-cationic surfactant structures 186
DNA-cosolute interactions 196
DNA-surfactant interactions 181, 189, 199
dodecyltrimethylammonium bromide (DTAB) 184
double layer distortion 74
drag force correction factor 271, 272
droplets 44
– determination 233
– size distributions 73, 234
– volumetric 234
drug molecules 24
– delivery applications 282–288
dynamic mobility 59, 60, 61
– formula 72
– spectra 73

e

electrical double layer repulsion/attraction 261
electro-acoustic spectroscopy 55
– particle characterization 55
electrokinetic sonic amplitude (ESA) effect 55, 56, 58, 59
– devices 59
– instruments 61
– measurements 59, 60, 69, 70
– method 75
– pressure wave 61
– signal 59, 61
– sound waves 55, 59
– voltage 61
electrolyte 61
– AcoustoSizer II 61
– induced aggregation processes 264
– induced coagulation processes 261, 264
– relaxation frequency 74
electro-osmotic flow 63
electro-rheological temperature device (ERD) 233
electrophoresis measurements 11
electrophoretic mobility 17, 75

electrostatic attraction 27
electrostatic destabilization drops 234
electrostatic interaction 103
electrostatic repulsion 102
emulsion films 97, 106
emulsion stability measurements 233
– Turbiscan analysis 233
emulsions 234
– droplet size distributions 234
– preparation 232
enzymatic assay 150
Epstein-Barr virus replication factor 192
esterification reactions 142–144
– acetylation 142
– carboxymethylation 143
– etherification 144–145
– reaction with anhydrides 142
ethylene glycol diglycidyl ether (EGDE) 196
ethylene oxide 41
– block-copolymers 41
Exerowa–Scheludko porous plate cell 99
exponential growth kinetics 265

f

fibers 10
– clay deposition 10
fibril bundles 138
fibril surface chemistry 141
flame aerosol reactors 273
– nanoparticle synthesis 273
floc fractal dimension 272
flocculation phenomenon 239
Florey–Huggins interactions 85, 88
fluid collection efficiency 271
fluid-solid transition 119
fluorinated organic liquid phases 41
– oil-in-water (o/w) emulsions 41
fluorinated systems 42
– interfacial adsorption layers 42
– rheological studies 42
– solid surfaces 51
fluoroorganic compounds 42
– structures 42
Fourier-transform infrared spectroscopy (FT-IR) 141
freeze-dried nanocrystals 145
freeze-dried suspensions 147
freeze-fracture electron microscopy (FF-EM) 282
freezing/melting in thin pores 156
frequency factor, *see* collision kernel
FS/HL systems 50
Fuchs' stability ratio 263

g

Gaussian coils 3
generalized free-volume theory (GFVT) 124
Gibbs free energy 32
Gibbs integral equation 158
Gibbs–Thomson equation 156
GL binodals 128, 129
globular polymers 5
glycidyl methacrylate (GMA) 146
gold-loaded poly(NIPAM-co-MAc) microgel particles 37
gold nanorods 38
graft polymeric surfactant 108

h

Hamaker constant 21, 48
Hamaker equation 111
Happel's cell model 272
Helmholtz energy 81, 89
– optimization 81
heteroflocculation clay, see paper machine
heteroflocculation latex, see paper machine
heterogeneous surfaces 203
high-affinity isotherm 31
high-internal-phase ratio (HIPRE) emulsions, see highly concentrated emulsions
high shear homogenization techniques 281
high shear mechanical treatment 138
highly concentrated emulsions 291
– introduction 291
Hildebrand theory of solubility parameters 48
HS/FL system(s) 46, 47, 49, 50
HS/HL system(s) 47, 49
hydrocarbon surfactants (HS) 42, 51
hydrocarbon systems 49
hydrodynamic correction factor 263
hydrogen-bonding interaction 179
hydrophilic polyfructose chains (PFC) 104
hydrophilic silica particles 234, 238
hydrophobic drugs 285
– diazepam 285
– griseofulvin 285
– propofol 285
– rifampicin 285
hydrophobic-hydrophilic balance 284
hydrophobic/hydrophilic domains 281
hydrophobic interactions 192, 198
hydrophobic silica particles 237
hydrophobized solid surfaces 47
– interaction 47
hydrotrope precursor method 285

i

industrial-scale coagulation/flocculation processes 269
inertia factor 64
inorganic electrolytes 261
– aggregation 264
insulin-loaded systems 285
interfacial adsorption layers (IAL) 42, 51
– block-copolymers 42
– rheological behavior 42
– rheological studies 42
interfacial dilational visco-elasticity 254–257
inulin polymeric surfactant(s) 103, 113
INUTEC SP1 molecule(s) 105, 113, 114
INUTEC SP1 polymeric surfactant 109
ionic additives 50
– SDS 50
ionic surfactants 79

k

Karl-Fischer titration 234

l

Langmuir kinetics 3
Langmuir-type adsorption isotherms 230
Laplace equation 166
large disk-shaped particle 14
latex particles, see white pitch
Lennard–Jones-type fluid 128
light scattering technique 262
linear polysaccharide, see cellulose
liquid crystal dispersions 281
liquid-solid interface 247
liquid-vapor equilibrium line 157
long-chain surfactant 199
loop-to-loop interaction 103, 104, 108
low-affinity isotherms 31
low-density macroporous materials 294
Lucassen's theory 256
lyotropic liquid crystalline phases 279

m

macropores 296
macroporous foams 295
macroscopic phase separation 182
magneto-electric dynamometer 47
Matec ESA 8000 device 59
Matec ESA 9800 device 62
maximum bubble pressure method (MBPM) 250
melting/freezing phase transitions 155
meso/macroporous materials 295
method of moments (MOM) 273

– PBM 273
methylated solid surfaces 51
micellar kinetics process 247, 248–251, 253, 254, 256
– impact 247, 254
– mechanisms 249–250
micellar solutions 248, 249, 255, 256
– direct measurements 248
– visco-elasticity modulus 255
micelle adsorption 248
micelle disintegration 248
micelle dissolution rate constant 251
micellosomes 287
– cryo-TEM image 287
microcrystalline cellulose and clay (MCC) 18
microelectronic devices 203
microfibril surfaces 141
– modification 141
microfibrillar cellulose (MFC) 135–137, 139, 147, 148
– cryo-TEM image 139
– preparation 137
microfluidic devices 203
microgel-coupled proteins 32
microgel particle 21, 22, 23, 24, 30, 34
– aspect of 23
– de-swelling behavior 21
– introduction 21
– semi-interpenetrating network 24
– sugar-based 24
– swelling behavior 21
– temperature-responsive 24
micro interferometric apparatus 98
microinterferometric method 98
microscopic thin liquid films 98
mildly hydrophilic particles 243
– Aerosil 7200 243
molecular fluids 119
molecular interaction 51
molecular model 82–83
– fundamentals 82
monodisperse polystyrene latex particles 214
monomers 248, 255, 256
– diffusion coefficient 255, 256
– diffusion flux 248
MTEG molecules 282
multidimensional population balances 274
multiple emulsification method 291

n

nanocrystal(s) 137
– rod-like 137

– suspensions 137
nanocrystalline cellulose 137
– preparation 137
nanocubosome 284
– FF-EM image 284
nanofibrillar cellulose 147
– applications 147
nanoparticles 73–75
– absorption 34
– dynamic mobility 73
nanosized cellulose 135
– preparation 135
– properties 135
nano-soft particles, see cubosomes
near-infrared (NIR) spectroscopy 230
net pressure force 56
neutron diffraction analysis 141
Newton black films (NBF) 108, 111, 116
Newton's second law 56
NIPAM-based microgel particles 36
non-aqueous environments 262
non-DLVO forces 116
non-DLVO interactions 108
non-equilibrium interfacial properties 248
non-ionic ethoxylated surfactants 251
non-ionic polymer theory 68
non-ionic polymeric surfactants 66, 97, 104, 111
– A-B-A triblock copolymers 97
– adsorption 66
– novel graft polymers 97
non-micellar solutions 254
nonpolar droplets 44
– schematic representation 44
nonpolar liquid phase 49

o

octanol-water partition coefficient 285, 286
oilfield emulsions 230
– hydrophilic characteristics 230
– hydrophobic characteristics 230
oilfield fluids 242
oil-in-water (o/w) emulsions 41, 229, 292
oil/water interfacial tension 230
one-component Yukawa system 122, 132
Ostwald ripening effect 50

p

paper machine 2
papermaking process 1–2
– orthokinetic heteroflocculation 1
papermaking suspension 13
paper surface treatments 1
– surface sizing/coating 1

particle bridging effect 229
particle-solution interface 266
PB-CFD models 273, 274, 275
PEO chains 30, 32
PEO molecules 29
PEO-PPO-PEO triblock copolymers 101
PFD drops 47
phase inversion temperature (PIT) emulsification method 291
phase transitions 172
photodynamic therapy 38
pH-responsive microgel particles 23
pH-responsive system 36
PIT emulsification method 292
plane-parallel quartz glass plate 99
Platikanov cell 99
Poisson-Boltzmann theory 85
polyacrylic acid (PAA) 267
– concentrations 267
poly(diethyl-dimethylammonium chloride) (PDADMAC) 1
polyelectrolytes 196
– DNA gels 196
– surfactant complex 193
– surfactant systems 185
polyelectrolytes/non-ionic polymers 65
polyethylene imine (PEI) 1, 4
poly(ethylene oxide) (PEO) chains 29
polymer adsorption 7, 10, 268
– dynamics models 275
– kinetics 4
polymer-coated spherical particles 268
– radius 267
polymer-covered particles 267
polymeric surfactants system 106, 116, 275
– hydrophobically modified inulin graft polymer 116
– PEO-PPO-PEO three-block copolymers 116
polymer-induced flocculation 262, 264, 267
polymer-induced orthokinetic heteroflocculation 2
polymer particles 21
– microgel 21
polymers 13, 29
– absorption 29
– chains 267
– coils 3
– desorption 5
– PEO 13
polymers aggregation 266–268
– charge patch neutralization 266
– polymer bridging 266
– polymer depletion 266
– simple charge neutralization 266
polymer transfer process 10–12
polyNIPAM-co-AAc microgel particles 27
polyNIPAM microgel particles 38
polystyrene foams 295
polystyrene latex aggregates 268, 271
polystyrene latex particles 8
polystyrene (PS) microgel particles 34
population balance equation (PBE) 262, 268, 269
population balance-fluid flow models 272–275
population balance (PB) framework 262
population balance models (PBMs) 267, 269, 270, 271, 274
– paradigm 263, 269
porous matrix 172
precipitated calcium carbonate (PCC) 15, 16, 17
propylene oxide 41, 42
– block-copolymers of 41
proteins 29
– absorption 29, 32
– drug-delivery systems 32
– immunoglobulin 284
– microgel derivatives 32
proteolytic enzymes 285
PSB latex 17
pseudo-ternary phase map 186

q
QCM-D technique 230
quadrature method of moments (QMOM) 273
– aggregation 273
– based PBMs 275
– nanoparticle synthesis 273
– population balance models 273
– precipitation 273
quantum dots (QDs) 34
– optical properties of 34
– tagged PS microgel particles 35
quasi-continuous surfaces 207
– deposition 207
quasi-homogeneous surfaces 215

r
random sequential adsorption (RSA) 205
– approach 205
– process 210
– simulation scheme 205, 208, 212
random site surfaces 207
– deposition 207
– regime 207

reactive sizing agents 1
- alkenyl succinic acid (ASA) 1
- alkyl ketene dimer (AKD) 1
Rheology test 239
- experiments 248
RheoPlus/32 software 233
RLA kinetics 265
rod-like crystallites 137
rotating bob 233
rotating vectors 74
RSS algorithm 210
RSS process 209
rupture plane 49

s
SARA components 231
SARA fractionation 231
scaling theory of de Gennes 101, 106
scanning electron microscopy (SEM) 140
SCF calculations 90, 92
SCF theory 80–83
- fundamentals 82
Scheludko–Exerowa cell 98
Schulze–Hardy rule 115
self-consistent field theory 79
- modeling 79
semipermanent (hydrolyzable) crosslinks 23
shear environments aggregation 269
shear-induced adsorption 3
short-chain alcohols 191
signal-triggered nanoparticle/microgel particle 37
silica nanoparticles 231, 232, 237, 242
- characterization 231, 232
- properties 231, 237
single component systems 274
single-crystal systems 282
SINTERFACE technologies 251
sinusoidal electric field 60
sinusoidal voltage pulse 56, 59
small-angle neutron scattering (SANS) 141
small-angle X-ray scattering (SAXS) 141
small-chain alcohols 22
small-molecule drugs 32
Smoluchowski kinetics 18
Smoluchowski's equation 265
Smoluchowski's theory 3
sodium polyacrylate (PAA) 69
solid-liquid interface 267
- potential 275
solid-vapor equilibrium line 157
solid-vapor interface 156
solute-interface interaction 268
somatostatin 285

- burst release process 285
spontaneous emulsification method 294, 295
spontaneous formation method, see phase inversion temperature (PIT) emulsification
stability measurements 236
stabilizing solids 229
- interaction effects 229
- Wetting properties 229
stirred tank compartments 272
- bulk zone 272
- dead space 272
- impeller zone 272
surface-active components 242
surface-charge groups 21
surface conductance 70–73
- effects of 71
surface features 212
surface forces 263
- models 270
- theories 269
surface phase transitions 157–165
surfactant molecules 25, 49, 50, 80
- absorption 25
surfactant sodium dodecyl sulfate (SDS) 25
- absorption 27
surfactant solutions 247
- interfacial properties 247
sustained release phenomenon 285
swelling-deswelling transitions 23

t
tangential field spectrum 71
- magnitude of 71
Taylor–Couette apparatus 273
TEMPO-mediated oxidation 139, 145
TGA-capped QDs 36
theory of Lucassen 255
theory of Smoluchowski 4
thin double layer systems 63
- dynamic mobility 63
- thickness 63
thin liquid films 98
- microinterferometric method 98
- pressure balance technique 99
thioglycolic acid (TGA)-capped 35
3D cubosomic growth 283
- bottom-up mechanism 283
transmission electron microscopy (TEM) 140
Turbiscan analyses 233, 236, 241, 242
Turbiscan EasySoft Formulaction software 233
Turbiscan lab 233
- backscattering data 233

– transmission 233
turbulent energy dissipation rate 272

u

unit cell 279
– aqueous channels 279

v

van der Waals attraction 107, 109, 261, 263, 266, 267, 268
van der Waals interactions 111
Vincent's expression 267
vinyl-functionalized acid-labile crosslinker 32
viscoelastic gel-like properties 147
visco-elasticity module 255
viscosity observations 234
viscosity ratio 233, 236, 240
visual process analyser (ViPA) 234
volume flux 57, 58
volume phase transition (VPT) 33

w

water-air interface 102
– manifestation 102
– peculiarity 102
water drops 233, 242
– agglomeration 233
water-in-crude oil emulsions 229, 230, 232, 236, 238, 241, 242, 292
– interface 239, 242
– processing 229
– production 229
– transportation 229
water-soluble nonionic amphiphiles 42
– hydrocarbon surfactants (HS) 42
water-soluble nonionic fluorinated surfactant (FS) 42
water-soluble polymeric surfactants 97
water-soluble polymers 190
wave damping methods 249
weakly-charged polymers 68
well-defined colloidal systems 119
well-dispersed starch 5
white pitch 13

x

X-ray diffraction analysis 141, 142
X-ray photoelectron spectroscopy (XPS) 141
X-ray scattering experiments 165

y

Yukawa attraction 122
Yukawa system 128